The
Superworld III

THE SUBNUCLEAR SERIES

Series Editor: **ANTONINO ZICHICHI**, *European Physical Society, Geneva, Switzerland*

Volume 1 was published by W. A. Benjamin, Inc., New York; 2-8 and 11-12 by
Academic Press, New York and London; 9-10, by Editrice Compositori, Bologna; 13-26
by Plenum Press, New York and London.

The
Superworld III

Edited by
Antonino Zichichi
European Physical Society
Geneva, Switzerland

PLENUM PRESS • NEW YORK AND LONDON

Library of Congress Cataloging-in-Publication Data

International School of Subnuclear Physics (26th : 1988 : Erice,
Italy)
    The superworld III / edited by Antonino Zichichi.
        p.   cm. -- (The Subnuclear series ; 26)
    "Proceedings of the Twenty-sixth course of the International
School of Subnuclear Physics on the Superworld III, held August
7-15, 1988, in Erice, Sicily, Italy"--T.p. verso.
    Includes bibliographical references.
    ISBN 978-1-4684-8871-5   ISBN 978-1-4684-8869-2 (eBook)
    DOI 10.1007/978-1-4684-8869-2
    1. Particles (Nuclear physics)--Congresses. 2. Superstring
theories--Congresses.   I. Zichichi, Antonino. II. Title.
III. Title: Superworld three. IV. Title: Superworld 3. V. Series:
Subnuclear series ; v. 26.
QC793.I555 1988
539.7'2--dc20
                                                        90-7154
                                                         CIP

Proceedings of the Twenty-Sixth Course of the International School
of Subnuclear Physics on The Superworld III,
held August 7-15, 1988, in Erice, Sicily, Italy

© 1990 Plenum Press, New York
Softcover reprint of the hardcover 1st edition 1990
A Division of Plenum Publishing Corporation
233 Spring Street, New York, N.Y. 10013

PREFACE

During August 1988, a group of 67 physicists from 45 laboratories in 17 countries met in Erice for the 26th Course of the International School of Subnuclear Physics. The countries represented were: Australia, Austria, Canada, China, Czechoslovakia, Denmark, France, Federal Republic of Germany, India, Italy, Poland, Portugal, Spain, Sweden, Switzerland, United Kingdom, and the United States of America.

The School was sponsored by the European Physical Society (EPS), the Italian Ministry of Public Education (MPI), the Sicilian Regional Government (ERS), and the Weizmann Institute of Science.

The interest in the Superworld is still very high. This is why, for the third year, the Erice School has been devoted, to a great extent, to review the many developments in Superstring, Supermembranes with their problems of quantization and compactification.

All these theoretical speculations are very far from the experimental frontier. In order to keep our feet on the ground, a series of lectures was included to cover the status of CP violation, of the Heavy Leptons, together with the projects for new physics at Gran Sasso and Fermi Lab. For completeness, Julian Schwinger reviewed the great problem of Anomalies in Quantum Field Theory and Shelly Glashow gave a closing lecture on the end of Superworld. If nothing new happens, next year there will be no Superworld in Erice.

I hope the reader will enjoy this book as much as the students enjoyed attending the lectures and the discussion sessions, which are the most attractive features of the School. Thanks to the work of the Scientific Secretaries, the discussions have been reproduced as faithfully as possible. At various stages of my work I have enjoyed the collaboration of many friends whose contributions have been extremely important for the School and are highly appreciated. I thank them most warmly. A final acknowledgement to all those who, in Erice, Bologna, Rome and Geneva, helped me on so many occasions and to whom I feel very much indebted.

Antonino Zichichi

CONTENTS

ANOMALIES IN QUANTUM FIELD THEORY

Julian Schwinger

University of California
Los Angeles, CA
U S A

ABSTRACT

The historically important decay of the neutral pion into two photons
is re-examined using the non-operator field theory techniques of Source
Theory. The equivalence of pseudoscalar and pseudovector couplings, or alternatively, the partial conservation of the axial current, is confirmed,
not only for the primitive interaction, but to the next order in α. In contrast, current algebra, which initially failed to account for the pion decay, uncovered, in the analysis of non-convergent integrals, an anomaly of
the axial current divergence. This saved the day at the primitive interaction level, but does not yield the α correction. The question, of vital
importance for non-Abelian gauge theories, is thereby raised but not answered: Do anomalies have physical significance, or are they mathematical
artifacts of a maladroit formulation?

The title of this lecture would be better stated as Anomalies in Quantum Field Theory: Physics? Mathematics? Or, to spell it out: Do anomalies
have physical meaning or are they artifacts of particular mathematical
schemes? In addition, I shall not talk about anomalies in general; the discussion to follow is restricted to the historically first one — the so-
called triangle anomaly.

I remind you of its context; the two-photon decay of the neutral π-me-
son, as modelled by pseudoscalar or pseudovector coupling to charged fermions. The Lagrange function (or density) for pseudoscalar coupling is

$$L = -\frac{1}{2}\,\psi\gamma^{\circ}[\gamma^{\mu}\Pi_{\mu} - g\gamma_5\phi + m]\,\psi \;,$$

$$\Pi_{\mu} = p_{\mu} - eq\,A_{\mu}\;, \quad q = \begin{pmatrix} 0 & -i \\ i & 0 \end{pmatrix}\;, \quad \gamma_5 = \gamma^{\circ}\gamma^1\gamma^2\gamma^3 \;\;,$$

which requires some additional comments on the notation. Corresponding to
the explicit appearance of the antisymmetrical charge matrix q, rather than
its ± 1 eigenvalues, the spin $\frac{1}{2}$ field $\psi(x)$ has $8 = 2 \times 4$ components. Also, the
matrices $\gamma^{\circ}\gamma^{\mu}$, $\mu = 0,1,2,3$, are symmetrical, whereas γ°, γ_5, and $\gamma^{\circ}\gamma_5$ are
antisymmetrical. (This is a Majorana representation.) Related algebraic pro-

perties are comprised in

$$-\frac{1}{2}\{\gamma_\mu,\gamma_\nu\} = g_{\mu\nu} \quad , \quad -g_{00} = g_{11} = g_{22} = g_{33} = 1$$

$$-\frac{1}{2}\{\gamma_\mu,\gamma_5\} = 0 \quad , \quad -\gamma_5^2 = 1 \quad .$$

Now, let us redefine ψ:

$$\psi \rightarrow e^{\frac{g}{2m}\gamma_5\phi}\,\psi = \psi\, e^{-\frac{g}{2m}\gamma_5\phi} \quad ,$$

so that

$$\frac{1}{2}\,\psi\gamma^\circ\gamma^\mu\Pi_\mu\,\psi \rightarrow \frac{1}{2}\,\psi\gamma^\circ\gamma^\mu\underbrace{e^{-\frac{g}{2m}\gamma_5\phi}\,\Pi_\mu\,e^{\frac{g}{2m}\gamma_5\phi}}\psi$$

$$\Pi_\mu - \frac{g}{2m}\,i\,\gamma_5\,\partial_\mu\,\phi \quad ,$$

where, despite the exponential forms, only terms linear in ϕ are of interest. That restriction becomes more relevant in

$$\frac{1}{2}\,\psi\gamma^\circ(m-g\gamma_5\phi)\psi \rightarrow \frac{1}{2}\psi\gamma^\circ m\,\underbrace{e^{\frac{g}{2m}\gamma_5\phi}(1-\frac{g}{m}\gamma_5\phi)\,e^{\frac{g}{2m}\gamma_5\phi}}_{\rightarrow 1}\psi$$

$$= \frac{1}{2}\,\psi\gamma^\circ m\,\psi \quad .$$

The outcome is the pseudovector coupling version of the Lagrange function:

$$L = -\frac{1}{2}\,\psi\gamma^\circ[\gamma^\mu\Pi_\mu - \frac{g}{2m}\,i\,\gamma^\mu\gamma_5\partial_\mu\phi + m]\psi \quad .$$

This brings us to the question whether the equivalence of pseudoscalar and pseudovector couplings is actually maintained in the specific problems of two-photon decay. The relevant pseudoscalar interaction term in the action W is produced as follows:

$$W_{int}^{ps} = \int(dx)\,\frac{1}{2}\,\psi\gamma^\circ\,g\,\gamma_5\phi\,\psi$$

$$= \frac{i}{2}\int(dx)\text{tr}\,g\,\gamma_5\phi(x)\underbrace{i\,\psi(x)\,\psi(x)\gamma^\circ}_{\rightarrow G_+(x,x)^A}$$

$$= \frac{i}{2}\,\text{Tr}[g\gamma_5\phi G_+^A] \quad ,$$

which uses the anticommutative character of fermion fields, and exhibits the vacuum replacement for the pair of spin $\frac{1}{2}$ fields in terms of Green's function, described symbolically by

$$(\gamma\Pi+m)G_+^A = 1 \quad , \quad G_+^A = \frac{m-\gamma\Pi}{\Pi^2+m^2-eq\sigma F} \quad ,$$

$$\sigma F = \frac{1}{2} \sigma^{\mu\nu} F_{\mu\nu} \quad , \quad \sigma^{\mu\nu} = \frac{i}{2}[\gamma^\mu,\gamma^\nu] \quad .$$

The trace symbol Tr refers to the space-time continuum, as well as to the discrete charge-spin variables, whereas tr signifies just the latter.

Inasmuch as an odd number of γ^μ's have zero trace,

$$W_{int}^{ps} = \frac{i}{2} \, Tr\left[g\gamma_5 \phi \, \frac{m}{\Pi^2+m^2-eq\sigma F} \right]$$

$$\rightarrow \frac{i}{2} \, g \, m \, Tr\left[\gamma_5 \phi \, \frac{1}{\Pi^2+m^2} \, eq\sigma F \, \frac{1}{\Pi^2+m^2} \, eq\sigma F \, \frac{1}{\Pi^2+m^2} \right] \quad ,$$

which exhibits the two $\sigma^{\mu\nu}$ factors needed to match γ_5 in the trace. In the following we use the assumption that

$$m_\pi \ll m$$

to justify treating the field $F_{\mu\nu}(x)$ as slowly varying. Then, from the trace

$$\frac{1}{8} \, tr\gamma_5 (\sigma F)^2 = -2 \, \vec{E}\cdot\vec{B} \quad ,$$

and the fact that the space-time trace of products of functions of the co-ordinates x with functions of the momenta p are given by four-dimensional phase space integrals, we get

$$W_{int}^{ps} = e^2 \, g \, m \, 8 \, I \, \int(dx)\phi \, \vec{E}\cdot\vec{B} \quad ,$$

$$I = \frac{1}{i} \int \frac{(dp)}{(2\pi)^4} \, \frac{1}{(p^2+m^2-i\epsilon)^3} = \frac{1}{32\pi^2} \, \frac{1}{m^2} \quad ,$$

from which it follows that

$$W_{int}^{ps} = \frac{\alpha}{\pi} \, \frac{g}{m} \int(dx) \, \phi \, \vec{E}\cdot\vec{B} \quad .$$

Before continuing, let us examine some evaluations of the momentum integral I that employ the Euclidean transformation

$$p_0 \rightarrow i \, p_4 \quad , \quad I = \int \frac{(dp)_E}{(2\pi)^4} \, \frac{1}{(p^2+m^2)^3} \quad .$$

We can use the surface area of a unit 4-sphere, $2\pi^2$:

$$I = \int_0^\infty \frac{2\pi^2 \frac{1}{2} p^2 dp^2}{(2\pi)^4} \, \frac{1}{(p^2+m^2)^3} = \frac{1}{16\pi^2} \, \frac{1}{m^2} \int_0^\infty dx \, \frac{x}{(x+1)^3}$$

$$= \frac{1}{32\pi^2} \, \frac{1}{m^2} \quad .$$

Or, we can employ the exponential representation of the denominator:

$$I = \int \frac{(dp)_E}{(2\pi)^4} \int_0^\infty ds \, \frac{s^2}{2} \, e^{-s(p^2+m^2)} = \int_0^\infty ds \, \frac{s^2}{2} \, e^{-sm^2} \underbrace{\int \frac{(dp)_E}{(2\pi)^4} \, e^{-sp^2}}_{(1/4\pi s)^2}$$

3

$$= \frac{1}{32\pi^2} \int_0^\infty ds\ e^{-sm^2} = \frac{1}{32\pi^2}\ \frac{1}{m^2}\quad .$$

The analogous pseudovector calculation begins with

$$W_{int}^{pv} = \frac{i}{2}\ Tr\left[\frac{g}{2m}\ i\ \gamma^\mu \gamma_5\ \partial_\mu \phi\ G_+^A\right]$$

$$= \frac{i}{2}\ Tr\left[\frac{g}{2m}\ i\ \gamma^\mu \gamma_5\ \partial_\mu \phi\ \frac{-\gamma\Pi}{\Pi^2+m^2-eq\sigma F}\right]$$

$$= -\frac{i}{2}\ \frac{g}{2m}\ Tr\left[\ i\ \gamma^\mu \gamma_5\ \partial_\mu \phi\ \gamma\Pi\frac{1}{\Pi^2+m^2}\ eq\sigma F\ \frac{1}{\Pi^2+m^2}\right]$$

where now it is the m-term that does not contribute to the trace. And, in contrast with the pseudoscalar version, the presence of two additional γ-factors makes a single $eq\sigma F$ factor suffice.

Suppose we test the equivalence of the two couplings at this stage? That involves the transfer of the gradient from the pion field. We do this in a gauge invariant manner by writing

$$\partial_\mu \phi = \frac{1}{i}\ [\phi,\Pi_\mu]\quad .$$

Then

$$W_{in}^{pv} \rightarrow -\frac{1}{2}\ \frac{g}{2m}\ Tr\left[i\gamma^\mu \gamma_5 \phi[\Pi_\mu,\gamma\Pi\ \frac{1}{\Pi^2+m^2}\ eq\sigma F\ \frac{1}{\Pi^2+m^2}]\right]\quad ,$$

which will supply the second field strength factor as a consequence of the commutator

$$[\Pi_\mu,\Pi_\nu] = ieq\ F_{\mu\nu}\quad .$$

The approximation of slowly varying fields allows one to write:

$$W_{int}^{pv} \rightarrow -\frac{1}{2}\ \frac{g}{2m}\ Tr\left[i\underbrace{\gamma^\mu \gamma_5 \phi\gamma^\nu eq\sigma F}[\Pi_\mu,\Pi_\nu\ \frac{1}{(\Pi^2+m^2)^2}]\right]$$
$$-\ \gamma_5 \sigma^{\mu\nu} eq\sigma F\phi\quad ,$$

along with

$$\left[\Pi_\mu,\Pi_\nu\ \frac{1}{(\Pi^2+m^2)}\right] = \underbrace{[\Pi_\mu,\Pi_\lambda]}_{ieqF_{\mu\nu}}\ \underbrace{\frac{\partial}{\partial\Pi_\lambda}\ \frac{\Pi_\nu}{(\Pi^2+m^2)^2}}_{\Pi\rightarrow p}\quad ,$$

which brings us to this point:

$$W_{int}^{pv} \rightarrow \frac{e^2}{2}\ \frac{g}{2m}\ \int(dx)tr(\gamma_5 \sigma^{\mu\nu}\sigma F)F_{\mu\lambda}\ \phi\ i\int\frac{(dp)}{(2\pi)^4}\ \frac{\partial}{\partial p_\lambda}\ \frac{p_\nu}{(p^2+m^2)^2}\quad .$$

The introduction of Euclidean metric, and the use of rotational symmetry, enables the momentum integral to be evaluated as a surface integral on a sphere of large radius P:

4

$$P \gg m \quad,$$

so that

$$i \int \frac{(dp)}{(2\pi)^4} \frac{\partial}{\partial p_\lambda} \frac{P_v}{(p^2+m^2-i\epsilon)^2} \to -\int \frac{(dp)_E}{(2\pi)^4} \frac{\partial}{\partial p_\lambda} \frac{P_v}{(p^2+m^2)^2} = -\delta_v^\lambda K \quad ;$$

$$K = \frac{1}{4} \int \frac{(dp)_E}{(2\pi)^4} \frac{\partial}{\partial p_\mu} \frac{P_\mu}{(p^2+m^2)^2} = \frac{1}{4} \frac{2\pi^2 P^3}{(2\pi)^4} \frac{P}{(P^2+m^2)^2} \bigg|_{P \gg m}$$

$$= \frac{1}{32\pi^2} \quad.$$

If this number seems familiar, it is not without reason. Indeed, on evaluating the divergence, K becomes

$$K = \int \frac{(dp)_E}{(2\pi)^4} \left[\frac{1}{(p^2+m)^2} - \frac{p^2}{(p^2+m^2)^3} \right]$$

$$= \int \frac{(dp)_E}{(2\pi)^4} \frac{m^2}{(p^2+m^2)^3} = m^2 I = \frac{1}{32\pi^2}$$

With all the ingredients in place, it is clear that the outcome will be just

$$W_{int}^{pv} \to W_{int}^{ps} \quad.$$

A surface integral in momentum space is a delicate instrument for maintaining the equivalence of pseudoscalar and pseudovector couplings, particularly if one converts it into a four-dimensional integral in which different orders of integration are possible. That is illustrated by giving an exponential representation for the differentiated version of K:

$$K = \int \frac{(dp)_E}{(2\pi)^4} \left[\frac{1}{(p^2+m)^2} - \frac{p^2}{(p^2+m^2)^3} \right]$$

$$= \int \frac{(dp)_E}{(2\pi)^4} \int_0^\infty ds\, s(1 - \frac{1}{2} sp^2)\, e^{-s(p^2+m^2)} \quad.$$

This s-integral can also be presented as

$$\int_0^\infty ds\, s\, e^{-sm^2} (1 + \frac{1}{2} s \frac{\partial}{\partial s})\, e^{-sp^2}$$

$$= \int_0^\infty ds\, e^{-sm^2} \frac{1}{2} \frac{\partial}{\partial s} [s^2 e^{-sp^2}]$$

Now, suppose one integrates first over the infinite momentum space, using

$$\int \frac{(dp)_E}{(2\pi)^4} e^{-sp^2} = \frac{1}{(4\pi)^2} \frac{1}{s^2} \quad?$$

That gives

$$K = \int_0^\infty ds\, e^{-sm^2} \frac{1}{2} \frac{\partial}{\partial s} \left[s^2 \frac{1}{(4\pi)^2} \frac{1}{s^2} \right]$$

$$= 0 \quad!$$

On the other hand, suppose the domain of momentum integration is a finite sphere of large radius $P \gg m$? Then we encounter

$$p < P : \int \frac{(dp)}{(2\pi)^4} \frac{1}{E} e^{-sp^2} = \int \frac{2\pi^2 p^2 \frac{1}{2} d\,p^2}{(2\pi)^4} e^{-sp^2}$$

$$= \frac{1}{(4\pi)^2} \left(-\frac{\partial}{\partial s}\right) \int_0^{P^2} dp^2 \, e^{-sp^2} = \frac{1}{(4\pi)^2} \left(-\frac{\partial}{\partial s}\right) \frac{1-e^{-sP^2}}{s}$$

$$= \frac{1}{(4\pi)^2} \left[\frac{1}{s^2} + \frac{\partial}{\partial s} \frac{e^{-sP^2}}{s}\right] \quad,$$

from which follows

$$K = \frac{1}{32\pi^2} \int_0^\infty ds \, e^{-sm^2} \frac{\partial}{\partial s} \left[s^2 \frac{\partial}{\partial s} \frac{e^{-sP^2}}{s}\right]$$

$$= \frac{1}{32\pi^2} \int_0^\infty ds \, s \, e^{-sm^2} \frac{\partial^2}{\partial s^2} e^{-sP^2} \quad,$$

just as with the radial part of the three-dimensional Laplacian operator. We now see directly that

$$K = \frac{1}{32\pi^2} \frac{P^4}{(P^2+m^2)^2} \bigg|_{P \gg m} = \frac{1}{32\pi^2} \quad,$$

exactly as with the surface integral, which also refers to a large but finite momentum space.

All right, what's going on? Which is to ask: What is the physics underlying this otherwise random mathematizing? It may help to draw the space-time triangle diagram, which has played no explicit role so far:

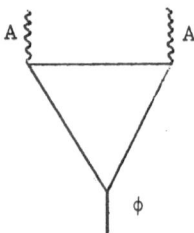

The mathematical problem associated with infinite momenta arises from the consideration of field products at the same point — with zero separation. The triangle diagram might suggest that one begin the calculation with a degree of space-time localizability of the pion field and of the electromagnetic field such that initially there is no overlap. The final stage of the calculation is one of space-time extrapolation to remove that restriction. The complementary description of this procedure is to begin the calculation with a bounded range of momenta and then extrapolate to infinite momenta. From this point of view, it is the surface integral in momentum space, or its equivalent finite volume integral, that is physically correct.

In order to get to the crux of the matter, which concerns the form of the pseudovector coupling when the pseudovector field is retained, I need some details of the initial non-overlap calculations. Here I lack the time to show all the steps and must be content to cite results, although there is reassuring contact with what has gone before. For pseudoscalar coupling, the action term is

$$W^{ps}_{int} = \int (dx)(dx') \frac{\alpha}{\pi} \frac{g}{m} \vec{E}\cdot\vec{B}(x) f(x-x')\phi(x')$$

$$f(x-x') = \int \frac{(dk)}{(2\pi)^4} e^{ik(x-x')} f(k)$$

$$f(k) = \int^{\infty}_{(2m)^2} d M^2 \frac{2m^2}{M^2} \log \frac{1+v}{1-v} \frac{1}{k^2+M^2} \quad , \quad v = \sqrt{1-\left(\frac{2m}{M}\right)^2}$$

$$f(k=0) = \int^{1}_{0} dv\, v \log \frac{1+v}{1-v} = 1 \quad .$$

In the situation under consideration,

$$-k^2 = m^2_\pi <\!<M^2, \quad f(k) \cong f(k=0) = 1 , \quad f(x-x') \cong \delta(x-x') \quad ,$$

and the known coupling for slowly varying fields reemerges.

For what follows, it is helpful to recall the dual field strength tensor.

$$*F^{\mu\nu} = \frac{1}{2} \epsilon^{\mu\nu\kappa\lambda} F_{\kappa\lambda} \quad : \quad \epsilon^{\mu\nu\kappa\lambda} \text{ totally anti, } \epsilon^{0123} = 1 \quad ,$$

$$-\frac{1}{4} *F^{\mu\nu} F_{\mu\nu} = \vec{E}\cdot\vec{B} \quad .$$

Then the pseudovector action term can be presented as

$$W^{pv}_{int} = \int (dx)(dx') \frac{\alpha}{\pi} \frac{g}{m} \frac{1}{2} *F^{\mu\nu} A_\nu(x) f(x-x') \partial'_\mu \phi(x') \quad ,$$

which, for slowly varying fields, reduces to

$$W^{pv}_{int} = \int (dx) \frac{\alpha}{\pi} \frac{g}{m} \frac{1}{2} *F^{\mu\nu} A_\nu \partial_\mu \phi \quad .$$

Despite superficial appearance, this structure is gauge invariant. Indeed, the addition to A_ν of the gradient of a scalar leads, on partial integration, to the combination:

$$(\partial_\nu *F^{\mu\nu})\partial_\mu \phi + *F^{\mu\nu}\partial_\mu\partial_\nu \phi = 0 \quad ,$$

where both terms vanish identically. Then, as might now be expected, the transfer of the derivative acting on ϕ produces

$$-\partial_\mu \left[\frac{1}{2} *F^{\mu\nu} A_\nu \right] = -\frac{1}{4} *F^{\mu\nu} F_{\mu\nu} = \vec{E}\cdot\vec{B} \quad ,$$

and the pseudoscalar coupling emerges.

So, what's the problem? As far as field theory is concerned there isn't any. The problem was created when the mind set and machinery of current algebra was introduced.

First, one should appreciate that the equivalence of pseudoscalar and pseudovector couplings, as expressed by

$$W_{int} = \int (dx) g\phi j_5 = \int (dx) \frac{g}{2m} \partial_\mu \phi \; j_5^\mu$$

$$j_5 = \frac{1}{2} \psi \gamma^0 \gamma_5 \psi \qquad j_5^\mu = \frac{1}{2} \psi \gamma^0 i \gamma^\mu \gamma_5 \psi \quad ,$$

implies the local partial conservation law,

$$- \partial_\mu j_5^\mu = 2m \; j_5 \quad ,$$

for the so-called axial current vector, as contrasted with the conservation law obeyed by the electric current vector:

$$j^\mu = \frac{1}{2} \psi \gamma^0 q \gamma^\mu \psi \quad , \qquad \partial_\mu j^\mu = 0 \quad .$$

Current algebra works with currents, not fields. Currents are supposed to be gauge invariant objects. Thus, it is not possible for a current algebra calculation to produce a structure like

$$W_{int}^{pv} = \int (dx)(dx') \frac{\alpha}{\pi} \frac{g}{m} \frac{1}{2} {}^*F^{\mu\nu}(x) A_\nu(x) f(x-x') \partial'_\mu \phi(x') \quad ,$$

with its gauge variant vector potential. Rather, that calculation will supply two additional derivatives, as required to take the curl of the vector potential, while maintaining a scalar form. The consequence of those derivatives for the low energy neutral-pion decay is a strong reduction of the decay rate. Indeed, the first such calculation gave a much too small result.

The anomaly turned up, as a solution to the problem, in studying the infinite momentum-space integral that represents the vacuum matrix element of the ordered operator product involving

$$\partial_\mu j_5^\mu + 2m j_5 \quad ; \quad j^\nu \quad ; \quad j^\lambda \quad .$$

This integral is linearly divergent, but consists of pairs of terms that would cancel if a displacement of the momentum variable were permissible. In fact they lead to finite surface integrals in momentum space. The result is the axial anomaly, as expressed by

$$- \partial_\mu j_5^\mu - 2 m j_5 = \frac{\alpha}{\pi} \left(-\frac{1}{4} {}^*F^{\mu\nu} F_{\mu\nu} \right) \quad ,$$

which reproduces the low energy, or slowly varying field, form of the coupling.

What is the general form, according to current algebra? We have only to return to the field theory version for pseudovector coupling and write

$$\frac{1}{k^2+M^2} = \frac{1}{M^2} - k^2 \frac{1}{M^2} \frac{1}{k^2+M^2}$$

$$f(k) = 1 - k^2 \bar{f}(k) \quad , \qquad \bar{f}(k) = \int_{(2m)^2}^\infty dM^2 \; \frac{2m^2}{(M^2)^2} \log \frac{1+v}{1-v} \frac{1}{k^2+M^2}$$

$$f(x-x') = \delta(x-x') + \partial^\lambda \partial_\lambda \bar{f}(x-x')$$

and get

$$W_{int} = [Anomaly] + [Non-Anomaly]$$

$$[A] = \int (dx) \frac{\alpha}{\pi} \frac{g}{m} \vec{E} \cdot \vec{B} \phi$$

$$[N-A] = \int (dx)(dx') \frac{\alpha}{\pi} \frac{g}{m} (-\partial^\lambda) \overline{\vec{E} \cdot \vec{B}}(x) \overline{f}(x-x') \partial'_\lambda \phi(x') \quad .$$

In short, the two versions are equivalent. But one has an anomaly, the other does not. It would seem that the anomaly is an aspect of the mathematical formulation, not of the physics.

There is, however, the possibility that, when pushed farther, the two formulations cease to be equivalent. This refers to the question of radiative corrections to the low energy pion-photon coupling

$$W_{int} = \frac{\alpha}{\pi} \frac{g}{m} \int (dx) \vec{E} \cdot \vec{B} \phi \quad .$$

Current algebraists have claimed that there are no such corrections because it is only the primitive triangle diagram that possesses the divergence needed for the anomaly. A field theory treatment, performed in several, concordant ways, arrives at a different conclusion. To the next order in α, there is a fractional change given by

$$\zeta_c - \zeta_{pv} \quad ,$$

$$\left\{ \begin{array}{c} \zeta_c \, \gamma^\mu \\ \zeta_{pv} \, i \, \gamma^\mu \gamma_5 \end{array} \right\}$$

$$= -i \, e^2 \int \frac{(dk)}{(2\pi)^4} \frac{1}{k^2} \gamma^\nu \frac{1}{\gamma(p-k)+m} \left\{ \begin{array}{c} \gamma^\mu \\ i \, \gamma^\mu \gamma_5 \end{array} \right\} \frac{1}{\gamma(p-k)+m} \gamma_\nu$$

$$\gamma p + m \rightarrow 0 \quad ,$$

where the parameters ζ_c, ζ_{pv} are associated with coupling-parameter renormalization, or maintenance of normalization, depending on the viewpoint.

The k-integral is logarithmically divergent. But the divergences:

$$\sim -i \, e^2 \int \frac{(dk)}{(2\pi)^4} \frac{1}{k^2} \frac{\gamma^\nu \gamma k}{k^2} \left\{ \begin{array}{c} \gamma^\mu \\ i \, \gamma^\mu \gamma_5 \end{array} \right\} \frac{\gamma k \gamma_\nu}{k^2}$$

are the same for both parameters. The difference, then, is finite and turns out to be

$$\zeta_c - \zeta_{pv} = \frac{\alpha}{2\pi} \quad .$$

Thus, the low energy coupling should be amended to

$$W_{int} = \frac{\alpha}{\pi} \frac{g}{m} \left(1 + \frac{\alpha}{2\pi} \right) \int (dx) \vec{E} \cdot \vec{B} \phi \quad .$$

All this is ancient history. I cannot tell you whether my story carries a message for today. It is you who must think about it, and, if necessary, act.

DISCUSSION

– *Volkas:*

Please expand on one of the last points you made. I am referring to the disagreement that you stated existed between a field theoretic and current algebraic calculation of radiative corrections to the triangle graph. In particular did you imply that the Adler–Bardeen theorem had only current algebraic validity? How is the theorem about high order corrections not contributing to the anomaly related to the difference between the calculations?

– *Schwinger:*

The difference between the calculations is to some extent a matter of attitude. The field theory calculation is a direct treatment of the phenomenon. The actual calculation is too elaborate to go into, but reduces to the difference between renormalization constants which were calculated with a non–zero answer. Whereas, in current algebra what is done is to isolate the renormalization terms, recognize that they are divergent in the same way, in that the ratio of the two renormalization constants is one (in the limit where cutoffs go to infinity) and they are therefore identified. The difference is in the philosophy of how one handles renormalization in current algebra, as compared to the field theoretic calculations which involves finite quantities throughout. I regard this as a deficiency of the current algebra approach.

– *Green:*

Your talk described the anomaly calculation for a situation in which there is no local gauge symmetry associated with the conserved axial current. Do you have any comments about the situation in which there is an anomaly in the conservation of a current coupling to a gauge field. Such anomalies are commonly considered disastrous for gauge theories although some people disagree.

– *Schwinger:*

I have none. The audience was invited to take up that challenge. I limited myself in my talk to the historical origin of anomalies and very specifically did not address the question of nonabelian groups. My curiosity is aroused however and perhaps I shall ultimately answer this question.

– *Sarid:*

What should we learn from the discussion on making contact between current algebra and field theory, i.e. splitting the field–theoretic result into an "anomaly" and "non–anomaly" part?

– *Schwinger:*

I was trying to suggest the form in which current algebra would give the answer as the anomaly plus an additional part which is negligible at low energies. Instead of picking it out of the current algebra calculation, which I did not have. I simply took the field theory calculation and rearranged it to have that form. Thus I picked out one part that is local, (which in one language is the anomaly, but not in field theory) plus an additional part with no low energy contribution.

– *Polychronakos:*

Do vacuum expectation values of currents expressed in the "original" and "chirally rotated" field coincide?

– *Schwinger:*

Yes, the axial current is invariant under the transformation.

– *Giannakis:*

Is there another manifestation of anomalies except from the violation of current conservation?

– *Schwinger:*

I know of an anomaly for spinor fields in the presence of an electromagnetic field. The trace of the stress tensor is proportional to the mass of the fermion and in the limit as the mass goes to zero (which to me is unphysical) you might think that the mass term would vanish, but nevertheless, the trace of the stress tensor does not vanish and that is considered an anomaly. If one explicitly studies the mass term one finds a finite limit as the mass goes to zero. I think that the stress anomaly is another situation where, when the calculation is done by field theory methods, there is no anomaly, although, at the moment, I have only a superficial knowledge of this problem.

SUPERSTRING THEORY: A SURVEY

Michael B. Green

Department of Physics
Queen Mary College
University of London
Mile End Road, London E1 4NS, U.K.

1. INTRODUCTION

Over the past few years the subject of superstring theory has exploded into a wideranging field covering many areas of modern mathematics, statistical mechanics, elementary particle physics and general relativity. It is appropriate that in this talk I should concentrate on the prospect that superstring theory may provide a unified quantum theory of all the fundamental forces, including gravity. Although it is this prospect that sparked the current interest in the particle physics community the links with mathematics and two-dimensional statistical mechanics have been very fruitful and exciting. This talk will be a very descriptive outline of the principles behind superstring theory and a survey of recent results - the interested reader may refer to any of a large number of references for technical details.[*]

Conventional relativistic quantum field theories may be described in terms of the dynamics of pointlike particles such as the photons and electric charges of quantum electrodynamics. String theory, on the other hand, is a relativistic quantum theory of stringlike particles - particles with one-dimensional extension moving through four-dimensional space-time. There are very obvious problems in describing extended objects in special relativity and it is the manner in which string theory overcomes these problems that is part of the fascination of the theory. The subject has its origins in the dual resonance model proposed by Veneziano in 1968 as a model of the strong interactions. This in turn was directly motivated by the experimental spectrum of hadronic resonances - the strongly interacting particles, namely, the mesons and baryons. This data suggested that the hadrons might be thought of as resonating states of an extended system. It is this extended nature of stringlike particles that leads to a radically new kind

---

[*] For example, for a compilation of recent work in string theory see *Proceedings of the 1988 ICTP Spring School on Superstrings*, M.B. Green, M. Grisaru, R. Iengo, E. Sezgyn and A.Strominger eds. (World Scientific, to be published).

The Superworld III
Edited by A. Zichichi
Plenum Press, New York, 1990

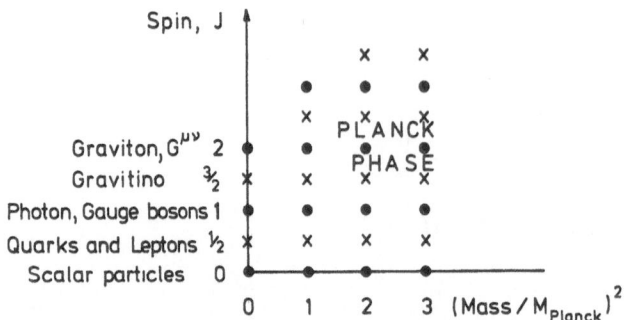

Figure 1.1. A typical superstring spectrum showing the angular momentum and mass of the states. The dots indicate bosonic states and the crosses indicate fermionic ones (each dot and cross generally represents a number of degenerate states). There are an infinite number of massive excitations corresponding to the infinite number of modes of higher and higher frequency. There an equal number of fermion and boson states at each mass level. The separation between successive levels is determined by the string tension $\Delta m^2 = T\pi h/c^3 \sim (10^{19} Gev/c^2)^2$. It is important that the massless states always include states with spins corresponding to the graviton (spin 2), gravitino (spin 3/2), photon and gauge bosons (spin 1), leptons and quarks (spin 1/2), scalar bosons (spin 0). These massless fields are typical of those required in conventional grand unified field theories with supersymmetry, which are approximations to string theory at low energy ($E \ll \sqrt{\pi h c T}$).

of theory since, unlike a point particle, a single string can oscillate and can therefore exist in any of an infinite number of excited quantum states.

The original hadronic string theories had certain inconsistencies which are overcome in the modern reincarnation in the form of superstring theories. The word *super* refers to the fact that these theories are supersymmetric in space-time, which is a necessary for their consistency. Fig. 1.1 shows a typical spectrum of the excitations of a superstring. One of the most striking facts about this spectrum is that it includes a massless spin-2 mode (as well as other massless modes) which is identified with the graviton, signifying that the theory describes the force of gravity and not simply the strong force. This means that this kind of string theory *necessarily* embodies general relativity and therefore should not be thought of simply as a theory of particles but also should determine the geometry of the space-time in which the string is moving.

One basic feature of string theory is that it contains a new dimensional constant, the string tension T, which is associated with a fundamental distance scale

$$l_S = \sqrt{\frac{ch}{\pi T}}. \tag{1.1}$$

This sets the scale of the frequencies of the infinite number of excited modes of vibration of the string. Since a mode of given frequency corresponds (via the de Broglie relation) to a quantum state of given energy and hence of given mass, the string describes an infinite number of quantum states with masses (and spins) that increase indefinitely (fig. 1.1). It is natural to identify l_S with the Planck distance,

$$l_{Planck} = \sqrt{\frac{Gh}{c^3}} \sim 10^{-35} meter \qquad (1.2)$$

(where G is Newton's constant) which is the distance scale that enters into any theory of quantum gravity. This means that $T \sim c^4/\pi G$ so the separation between the masses of the excited string states is huge $\Delta m^2 \sim T\pi h/c^3 \sim (10^{19} Gev/c^2)^2$. The infinitely rich spectrum of states in string theory is correlated with an infinite extension of the conventional Yang–Mills and gravitational gauge invariances. In the "low energy" approximation (at energies much less than the Planck energy, $E \ll \sqrt{\pi chT} \sim 10^{19} Gev$) the massive states decouple, leaving an effective theory of massless pointlike particles corresponding to the massless states in fig. 1.1. Since the massless states include the spin-2 graviton and spin-1 Yang–Mills particles (as well as the spin-3/2 gravitino characteristic of supersymmetry), superstring theory reduces in this low energy limit to a conventional supersymmetric Yang–Mills theory interacting with supergravity. Different string "theories" (i.e., theories with different choices of background) differ in the details of the spectrum of massless states (such as the Yang–Mills gauge group and the representations of the quarks and leptons) and so correspond to different possible unifications of the non-gravitational forces. Certain of these low energy limits come remarkably close to describing the observed world of elementary particles and their forces.

The present formulation of string theory is one in which the quantum-mechanical strings are considered to be moving through a classical space-time background. This separation of the particle dynamics from space-time is analogous to the perturbative description of Einstein's theory in which the pointlike graviton corresponds to a small fluctuation of the metric of space-time around some classical solution of Einstein's equations. In conventional quantum gravity this perturbation expansion has terrible problems with nonrenormalizable infinities. In superstring theory, however, the individual terms in this perturbation expansion (the string version of Feynman diagrams) have certain extraordinary features which suggest that the usual problems of conventional perturbative quantum gravity are avoided. In particular, there are no infinities or chiral anomalies when the classical background is chosen consistently.

There are many possible consistent choices for the space-time background in string perturbation theory which, for brevity, I shall refer to as different string "theories" or "models". They are more properly described as different semiclassical expansions of the same theory as I will describe later.

Despite these attractive features of the individual terms in string perturbation theory there should be more to a quantum theory of gravity than a semiclassical expansion around a fixed space-time background so our understanding of the theory is far from complete. At present we are in the rather extraordinary situation of knowing these very interesting perturbation expansions without knowing the fundamental structure that is being expanded (i.e., the analogue of the curvature scalar in the Einstein–Hilbert theory). One would expect the perturbation expansion to be inadequate for understanding many issues in superstring theory (as it is in many ordinary field theories, such as QCD) In particular, in a quantum theory of gravity space-time coordinates should not be introduced into the theory as fundamental input, as in string perturbation theory, but rather should emerge from the dynamics unified with the particles and forces.

One might expect that the semi-classical approximation to the space-time background breaks down at the Planck scale and therefore that the string perturbation may be inconsistent even though it appears consistent at any finite order. In other words, just at the scale at which the extended nature of strings becomes important - leading, for example, to the good ultraviolet properties of individual string Feynman diagrams - the simple picture of a single string moving in a fixed background may not be sensible. This suggests that strings are incorrect variables with which to describe string theory. Indications of a phase transition in the theory around the Planck scale* support the view that new coordinates are needed to describe the theory in the "Planck phase" (fig. 1.1). In the perturbative description of string theory the background metric of space-time, $G_{\mu\nu}$, takes a fixed nonzero value (approximately the Minkowski metric) characteristic of the spontaneous breakdown of general coordinate invariance. A more fundamental treatment would describe the theory in the unbroken phase in which the spacetime metric would not be singled out from the rest of the string states (Witten's topological models of gravity are possible examples of this kind of theory). Needless to say an intensive search is being undertaken for such a fundamental setting to string theory. Many paths are open since the only criterion for deciding on a correct approach (apart from aesthetic ones) is that it should reproduce the (presumably divergent) string perturbation expansion in a (possibly inconsistent!) approximation.

I will begin with a survey of the developments in string perturbation theory and attempts to classify consistent string theories. I will then comment on a few of the ideas that have been suggested about how to formulate a nonperturbative theory that reproduces string perturbation theory.

2. STRING FEYNMAN DIAGRAMS

The quantum mechanical description of string theory is based closely on the Feynman path integral formulation of quantum mechanics. Recall that for relativistic point particles the amplitude for a particle to move from a point labelled x_1^μ (where $\mu = 0, 1, \ldots, D-1$ is the space-time vector index) to a point x_2^μ is given by an average over all world lines $x^\mu(\tau)$ connecting these points (where τ is the parameter labelling the path),

$$A(x_1 \to x_2) = \sum_{world\ lines} e^{\frac{i}{\hbar} S[x(\tau)]}. \tag{2.1}$$

The action S is the length of the world line, a parametrization-independent quantity.

---

* For hadronic strings the string tension has a value of around $10^{-19} M_{Planck}$ and there is a phase transition at a correspondingly low temperature to a phase of deconfined quarks and gluons, the constituents of QCD. Nowadays, the stringlike spectrum observed in hadronic experiments is supposed to be explained by QCD.

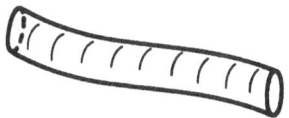

Figure 2.1. A world surface joining initial and final closed strings. The path integral includes sums over all possible surfaces joining initial and final surfaces.

The analogous quantity in string theory describes the amplitude for a string to move from a given initial curve $X_1^\mu(\sigma)$ to a curve $X_2^\mu(\sigma)$ and is given as a sum over all surfaces connecting the curves These two-dimensional surfaces may be parametrized by σ and τ but no physical quantity should depend on the parametrization used. The simplest parametrization-independent action is the *area* of the world-sheet embedded in space-time - this is the 'Nambu' action. It proves to be very convenient to describe any world sheet not only its embedding in space time, $X^\mu(\sigma, \tau)$, but also by an intrinsic metric, $g^{\alpha\beta}(\sigma, \tau)$ (where $\alpha, \beta = 0, 1$ label the worl-sheet coordinates so that g is a symmetric 2×2 matrix with three independent components). This leads to the so-called 'Polyakov' method of calculating string theory amplitudes in which the sum over histories involves an integral over all intrinsic metrics as well as all embeddings of the world sheet, so that (2.1) is replaced by

$$A(X_1^\mu(\sigma) \to X_2^\mu(\sigma)) = \int_1^2 \mathcal{D}g^{\alpha\beta} \mathcal{D}X^\mu \mathcal{D}\psi \, e^{-S/\hbar}. \qquad (2.2)$$

The variables $\psi(\sigma, \tau)$ in this expression represent the additional fermionic variables that are present in the various kinds of superstring theories. It has also been assumed that it is sensible to continue τ to imaginary value, $\tau \to i\tau$ (a 'Wick rotation') so that $e^{iS/\hbar} \to e^{-S/\hbar}$. An obviously reparametrization invariant action is given by

$$S = -\frac{T}{2} \int d\sigma d\tau \sqrt{g} g^{\alpha\beta}(\sigma, \tau) \partial_\alpha X^\mu(\sigma, \tau) \partial_\beta X^\nu(\sigma, \tau) G_{\mu\nu}(X). \qquad (2.3)$$

where $g \equiv \det g_{\alpha\beta}$ and $G_{\mu\nu}(X)$ is the (symmetric) $D \times D$ metric tensor of the curved space-time in which the string is moving. The expression (2.3) can be viewed as the action for a *two-dimensional* field theory. The coordinates $X^\mu(\sigma, \tau)$ are simply D scalar two-dimensional fields (the Lorentz index μ playing the rôle of an internal symmetry index). In fact, the action S is the action of two-dimensional gravity coupled to D scalar fields and hence it is manifestly reparametrization invariant. The variables $g_{\alpha\beta}$ are auxiliary fields which can be eliminated from (2.3) by using their equations of motion, in which case the resulting action is simply the area of the world sheet.

17

It is intriguing that while string theory is based on the principle of two-dimensional coordinate invariance of the world sheet it turns out to contain the theory of general relativity in the embedding space - the space-time in which the string is moving.

In addition to (2.3) a number of other terms may also be considered in the action. One of these involves an antisymmetric background field $B_{\mu\nu}(X)$, and another a scalar field, $\Phi(X)$ (which corresponds to a 'dilaton' field in the embedding space). Other terms involving fermionic background fields may also be added to the action.

Conformal Field Theory

Reparametrization invariance means that the action S_0 is invariant under arbitrary redefinitions of the world-sheet coordinates $\tau \to \tau + \xi^0(\sigma,\tau)$ and $\sigma \to \sigma + \xi^1(\sigma,\tau)$, where ξ^α is an arbitrary world-sheet vector. Furthermore, the action (2.3) is also invariant under Weyl transformations, $g^{\alpha\beta} \to e^{\lambda(\sigma,\tau)}g^{\alpha\beta}$ where λ is an arbitrary scalar function. These three invariances can be used to gauge away all three degrees of freedom in $g_{\alpha\beta}$ so that we can choose the gauge (at least locally)

$$g^{\alpha\beta} = \eta^{\alpha\beta} \tag{2.4}$$

where $\eta_{\alpha\beta} = \text{diag}(-1,1)$ is the Minkowski world-sheet metric which becomes the unit matrix after a Wick rotation to imaginary τ. Although this choice of gauge is possible in the classical theory it is generally obstructed by quantum mechanical violations of the symmetries of the classical action *i.e.*, by *quantum anomalies*. Evidently the action S_0 simplifies considerably in this gauge so that, defining the complex coordinates

$$z = e^{\tau+i\sigma}, \qquad \bar{z} = e^{\tau-i\sigma}, \tag{2.5}$$

we have

$$S_0 = -\frac{T}{2} \int d^2z \, \partial_\alpha X^\mu \partial^\alpha X^\nu G_{\mu\nu}. \tag{2.6}$$

In a flat space-time background $(G_{\mu\nu}(X) = \eta_{\mu\nu})$, S_0 is simply the action for D free two-dimensional boson fields. As usual, the choice of such a gauge requires the imposition of constraints which are simply the $g^{\alpha\beta}$ equations of motion evaluated in this gauge (the two-dimensional Einstein equations),

$$T_{\alpha\beta} \equiv \frac{-2}{\sqrt{g}T} \frac{\partial S_0}{\partial g_{\alpha\beta}} = 0, \tag{2.7}$$

where $T_{\alpha\beta}$ is the two-dimensional energy-momentum tensor. These constraints generate the residual symmetries of the theory (2.6) that preserve the gauge conditions (2.4). These residual symmetries are the conformal transformations $z \to \tilde{z}(z)$ and $\bar{z} \to \tilde{\bar{z}}(\bar{z})$ which is why this gauge is called a *conformal gauge*. It is convenient to express the two independent components of $T_{\alpha\beta}$ in terms of their modes, L_n and $\bar{L}_n$. These satisfy the Virasoro algebra

$$[L_m, L_n] = (m-n)L_{m+n} + \frac{c}{12}\delta_{m+n,0}(m^3 - m), \tag{2.8}$$

where c is an important constant. The modes of $\bar{T}$ define another set of generators, $\bar{L}_n$, which satisfy a similar algebra and commute with L_n. The *central extension* with coefficient c is a crucial quantum mechanical term which corresponds to an anomaly in the classical algebra of the generators of conformal transformations. The Virasoro algebra, the algebra of the constraints, plays a central rôle in understanding string theory.

For closed strings in the simplest string theory, defined by the action S_0 (with $G_{\mu\nu} = \eta_{\mu\nu}$) with no other terms, the X^μ equation of motion (which is simply the two-dimensional wave equation, $\partial^2 X^\mu = 0$) has the general solution

$$X^\mu(z, \bar{z}) = (X_L^\mu(\bar{z}) + X_R^\mu(z))/2, \qquad (2.9)$$

where the left and right handed polarizations, X_L and X_R, are referred to as *left-movers* and *right-movers*. Each component of X^μ contributes 1 to the value of c in (2.8) so that the total value is $c = D$, the dimension of space-time. It turns out that the quantum theory is only free of troublesome anomalies if $c = 26$, which means that the simplest theory only makes sense in $D = 26$ space-time dimensions.

The simplest generalization of the bosonic theory includes free world-sheet fermionic fields in addition to the bosonic $X^\mu(\sigma, \tau)$. A single Majorana fermion contributes $c = \frac{1}{2}$ to the Virasoro anomaly. In the *spinning string* model the fermions also have a *space-time vector* index, *i.e.*, $\psi_a^\mu(\sigma, \tau)$ (where $a = 1, 2$ is the world-sheet spinor index) describes D Majorana *world-sheet spinors*. The fields ψ^μ contribute Dirac terms to the action S which make it supersymmetric on the world sheet ($S_\psi \sim \int d\sigma d\tau \bar{\psi}^\mu \rho \cdot \partial \psi^\nu \eta_{\mu\nu}$ in the conformal gauge). The total contribution to c from X^μ and ψ^μ is now $c = 3D/2$. In this case $c = 15$ is required for a consistent theory so that the theory only makes sense in $D = 10$ space-time dimensions. A truncated version of this theory, the GSO projection, not only has world-sheet supersymmetry but also turns out to be supersymmetric in the space-time in which the string is moving. This means that there are an equal number of fermion and boson states at every mass level. Theories with space-time supersymmetry are superstring theories and it is such theories which are remarkably consistent.

The string theories I have described so far are the simplest ones, defined by free two-dimensional field theories. More generally, there is a string theory corresponding to any conformal field theory with $c = 26$ or $c = 15$ (in the case of world-sheet supersymmetry) where the fields may have internal symmetry indices as well as space-time indices. Such general theories are classified by studying the representations of the Virasoro algebra without the necessity of starting from any particular action.

The critical behaviour of a two-dimensional statistical mechanical system at a continuous phase transition (*i.e.*, a second-order or higher-order critical point) is also determined by a conformal field theory. Therefore the classification of representations of the Virasoro algebra for all values of c amounts to a classification of possible critical behaviour in two-dimensional statistical systems. In the description of critical systems there is no integration over the geometry of the two-dimensional surface so that there is no need to impose the gauge constraints, (2.7). Consequently, there is no constraint on the value of c.

Any of these generalized string (superstring) theories with $c = 26$ ($c = 15$ for theories with supersymmetry) are special because in these cases the physical states are transversely polarized in the same way that the polarization states of a photon are transverse in quantum electrodynamics. These are the string theories with massless gauge particles which are of interest as possible unifying theories of all the forces. It is probable that consistent "subcritical" string theories can also be defined with smaller values of c, following the work of Polyakov and many others. It may be that such theories describe hadrons as well as being related to statistical systems.

Curved space ($G_{\mu\nu} \neq \eta_{\mu\nu}$) - the connection with general relativity

So far we have seen that certain (unreasonably large) dimensions of space-time are singled out for the simplest theories describing quantum mechanical strings propagating in a flat space-time background. More general situations involve curved backgrounds. This, for example, is one way of obtaining theories in lower dimensions starting from one of the above flat-space theories. The extra spatial dimensions may be curled up so that the effective dimensionality of the theory may then be more realistic. When $G_{\mu\nu} \neq \eta_{\mu\nu}$ the action (2.6) is that of an interacting two-dimensional theory - a *nonlinear sigma model*. Although the action is conformally invariant as a classical two-dimensional theory (*i.e.*, $T^\alpha_\alpha = 0$) there will generally be quantum anomalies in the conformal symmetry. Although the theory cannot be solved exactly with a general $G_{\mu\nu}$ it can be studied in perturbation theory in powers of the inverse string tension, $\alpha' = 1/\pi T$. The condition for the absence of these anomalies is more or less equivalent to the statement that the renormalization group β functions of the two-dimensional field theory vanish. In a nonlinear sigma model, such as that defined by the action $S_0 + S_1 + S_2$ in the gauge $g_{\alpha\beta} = \eta_{\alpha\beta}$, there is a beta *functional* that describes the renormalization of each of the background fields, $G_{\mu\nu}$, $B_{\mu\nu}$, Φ, ... The conditions for conformal invariance are therefore firstly that $c = 26$ (or $c = 15$ for superstring theories) and

$$\beta^G[G_{\mu\nu}, B_{\mu\nu}, \Phi, \ldots] = 0, \quad \beta^B[G_{\mu\nu}, B_{\mu\nu}, \Phi, \ldots] = 0, \quad \beta^\Phi[G_{\mu\nu}, B_{\mu\nu}, \Phi, \ldots] = 0, \quad (2.10)$$

with further equations corresponding to other background fields (such as the fermionic ones in superstring theories). The perturbative evaluation of the β functions therefore leads to a set of consistency conditions on the background fields. For example, ignoring all fields other than $G_{\mu\nu}$ and Φ the $\beta^G = 0$ equation gives

$$R_{\mu\nu} - \nabla_\mu\nabla_\nu\Phi + \frac{\alpha'}{2}R_{\mu\rho\omega\gamma}R^{\rho\omega\gamma}{}_\nu + \ldots = 0. \quad (2.11)$$

The ... indicate an infinite number of terms of higher order in α', including fermionic terms that are characteristic of a supersymmetric theory. In this equation $R_{\mu\rho\omega\nu}$ is the Riemann curvature of the embedding space formed out of the metric $G_{\mu\nu}$ in the usual way and $R_{\mu\nu}$ is the Ricci tensor. Equation (2.11) is a generalization of Einstein's equation for the space-time in which the string is moving! At energies much smaller than $1/\sqrt{\alpha'}$ only the lowest-order term is relevant and Einstein's equation is recovered. The higher order terms alter Einstein's equation at very high energies (short distances). This is one of the clearest ways of seeing that string theory is a theory that contains general relativity.

Remarkably, we see that the quantum mechanical description of a string moving through a curved space-time *requires* that space-time to satisfy a generalization of Einstein's equation.

The connection between string theory and general relativity goes beyond this. It is possible to identify the massless spin-2 particle excitation of the string with fluctuations of the background metric, $G_{\mu\nu}$. This connection shows that string theory is not simply a theory of extended particles moving through an inert space-time, but determines geometrical aspects of space-time.

Although (2.11) is given as a perturbation expansion in powers of α' there are classes of curved spaces which are known to be solutions of this equation (together with the other $\beta = 0$ equations) to all orders in α'. Consider, for example, a superstring theory which is initially defined in ten space-time dimensions and assume that four dimensions are flat and six spatial dimensions are curved. In this case the solutions to the $\beta = 0$ equations are special six-dimensional spaces known as *Calabi–Yau* spaces[*]. The resulting four-dimensional theory has space-time supersymmetry as well as other features which suggest that it may describe a kind of Grand Unified Theory.

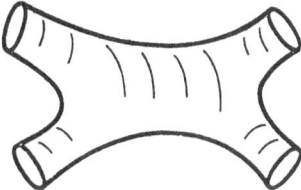 

Figure 2.2. (a) A world sheet for the scattering of two incoming strings to give two outgoing strings. (b) The conformal symmetry of string theory means that the parameter-space of (a) can be mapped to a standard sphere, with the external particles mapped onto the dots.

String tree amplitudes

The path history method can be used to evaluate string scattering amplitudes. For example, the four-string on-shell tree amplitude is associated with a sum over world-sheets of the form shown in fig. 2.2(a) which link the incoming and outgoing particles. The parameter-space of this world sheet has the topology of a sphere and the expression for the amplitude turns out to be given by an integral over the positions of the external particles attached to the sphere in fig. 2.2(b). In superstring theory the world sheet becomes a *super*-world sheet with ordinary coordinates z, $\bar{z}$ and anticommuting coordinate θ. Superstring amplitudes are then given by an integral over the positions

---

[*] Technically, these are Kähler spaces with vanishing first Chern class.

of the M external states over the super world sheet,

$$A(k_1,\ldots,k_M) = g^{M-2} \int \prod d^2 z_r d\theta_r F(\{k_r, z_r, \theta_r\}). \qquad (2.12)$$

When the on-shell external scattering particles are massless this expression reduces at low energy to the sum of the Feynman diagrams for the scattering of these states in a standard field theory of (super)-Yang–Mills coupled to (super)gravity. It is one of the attractive features of string theory that all the Feynman diagrams of any given order are packaged together into one expression.

Loop amplitudes - modular invariance

The study of one-loop scattering amplitudes raises some important new issues. These are the lowest order radiative corrections to the tree amplitudes.

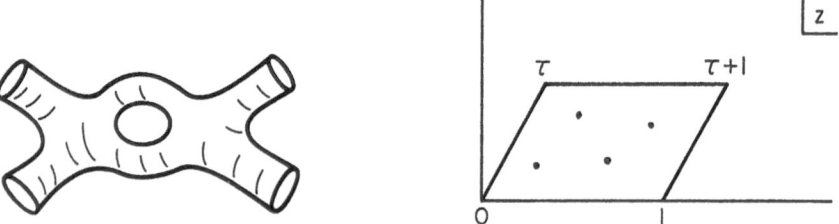

Figure 2.3. (a) A toroidal world sheet representing the one-loop correction to the four-string scattering amplitude. (b) Representation of the torus as a parallelogram with opposite edges identified. The complex *modulus* τ labels tori which are not equivalent under reparametrizations and conformal transformations. The external particles are mapped onto the dots.

In this case the world-sheet has the topology of a torus (fig. 2.3(a)). Unlike in the case of the sphere it is not true that all tori are equivalent up to reparametrizations and Weyl transformations. In other words the integration over all intrinsic metrics $g_{\alpha\beta}(z,\bar{z})$ cannot simply be performed by choosing a particular gauge. There is a one (complex) parameter class of gauge-inequivalent tori which are labelled by the modulus, τ (fig. 2.3(b)). The expression for the amplitude is therefore given by an integral over τ as well as over the positions of the external particles over the parallelogram in fig. 2.3. In the case of a torus there are also *large* reparametrizations which are not continuously deformable to the infinitesimal ones considered earlier. For example, if the torus in fig. 2.3 (a) is cut along a circle and the two edges reglued after a relative rotation of 2π, a twist is introduced into the parametrization which cannot be undone continuously. All transformations of this type (known as *modular* transformations) are expressed in terms of τ as combinations of the two transformations

$$\tau \to \tau + 1, \qquad \tau \to -\frac{1}{\tau}. \qquad (2.13)$$

Reparametrization invariance of string theory requires invariance under modular transformations as well as infinitesimal reparametrizations imposes important extra restrictions on possible theories. These comments generalize to multi-loop amplitudes which are associated with more moduli analogous to τ.

The simplest one-loop amplitude is the lowest-order contribution to the cosmological constant, Λ, which is a measure of the curvature of the universe in the absence of matter. It is determined by the vacuum energy density which is given by loop diagrams with no external particles. Experimentally the cosmological constant is known to be zero or extremely close to zero, $\Lambda < 10^{-120} M_{planck}^4$, when expressed in the natural units. In conventional quantum field theory the characteristic value of Λ generated by quantum fluctuations of the vacuum is $\Lambda \sim 1 \times M_{planck}^4$, one of the worst mistakes in the history of science!

In the simplest string theory, the 26-dimensional bosonic string theory, Λ is given, at one loop, by

$$\Lambda = \int_F \frac{d^2\tau}{(\text{Im}\tau)^2} (\text{Im}\tau)^{-12} e^{4\pi \text{Im}\tau} \left[\prod_{n=1}^{\infty} (1 - q^{2n})(1 - \bar{q}^{2n}) \right]^{-24}, \qquad (2.14)$$

where $q = e^{i\pi\tau}$ and F indicates that the τ integration region is taken to contain only one copy of τ's related by (2.13). The measure $d^2\tau/(\text{Im}\tau)^2$ is modular invariant as is the rest of the integrand – but this depends on the precise factors such as the powers of -24 which arise only in $D = 26$ dimensions. In a more general theory the corresponding expression is of the form

$$\Lambda = \int \frac{d^2\tau}{(\text{Im}\tau)^2} \sum \mathcal{N}^{ij} \chi^i(\tau) \chi^j(\bar{\tau}), \qquad (2.15)$$

where χ^i is the character of the highest-weight representation of the Virasoro algebra associated with conformal weight h^i.

Several features of string loop amplitudes are illustrated by (2.14) and (2.15):

It might have seemed plausible that Λ in (2.14) should equal the sum of the cosmological constants of the infinite number of pointlike fields propagated by the string. In fact, the string theory result is effectively the field theory result divided by an infinite factor. The τ integral in (2.14) over the region F has *no ultraviolet divergences*. This comment generalizes to loop amplitudes with external particles.

In superstring theories the integrand of Λ vanishes point by point in τ space. We do not, however, expect that the experimental vanishing of the cosmological constant is explained by this mechanism since this would require supersymmetry to be unbroken in the real world.

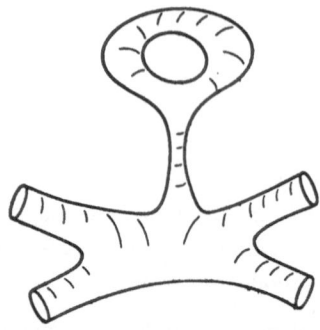

Figure 2.4. Divergences may arise in loop amplitudes due to the sum over worldsheets in which a toroidal tadpole disappears into the vacuum with a very thin neck. The propagation of a massless scalar (dilaton) particle in the neck causes the divergence. The coefficient of the divergence is proportional to the dilaton vacuum expectation value (which is also the cosmological constant).

Moore has pointed out that even in non-supersymmetric string theories there may be a symmetry (known as Atkin–Lehner symmetry) of the integrand of Λ which is nonlocal in τ such that the integrated cosmological constant vanishes.

Even though there are no ultraviolet divergences, string loop amplitudes may still diverge due to the emission of a closed-string massless scalar state (the 'dilaton') into the vacuum with zero momentum. This infinity arises from the propagtor $1/k^2$ (with $k^\mu = 0$) in the neck of the tadpole in fig. 2.4. Divergences of this type (which are proportional to the dilaton vacuum expectation value) suggest that the theory has been defined around an incorrect vacuum state. Fischler and Susskind and others have shown how these one-loop infinities may be cancelled by modifying the string vacuum state to include a condensate of dilatons. In superstring theories space-time supersymmetry ensures that the dilaton expectation value vanishes so that the one-loop amplitudes are finite.

An important feature of (2.15) is that it is given by a sum of terms which are products of holomorphic and antiholomorphic functions of the modulus (*i.e.*, functions of τ and functions of $\bar{\tau}$). This 'holomorphic factorization' is a very general feature that originates from the fact that the left-polarized and right-polarized modes contribute independent factors to the amplitudes (apart from subtleties with the zero modes). This generalizes to multiloop amplitudes and has deep geometrical significance. The question of modular invariance is also of great significance in critical behaviour of two-dimensional statistical mechanical systems. In an important recent development E. Verlinde has discovered that the conditions for modular invariance are related to local features of the conformal algebra.

Multiloop amplitudes

The multiloop amplitudes for closed string theories are defined by integrating over world sheets of higher genus. The fact that there is precisely one string diagram at each order in perturbation theory is an incredible simplification compared to standard pointlike quantum field theories, in which there is a dramatic escalation of diagrams as the order increases.

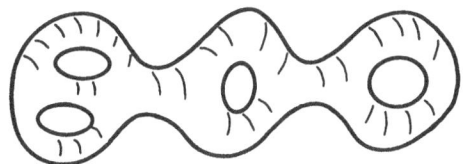

Figure 2.5. A l-loop amplitude is described by a world sheet of genus l.

The multiloop diagrams are orientable Riemann surfaces of nontrivial topology. Any surface is characterized by its genus which is equal to the number of handles. The integration over all possible intrinsic metrics, $g_{\alpha\beta}(\sigma,\tau)$, now reduces to a finite-dimensional integral over $3l-3$ complex moduli (analogous to the one-loop modulus, τ) when $l > 1$ (three extra moduli for each handle added). The l-loop cosmological constant for the bosonic theory in a flat space-time background has a form that generalizes (2.14). Multiloop amplitudes have also been constructed.

There has been much effort invested in generalizing this analysis to the case of superstring amplitudes. In this case there are $(2l-2)$ complex *super*moduli which are anticommuting (Grassmann) coordinates - the supersymmetric analogues of the moduli. The cosmological constant is now expressed as an integral over a density in supermoduli space. There have been some problems in analysing the integration over the supermoduli.

Finiteness of superstring scattering amplitudes to all orders

The coefficient of the divergence of a string loop amplitude is proportional to the cosmological constant so it is necessary, and possibly sufficient, to have $\Lambda = 0$ for the amplitude to be finite. Atick, Moore and Sen have given a somewhat indirect proof of the vanishing of the cosmological constant to all orders in the case of heterotic theories.

Use of the light-cone parametrization (in which $\tau = X^0 + X^{D-1}$) leads to a very useful parametrization for the supermoduli. Mandelstam has used this light-cone gauge to construct the amplitudes for the ten-dimensional type 2 theories and he claims that they are finite to all orders in perturbation theory. Mandelstam's expression for the amplitudes is similar to the expression derived from another light-cone analysis based on the space-time supersymmetric formalism by Restuccio and Taylor (but the presence of contact interactions in that formalism makes the analysis rather complicated). The analysis of Restuccio and Taylor as well as that of Kallosh and Morozov also point to the finiteness of the heterotic theories.

It therefore seems that the hope that superstring theory leads to a well-defined perturbative quantum theory of gravity is in good shape.

3. MODEL BUILDING

We have seen that consistent string theories can be built from representations of the Virasoro algebra with $c = 26$ or $c = 15$. In closed string theories the left polarized modes are to some extent independent of the right polarized modes. Denoting the left-moving and right-moving Fock spaces by $|L, p\rangle$ and $|R, p\rangle$, respectively, a general closed-string state has the form

$$|L\rangle \otimes |R\rangle. \tag{3.1}$$

The only link between the left-movers and right-movers is that they have the same space-time momentum, p^μ. Otherwise $|L\rangle$ and $|R\rangle$ are independent states that generally belong to different representations of the Virasoro algebra which might have different values of c. This means that (neglecting open string theories[*]) there are two classes of closed supersymmetric string theories.

The first class is that of **type 2** theories in which both left and right-movers have supersymmetry and hence $c = 15$. There are two possible models of this type one of which is non-chiral (type 2a) while the more interesting one is chiral (type 2b).

The second class of closed superstring theories has supersymmetry, and hence $c = 15$, for the right polarized modes and $c = 26$ for the left polarized modes which therefore correspond to the bosonic theory. This is the class known as **heterotic** theories. Since the left-moving and right-moving modes of the heterotic string are different, theories of this type are chiral with respect to the world sheet.

The value of c may be composed from smaller values. In particular, if we wish to consider string theories in D flat space-time dimensions (with $D = 4$ being of particular interest) the total c is given by

$$26 = c_i + D, \qquad \text{or} \qquad 15 = c_i + \frac{3}{2}D, \tag{3.2}$$

where the residual c_i represents the part of c which must be provided by 'internal' symmetry.

The original superstring theories were defined in $D = 10$ space-time dimensions. In this case the type 2 theories have no internal symmetry since $c_i = 0$ for both left-movers and right-movers. However, the ten-dimensional heterotic theories have left-movers with $c_i = 16$ ($26 = 16+10$). Modular invariance of the loop amplitudes constrains the internal gauge groups to those with even, rank 16, self-dual root lattices. This means that there are only two possible ten-dimensional heterotic theories - those with the gauge groups $E_8 \times E_8$ and Spin $32/Z_2$.

In $D = 4$ space-time dimensions c_i must obviously take the values

$$c_i = 22, \qquad \text{or} \qquad c_i = 9. \tag{3.3}$$

These values may be obtained from tensor products of representations of Virasoro algebras with smaller values of c such that the value of the sum of the c's add up to 22 (for

---

[*] There is one consistent open string, or type 1, model (associated with the gauge group $SO(32)$).

bosonic strings) or 9 (for superstrings). The particular combinations of representations that are permitted are highly constrained by requiring modular invariance of the loop amplitudes.

The earliest four-dimensional string models were those constructed (by Candelas, Horowitz, Strominger and Witten) by curling up six dimensions in the ten-dimensional heterotic model. There are powerful restrictions on these six compactified dimensions which require the resulting four-dimensional theory to be supersymmetric in space-time. The known spaces of this type are the Calabi–Yau spaces referred to earlier, as well as spaces which are flat everywhere except at points. The latter spaces are known as orbifolds. There are many attractive features of these models as far as their possible relevance to observed physics.

A more general study of all possible string theories makes use of the classification of all possible modular invariant unitary conformal field theories with arbitrary c. This also amounts to a classification of all possible (unitary) critical two-dimensional statistical mechanical models. This is a subject of intense research and there has been very much recent progress. since the original work by Belavin, Polyakov and Zamolodchikov (which was based on Polyakov's conformal bootstrap of the early 1970's) and Friedan, Qui and Shenker. At present the classification is complete for theories with $c < 1$ but not for $c \geq 1$ which include the values of c needed for consistent string theories. Even with the incomplete classification there are very many known consistent string models.

It is somewhat extraordinary that the analysis of two-dimensional statistical mechanical systems at a phase transition leads to the construction of generic string theories, including some of those that might describe realistic four-dimensional physics. Explicit examples have been given by Gepner and others of how certain curled-up six-dimensional Calabi–Yau spaces can be described in terms of two-dimensional statistical models.

The construction of particular background space-time geometries by assembling different two-dimensional statistical mechanical models again illustrates the interplay between properties of the world sheet and the space-time in which it is embedded. It seems that many of the properties that we need for an effective space-time are also incorporated in the world-sheet. These include: coordinate invariance, chirality and supersymmetry. This interplay between the geometry of the world sheet and space-time ought to be a clue for understanding a more fundamental formulation of string theory.

Comments

There are very many string theories in four dimensions. which are defined by a series of string diagrams analogous to ordinary Feynman diagrams. This classification is based on considering properties of the free theory along with the modular constraints deduced from the one-loop amplitudes. In addition, finiteness of one-loop amplitudes (or vanishing of the one-loop cosmological constant) singles out the models with space-time supersymmetry (*i.e.*, superstrings) as the only possibly consistent ones. There are good arguments which suggest that these conditions are sufficient to give consistent amplitudes to all orders in string perturbation theory although certain problems can arise.

Many methods are used to construct such string theories. Among these are: Calabi–Yau spaces; orbifolds; world-sheet fermions; bosonized fermions; current algebra; asymmetric orbifolds; and many others.

For each model there is a determined spectrum of massless particles and couplings between them. There is therefore some hope of selecting a physically interesting string theory on the basis of experimental input, such as the number of generations of quarks and leptons, the absence of very rapid proton decay and other experimental input. For the moment models of the heterotic type ($c = 15$ for the right-movers and $c = 26$ for the left-movers) are the only ones which are able to accomodate the standard model with the observed particle representation content. Type 2 models ($c = 15$ for both left-movers and right-movers) come tantalizingly close but, in our present way of thinking about phenomenology (which might be wildly wrong), they are unable to accomodate the complete standard model.

Some of the consistent heterotic models have a spectrum of massless states and interactions that corresponds quite closely to that of the standard model $SU(3) \times SU(2) \times U(1)$ (together with other low-energy symmetries in many examples). The way in which such semi-realistic models work appears to be very nontrivial since it requires the presence of very particular anomaly-free complex representations as well as dynamical symmetry breaking of the appropriate type.

Unfortunately, there are several arguments that sensible detailed phenomenology can never emerge from string perturbation theory. For example, supersymmetry is unbroken to all orders - this is crucial for the string diagrams to be finite. However, if supersymmetry is of any relevance in the physical world it must very obviously be a broken symmetry. As long as supersymmetry is unbroken the lowest-mass string states (which are supposed to describe the observed elementary particles) have precisely zero mass. In order to understand the observed non-zero masses, which are tiny when expressed in Planck units, it is important to understand the non-perturbative mechanism that breaks supersymmetry. Furthermore, since any familiar symmetry-breaking mechanism generates an unacceptably large cosmological constant it is important to understand why the observed cosmological constant is so small. Similarly, the implications of string theory for the early history of the universe and other cosmological phenomena must await a deeper understanding of nonperturbative effects.

The apparently vast array of string "theories" should be viewed as different semi-classical approximations to a more fundamental theory. This means that they are different solutions of the same set of equations, such as the equations for (super)conformal invariance, (2.10). On purely aesthetic grounds it is unsatisfactory to be forced to choose one particular conformal field theory out of this huge number of possibilities as the semi-classical ground state of the string perturbation expansion. If such an approximation is of relevance to real physics it is essential to understand the theoretical principle that selects one particular ground state. Furthermore, the choice of a particular fixed background space-time appears to be at variance with the logic of the uncertainty principle at the Planck distance or smaller scales. At larger length scales strings behave like point particles, so there appears to be no scale at which strings should be the relevant degrees of freedom. This suggests that the perturbative expansion of string theory should diverge.

Bosonic string perturbation theory has indeed been shown to be divergent by Gross and Periwal and is at best an asymptotic series. The same is probably true for superstring theories. This means that nonperturbative effects are certainly of importance.

4. BEYOND PERTURBATION THEORY

As yet there is no single compelling approach to a formulation of string theory that goes beyond the perturbation expansion. The only clues in discovering such an underlying formulation are that it should reproduce the perturbation expansion in some approximation (maybe a poor approximation) and it should presumably be based on an elegant principle. There are a number of different approaches.

String field theory

One rather obvious strategy, by analogy with second-quantization in relativistic point-particle physics, is to introduce string fields which are functionals of the string coordinates, $\Phi[X^\mu(\sigma)]$. In string quantum field theory such fields create and destroy complete strings. In the case of superstring theories the fields also depend on the fermionic world-sheet coordinates. Such fields can be expanded as a sum over a complete basis with coefficients which are fields depending only on the zero modes, x^μ. These coefficients are the ordinary fields associated with each of the infinite number of states described by the string.

Several different formulations of open string field theory have been proposed. The most geometrical of these is the open-string theory of Witten based on an action that generalizes the "Chern–Simons" form of three-dimensional electromagnetism - this is a topological invariant made out of the vector potential. It has been shown (by D'Hoker, Giddings, Martinec and Witten) that the perturbation expansion of this string field theory reproduces all the correct string Feynman diagrams. This string field theory is an explicit generalization of the Yang–Mills theory action in which conventional gauge invariance is seen to be just one out of an infinite number of stringy gauge invariances.

One of the unappealing features of this approach is that it it selects a particular parametrization for the world sheet which obscures the duality features of the complete string diagrams. Furthermore, the field theory depends on a fixed background space-time metric which appears explicitly in Witten's action. It is therefore only appropriate for describing an expansion around a particular classical background. There are also particular problems associated with closed-string field theory as well as with the generalization to superstrings.

World-sheet dynamics

A different kind of approach towards finding a fundamental formulation of string theory is based on generalization of the geometrical properties of conformal field theories on (super)Riemann surfaces. Recall that any l-loop string amplitude is determined as an integral over the moduli (and supermoduli) of the surface together with the points corresponding to the positions of the external on-shell states. The full perturbation series is given as a sum of integrals over the moduli spaces of the surfaces of arbitrary genus. This means that all of the properties of the string theory are encoded in the properties of (super)moduli space. The analytic properties of the l-loop partition function as a density on (super)moduli space are therefore of vital significance in determining properties of the theory, and in some sense they determine the nature of the embedding space-time. This is one of the most attractive features of string theory. This motivated Friedan and Shenker to suggest a setting for string theory based on an infinitely large space - 'Universal Moduli Space' - which in some sense contains the union of the moduli space of all Riemann surfaces of arbitrary genus, including disconnected surfaces. The hope is that, at least in its supersymmetric version, the geometry of universal moduli space is so constrained that there may be few consistent theories (a unique theory?) defined in this way.

A rather different approach towards the search for a fundamental setting for string theory is based on the fact that the set of conformal field theories is a subspace of the space of all possible two-dimensional field theories. The conformal field theories are determined by the fixed points of the renormalization group in this large space (the points at which the β functions vanish, given by (2.10)). These points are connected by renormalization group trajectories. This suggests that this larger space may be an appropriate fundamental setting for quantum string theory, reducing to the usual conformal field theory formulation only in semiclassical approximation. This suggestion would incorporate quantum tunnelling between the different ground states (the different string "theories" described earlier). A prerequisite to studying this idea is to develop an understanding of the connectivity of conformal field theories in this space, a programme pioneered by Zamalodchikov.

In this and other approaches the Riemann surface picture of world sheets should emerge as an approximation to the more fundamental underlying theory.

Ideas about the Planck phase

Earlier I indicated that strings are probably not the appropriate coordinates for a fundamental description of string theory. This point of view is supported by the evidence for the existence of a phase transition at the "Hagedorn" temperature, T_0, which is around the Planck temperature ($T_{Planck} = M_{Planck} c^2 / k$). Such a phase transition is a very general thermodynamic feature of a system (such as string theory) with a density of string states which increases exponentially with mass. In fact, Hagedorn had described transitions of this type in the context of his hadronic bootstrap which was a precursor to the original string theory. Originally the Hagedorn temperature was thought to be an "ultimate" temperature, reached only at infinite energy density. Recently Kagan and Sathapalian have given a very simple explanation of this transition and pinpointed a relevant order parameter.

Atick and Witten have stressed the strong analogy between this transition and the deconfining phase transition in four-dimensional QCD. They argue that certain features of the high temperature phase may be deduced from the asymptotic behaviour of the series of string perturbation diagrams that describe the low temperature phase. In particular, they argue that the Hagedorn transition is strongly first order and that the free energy density in the high temperature phase increase only like T^2. This analysis is problematic in detail since the very notion of equilibrium and the definition of temperature can only be approximate in a theory of gravity due to the instability of an infinite gravitating system. Nevertheless it has pinpointed some dramatic ways in which string theory differs from any conventional field theory. Starting from an entirely different viewpoint Klebanov and Susskind have also concluded that there is evidence that string theory arises from a theory with the number of degrees of freedom appropriate to a 2-dimensional field theory (one space and one time). Gross and others have emphasized the fact that there are indications from the asymptotic behaviour of string perturbation theory diagrams that there should be a phase with infinitely larger symmetry. In this case the masses of all the string states may in some sense be generated by spontaneous breakdown of this huge symmetry, in the same way as gauge boson masses are generated in the electro-weak theory.

In a separate development Witten has interpreted the mathematical results of Donaldson concerning the topology of four-dimensional manifolds in terms of relativistic quantum field theory (generalizing non-relativistic results of Atiyah) in a phase in which general covariance is unbroken. These mathematically motivated ideas may eventually turn out to be connected to the ideas about the high temperature phase of string theory which should also be a phase in which general covariance is unbroken. It would then be a challenge to understand the mechanism that breaks this symmetry to give the observed world with particles moving in approximately Minkowski space-time.

5. SUMMARY

I have tried to emphasize the scope of the current activity in research into string theory.

The main activity has been in the context of string perturbation theory - conformal field theory. This has lead to a partial classification of possible consistent string theories in four space-time dimensions. Among the large number of apparently consistent models there are some which come remarkably close to explaining the observed phenomenology of elementary particle physics - the standard model with the correct anomaly-free assignment of representations for the quarks and leptons. This success indicates one path by which string theory might provide the framework for a unified theory of all the forces.

However, non-perturbative effects are essential in breaking space-time supersymmetry. Since all observable particles are predicted to be massless in the initial supersymmetric approximation it is essential to have an understanding of supersymmetry-breaking effects in order to determine the observed non-zero masses.

It is possible that there are instanton-like effects (analogous to those in QCD) which could lead to a controlled estimate of non-perturbative symmetry breaking. However, it is quite likely that nonperturbative effects will only be properly understood in terms of a reformulation of string theory - possibly along the lines of one of the proposals outlined earlier. Certainly, such a reformulation is essential for understanding the logic of string theory and how it gives rise to a dynamical theory of space-time unified with the elementary particles.

I have indicated that string theory is in a very vigorous phase, encompassing some of the most original research in several areas of physics and mathematics. It has suggested avenues for exploration that were not dreamed of a few years ago - avenues which unite the geometrical features of general relativity and Yang–Mills theories with quantum mechanics. Clearly, there is some way to go before a detailed understanding is developed of present experimental particle phenomena and obviously experimental verification will be essential once the theory is properly formulated. Since string theory represents such a radical departure from the traditional approach to relativistic dynamics it is to be expected that the theory will have experimental predictions of a qualitatively different nature from those of conventional point-particle models.

Chairman Michael B. Green

Scientific Secretaries: I. Giannakis and B. Mitra

DISCUSSION

– *Volkas:*

Do you know of any phenomena or particles that we have a hope of seeing experimentally in the near future and which can be incorporated in standard gauge field theory, but whose discovery would rule out string theory?

– *Green:*

No.

– *Horne:*

Do you have any idea about how one can choose between the different semi–classical solutions of string theory?

– *Green:*

Not really. The situation at present is that we apparently have a very large number of choices. Each choice appears superficially to correspond to a different string theory with a different particle spectrum and different low energy consequences. However, we know that these different "theories" should really be described as different semi–classical expansions of some more fundamental structure. There may therefore be an analogy with the situation in conventional quantum field theory. When the action has a large number of degenerate minima it is often incorrect to consider only the small fluctuations around one particular minimum. It is important to allow for quantum tunnelling between them. In order to do that you have knowledge of the theory which goes beyond the perturbative approximation. So at the very least you have to know what the configuration space of the fields is and you have to know much more. Unfortunately, in string theory we only have knowledge of the minima (maybe we will eventually know them all) but do not yet know in what larger space they lie. Each one of these minima corresponds to a particular two–dimensional conformal field theory and therefore the classification of string theories is contained in the classification of two dimensional conformal field theories. But we do not yet know how to go beyond this semi–classical aproximation. There are, however, a large number of ingenious suggestions, some of which I described in my talk.

– *Warr:*

In the string perturbation theory what is it that is finite to all orders? Surely not the scattering amplitudes for arbitrary modes of the strings?

– *Green:*

The scattering amplitudes which are finite are the on–shell S-matrix elements for the scattering of the massless states of the theory. The amplitudes for massive states have trivial on-shell singularities just as they do in ordinary field theory. The finiteness results claimed so far relate to the ten-dimensional type 2 and heterotic theories - remember that normal field theories are horribly ultra-violet divergent in ten dimensions. In four-dimensional string theories one will obviously faced by the usual infrared infinities of non-abelian gauge theories associated with the massless string states.

– *Rahal:*

Is there any progress towards a physical principle from which string theory would follow for example general relativity follows from general coordinate covariance?

– *Green:*

The problem with string theory is that it is not based on any logical physical principle analogous to Einstein's principle of relativity. We have discovered certain features of what is clearly a very interesting structure which has a strong possibility of being of relevance to physics. The present theme of much research is the search for this principle. We know that there must be some very interesting principle lurking there because we know the theory contains general relativity and Yang–Mills as a small piece of it.

– *Cappiello:*

Using conformal field theories as building blocks one can construct different string models. The only requirement in that case is the cancellation of the total central charge of the Virasoro algebra. If this is not the case the theory is not a string theory because of the appearance of the Liouville mode. My question concerns this theory: what is the status of Liouville theory? Do you think that models containing Liouville modes could be useful in string thery?

– *Green:*

The existence of an anomaly in the trace of the two-dimensional energy-momentum tensor for subcritical theories (*i.e.*, theories with $c < 26$ in the bosonic case or $c < 10$ in the supersymmetric case) means that one of the components of the intrinsic metric is not a pure gauge but becomes an extra dynamical mode in the string quantum theory. This is the so-called Liouville mode, introduced in this context by Polyakov. In this case the string is no longer described by purely transverse excitations as it is in the critical dimension ($D = 26$ in the simplest theory). The Liouville theories are not yet completely understood, but it seems very likely that they do not contain massless gauge particles, so they are not of very obvious relevance in our present understanding of unified string theories. However, they may be important in understanding how to go beyond the present semi-classical approximation to string theory. They may also turn out to be relevant as effective string theories of hadrons in four dimensions. Furthermore Polyakov has emphasised that the three-dimensional supersymmetric version describes the critical behaviour of the three-dimensional Ising model. It seems unlikely that sense can be made of bosonic theories with $c > 26$ (or $c > 10$ in the supersymmetric case).

– Bobbink:

You mentioned that the strings are not point like. What does it mean for the charge and spin distribution not to be point like?

– Green:

Loosely speaking the charge is smeared out over the world sheet swept out by the string. From this two-dimensional point of view one has a density of charge on the world sheet. Although it sounds as though introducing such a charge distribution is arbitrary, it really is not. There are strong constraints on the way in which the charges enter into the theory. For example in the ten-dimensional heterotic theories modular invariance constrains the charges so that the symmetry groups are restricted to those associated with an even, rank 16, self-dual root lattice. There are only two groups with such root lattices, namely, $E_8 \times E_8$ and $\mathrm{Spin}32/Z_2$. In other words, the charge assignment is restricted, although I should emphasize again that in lower dimensions there is still a very large number of possibilities.

– Plychronakos:

Is there an intuitive understanding of why the number of the degrees of freedom of the string is so drastically reduced at high temperature? Could it mean that say space–time as emerges from string theories has microscopic dimension 2 and macroscopic dimension 10?

– Green:

The arguments of Atick and Witten are very interesting but nobody really understands yet how string theory manages to have so few fundamental degrees of freedom compared with any conventional field theory. They talk about "quantum" Riemann surfaces but it is not clear that they really know what that means yet.

– Mincer:

Is there any reason not to generalize from strings to higher dimensional objects. In particular, for the question about a physical principle, why not go to three dimensional objects?

– Green:

I cannot give an intuitive reason why it is unreasonable to consider higher dimensional objects. However, there are some very obviously special mathematical features of strings and their associated two-dimensional Riemann surface world sheets which make these theories tractable. There appear to be terrible obstacles in the case of membranes or higher-dimensional objects. But that is the subject of Mike Duff's lectures.

CLASSICAL AND QUANTUM SUPERMEMBRANES

Michael Duff

Center for Theoretical Physics
Physics Department
Texas A&M University
College Station, TX 77843

"Our universe may be only an atom in the leg of some
superdog, barking in a superworld." Anatole France

1. CLASSICAL SUPERMEMBRANES

(1.1) Introduction

I am very pleased to be invited to join the other superdogs barking
in Superworld III. In Superworld II I discussed the exciting new discovery of supermembranes [1] and in these lectures I would like to review
what has been achieved over the last 12 months. The subject naturally
divides into classical supermembranes, which we now understand quite
well, and quantum supermembranes, which we now understand much better
than a year ago, but which still present many unanswered questions.

(1.2) d-dimensional objects in D-dimensional spacetime

Consider some extended object with 1 time and (d-1) space dimensions
moving in a spacetime with 1 time and (D-1) space dimensions. We shall
demand that its dynamics is governed by minimizing the worldvolume which
the object sweeps out

$$S = -T \int d^d\xi \left[- \det \partial_i x^\mu \partial_j x^\nu \eta_{\mu\nu} \right]^{1/2} \tag{1.1}$$

where we have introduced worldvolume coordinates ξ^i (i = 1, ..., d) and
spacetime coordinates x^μ (μ = 1, ..., D). To begin with, we assume
spacetime is flat with Minkowski metric $\eta_{\mu\nu}$ and signature (-, +,...,+).
The tension of the object is given by the constant T which renders the
action S dimensionless. This action was first introduced by Dirac [2] in
the case of a membrane (d = 3) and later by Nambu and Goto [3] in the
case of a string (d = 2). We shall frequently use the word "membrane" to
describe extended objects with any d.

The Superworld III
Edited by A. Zichichi
Plenum Press, New York, 1990

The classical equations of motion that follow from (1.1) may equivalently be obtained from the action

$$S = -\frac{T}{2} \int d^d\xi \left[\sqrt{-\gamma}\, \gamma^{ij}\, \partial_i x^\mu\, \partial_j x^\nu\, \eta_{\mu\nu} - (d-2)\sqrt{-\gamma} \right] \qquad (1.2)$$

where, following Howe and Tucker [4] and Polyakov [5], we have introduced the auxiliary field γ_{ij}. γ denotes its determinant and γ^{ij} its inverse. Varying with respect to γ_{ij} yields the equation of motion

$$\frac{1}{2} \sqrt{-\gamma}\, \gamma^{ij}\, \gamma^{k\ell}\, \partial_k x^\mu\, \partial_\ell x^\nu\, \eta_{\mu\nu} - \sqrt{-\gamma}\, \partial_k x^\mu\, \partial_\ell x^\nu\, \gamma^{ik}\, \gamma^{j\ell}$$

$$= \frac{1}{2}\,(d-2)\,\sqrt{-\gamma}\,\gamma^{ij}. \qquad (1.3)$$

Taking the trace, we find for $d \neq 2$, that

$$\gamma^{k\ell}\, \partial_k x^\mu\, \partial_\ell x^\nu\, \eta_{\mu\nu} = d \qquad (1.4)$$

and hence that γ_{ij} is just the induced metric on worldvolume

$$\gamma_{ij} = \partial_i x^\mu\, \partial_j x^\nu\, \eta_{\mu\nu} . \qquad (1.5)$$

Varying (1.2) with respect to x^μ yields

$$\partial_i \left(\sqrt{-\gamma}\, \gamma^{ij}\, \partial_j x^\nu\, \eta_{\mu\nu} \right) = 0 . \qquad (1.6)$$

Thus equations (1.5) and (1.6) are together equivalent to the equation of motion obtained by varying (1.1) with respect to x^μ.

Note that the case $d = 2$ is special. Here, the worldvolume cosmological term drops out and (1.2) displays a conformal symmetry

$$\gamma_{ij}(\xi) \rightarrow \Omega^2(\xi)\gamma_{ij}(\xi)$$

$$x^\mu(\xi) \rightarrow x^\mu(\xi) \qquad (1.7)$$

where Ω is some arbitrary function of ξ. In this case γ_{ij} and $\partial_i x^\mu \partial_j x^\nu \eta_{\mu\nu}$ are related only up to a conformal factor. The actions (1.1) and (1.2) are, however, equivalent for all d, at least classically. As discussed by Deser, Duff and Isham [6], it is possible to construct a conformally invariant action for all d, by the simple exedient of raising the usual Lagrangian to the power d/2

$$S = - T \int d^d\xi \sqrt{-\gamma} \left(\frac{1}{d} \gamma^{ij} \partial_i x^\mu \partial_j x^\nu \eta_{\mu\nu} \right)^{d/2} . \tag{1.8}$$

Now the equations of motion read

$$\frac{1}{2} \sqrt{-\gamma} \gamma^{ij} \left(\frac{1}{d} \gamma^{k\ell} \partial_k x^\mu \partial_\ell x^\nu \eta_{\mu\nu} \right)^{d/2}$$

$$= \frac{1}{2} \left(\frac{1}{d} \gamma^{k\ell} \partial_k x^\mu \partial_\ell x^\nu \eta_{\mu\nu} \right)^{d/2 - 1} \gamma^{im} \gamma^{jn} \partial_m x^\rho \partial_n x^\sigma \eta_{\rho\sigma} \tag{1.9}$$

and

$$\partial_i \left[\sqrt{-\gamma} \, (\gamma^{k\ell} \partial_i x^\mu \partial_j x^\sigma \eta_{\rho\sigma})^{d/2 - 1} \gamma^{ij} \partial_j x^\nu \eta_{\mu\nu} \right] = 0 \tag{1.10}$$

which are again equivalent to the Dirac and Howe-Tucker equations of motion.

  There are two useful generalizations of the above. The first is to go to curved space by replacing $\eta_{\mu\nu}$ by $g_{\mu\nu}(x)$; the second is to introduce an antisymmetric tensor field $B_{\mu\nu...\rho}(x)$ of rank d which couples via a Wess-Zumino term. The action (1.2) then becomes

$$S = - \frac{T}{2} \int d^d\xi \left[\sqrt{-\gamma} \gamma^{ij} \partial_i x^\mu \partial_j x^\mu g_{\mu\nu}(x) - (d-2)\sqrt{-\gamma} \right.$$

$$\left. + \frac{2}{d!} \epsilon^{i_1 i_2 .. i_d} \partial_{i_1} x^{\mu_1} \partial_{i_2} x^{\mu_2} ... \partial_{i_d} x^{\mu_d} B_{\mu_2 \mu_2 .. \mu_d}(x) \right] \tag{1.11}$$

and the equations of motion are

$$g_{\mu\rho} \left(\Box x^\rho + \Gamma^\rho_{\kappa\lambda} \partial_i x^\kappa \partial_j x^\lambda \gamma^{ij} \right) - \frac{1}{d!} F_{\mu\nu\tau...\sigma} \epsilon^{ij...k} \partial_i x^\nu \partial_j x^\tau ... \partial_k x^\sigma$$

and

$$\gamma_{ij} = \partial_i x^\mu \partial_j x^\nu g_{\mu\nu}(x) \tag{1.12}$$

where the field-strength F is given by

$$F = dB \tag{1.13}$$

and hence obeys the Bianchi identity

$$dF = 0 \ . \qquad\qquad (1.14)$$

The virtue of these generalizations is that they now permit a straight-forward transition to the supermembrane.

Our experience with string theory suggests that there are two ways of introducing supersymmetry into membrane theory. The first is to look for a "supermembrane" for which has manifest spacetime supersymmetry but no supersymmetry on the worldvolume. The second is to look for a "spinning membrane" which has manifest worldvolume supersymmetry but no supersymmetry in spacetime. An early attempt at spinning membranes by Howe and Tucker [4] encountered the problem that the worldvolume cosmological term does not permit a supersymmetrization using the usual rules of $d = 3$ tensor calculus without the introduction of an Einstein-Hilbert term [7]. Indeed, these objections have recently been elevated to the status of a "no-go theorem" for spinning membranes [8]. As discussed in Section (1.6), however, these have recently been circumvented by Lindstrom and Rocek [9] starting from the conformally invariant action (1.8). Progress in supermembranes, on the other hand, was hampered by the belief that κ symmetry, so crucial to Green-Schwarz superstrings, could not be generalized to membranes [10]. The breakthrough came when Hughes, Liu and Polchinski [11] showed that it in fact could, and constructed explicitly a $d = 4$ object displaying an explicit $D = 6$ spacetime supersymmetry and κ invariance on the worldvolume. Shortly afterwards, Bergshoeff, Sezgin and Townsend [12] constructed the $d = 3$, $D = 11$ supermembrane. Let us first consider supermembranes.

(1.3) Supermembranes

Let us introduce the coordinates Z^M of a curved superspace

$$Z^M = (x^\mu, \ \theta^\alpha) \qquad\qquad (1.13)$$

and the supervielbein $E_M{}^A(Z)$, where $M = \mu,\alpha$ are world indices and $A = a,\alpha$ are tangent space indices. We also define the pull-back

$$E_i{}^A = \partial_i Z^M E_M{}^A \ . \qquad\qquad (1.14)$$

We also need the super-d-form $B_{MN...P}(Z)$. Then the supermembrane action is (with $T = 1$)

$$S = \int d^d\xi \left[-\frac{1}{2} \sqrt{-\gamma}\ \gamma^{ij}\ E_i{}^a E_j{}^b\ \eta_{ab} + \frac{1}{2} (d-2) \sqrt{-\gamma} \right.$$

$$\left. + \frac{1}{d!}\ \epsilon^{i_1 \cdot \cdot i_d}\ E_{i_1}{}^{A_1} \ldots E_{i_2}{}^{A_d}\ B_{A_d \cdot \cdot A_1} \right] . \tag{1.15}$$

As in (1.11) there is a kinetic term, a worldvolume cosmological term, and a Wess-Zumino term. The action (1.15) has the virtue that it reduces to the Green-Schwarz superstring action when d = 2.

The target-space symmetries are superdiffeomorphisms, Lorentz invariance and d - form gauge invariance. The worldvolume symmetries are ordinary diffeomorphisms and "Siegel symmetry". This is the fermionic κ - invariance referred to earlier which is known to be crucial for superstrings, so let us examine it in more detail. The transformation rules are

$$\delta Z^M\ E_M{}^a = 0, \quad \delta Z^M\ E_M{}^\alpha = \kappa^\beta (1 + \Gamma)^\alpha{}_\beta \tag{1.16}$$

where $\kappa^\beta(\xi)$ is an anticommuting spacetime spinor but worldvolume scalar, and where

$$\Gamma^\alpha{}_\beta = \frac{(-1)^{d(d-3)/4}}{d!\sqrt{-\gamma}}\ \epsilon^{i_1 \cdot \cdot i_d}\ E_{i_1}{}^{a_1}\ E_{i_2}{}^{a_2} \ldots E_{i_d}{}^{a_d}\ \Gamma_{a_1 \cdot \cdot a_d} . \tag{1.17}$$

Here Γ_a are the Dirac matrices in spacetime and

$$\Gamma_{a_1 \cdot \cdot a_d} = \Gamma_{[a_1} \ldots \Gamma_{a_d]} . \tag{1.18}$$

This κ - symmetry has the following important consequences:

1) The symmetry is achieved only if certain constraints on the antisymmetric tensor field strength $F_{MNP \cdot \cdot Q}(Z)$ and the supertorsion are satisfied. In particular the Bianchi identity dF = 0 then requires the Γ matrix identity

$$\left(d\bar\theta\ \Gamma_a d\theta \right) \left(d\bar\theta\ \Gamma^{ab_1 \ldots b_{d-2}}\ d\theta \right) = 0 \tag{1.19}$$

for a commuting spinor $d\theta$. As shown by Achucarro, Evans, Townsend and Wiltshire [13] this is satisfied only for certain values of d and D. Specifically,

$$
\begin{aligned}
&d = 2; && D = 3, 4, 6, 10 \\
&d = 3; && D = 4, 5, 7, 11 \\
&d = 4; && D = 6, 8 \\
&d = 5; && D = 9 \\
&d = 6; && D = 10 \ .
\end{aligned}
\tag{1.20}
$$

Note that we recover as a special case the well-known result that Green-Schwarz superstrings exist <u>classically</u> only for $D = 3$, 4, 6 and 10. Note also $d_{max} = 6$ and $D_{max} = 11$. The upper limit of $D = 11$ is already known in supergravity but there it is necessary to make extra assumptions concerning the absence of consistent higher spin interactions. In supermembrane theory, it follows automatically (see, however Section (2.5)). We shall return to the upper limit of $d = 6$ in Section (1.4).

 2) The matrix Γ of (1.17) satisfies

$$
\Gamma^2 = 1
\tag{1.21}
$$

when the equations of motion are satisfied and hence the matrices $(1 \pm \Gamma)/2$ act as projection operators. The transformation rule (1.16) therefore permits us to gauge away one half on the fermion degrees of freedom. This gives rise to matching of physical boson and fermion degrees of freedom on the worldvolume. One finds

$$
D - d = \frac{nN}{4} \qquad d > 2
\tag{1.22}
$$

where N is the number of supersymmetries and n is the real dimension of the minimal spinor in D dimensions. The equation is satisfied only for $N = 1$ and then only for those values of d and D given in (1.20). The $d = 2$ case is special because of the possibility of having left and right movers on the worldsheet. Here we can have either

$$
D - 2 = \frac{nN}{4}
\tag{1.23}
$$

which is satisfied only for $N = 2$ or

$$
D - 2 = \frac{nN}{4}
\tag{1.24}
$$

which is satisfied only for $N = 1$. In all cases, these conditions are precisely the requirement of equality of bose and fermi degrees of freedom on the worldvolume. Starting from D bosonic coordinates and taking

42

into account d worldvolume reparameterisations yields D - d boson degrees of freedom. Whereas nN/4 is the number of fermionic degrees of freedom when one takes into account the fermionic gauge invariance (which removes half of the fermionic coordinates) and the fact that they obey first order field equations (which removes half again if d > 2). There are four types of solution as shown in Table I with 8 + 8, 4 + 4, 2 + 2 or 1 + 1 degrees of freedom respectively. Since the numbers 1, 2, 4 and 8 are also the dimension of the four division algebras.

Table I. Bose-Fermi matching

Bose D-d	Fermi nN/4	Algebra
8	8	O
4	4	H
2	2	C
1	1	R

these four types of solution are referred to as real, complex, quaternion and octonion respectively. The connection with the division algebras can in fact be made more precise [14,15]. The possible values of D and d given in (1.20) are displayed on the, by now familiar, "brane-scan" of Fig. 1. As described in [16] and [13], the diagonal lines relate different extended objects by the process of "simultaneous dimensional reduction". In particular, we can derive the Type IIA superstring in D = 10 starting from the supermembrane in D = 11. The horizontal lines correspond to swapping the 3-index antisymmetric tensor field strength by its dual.

In summary, another way to obtain (1.20) is to list those values of d and D which allow equal numbers of bose and fermi degrees of freedom. This may appear puzzling because we are here counting worldvolume degrees of freedom, yet the supermembrane displayed spacetime supersymmetry! The explanation is that, after going to a physical gauge for the worldvolume diffeomorphisms and fermionic gauge invariance, there is a residual fermionic symmetry which is nothing but an worldvolume supersymmetry with N = 1, 2, 4 or 8 for R, C, H and O.

3) In the case of the eleven-dimensional supermembrane, it has been shown [16] that the constraints on the background fields [12] $E_M{}^A$ and B_{MNP} are nothing but the equations of motion of eleven-dimensional supergravity. Thus the supermembrane has revived an interest in compactifications of supergravity [17].

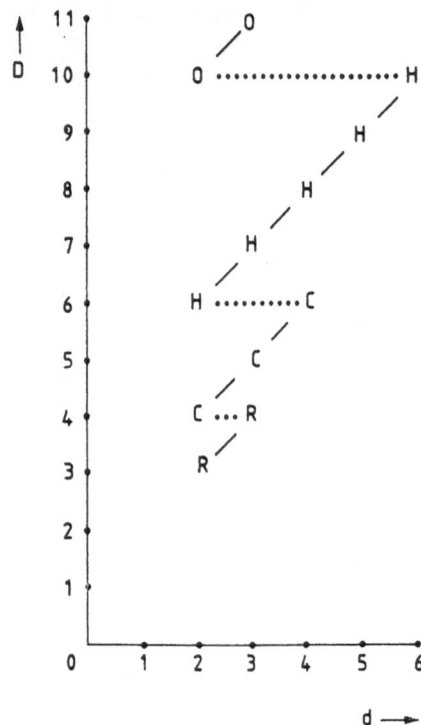

Fig. 1.   The brane-scan.

(1.4) Spontaneous compactification

Of particular interest are those of the Freund-Rubin [18] type for which the bosonic four-index field strength is given by

$$F_{\mu\nu\rho\sigma} = \frac{3a}{2} \sqrt{-g} \; \epsilon_{\mu\nu\rho\sigma}, \qquad a = \text{constant} \qquad (1.25)$$

and spacetime is given by four-dimensional anti-de Sitter space. In co-ordinates $x^{\mu} = (t, \theta, \phi, r)$ the line-element is

$$ds^2 = -(1 + a^2 r^2) dt^2 + r^2 (d\theta^2 + \sin^2\theta d\phi^2) + (1 + a^2 r^2)^{-1} dr^2. \quad (1.26)$$

The internal space is some compact seven-dimensional Einstein space, satisfying $R_{mn} = 3a^2 g_{mn}/2$. Recently, Bergshoeff, Duff, Pope and Sezgin [19] constructed new vacua for the eleven-dimensional supermembrane for which the membrane is a sphere of non-zero radius in the anti-de Sitter space. In one class of solution, the radius is a specific multiple of 'a', the de Sitter scale parameter while in a second class the radius is arbitrary. Remarkably, only two Freund-Rubin compactifications admit vacua of the latter class: the round seven-sphere with N = 8 spacetime supersymmetry [17] and the N (0,1) space which is an SO(3) bundle over CP^2 yielding N = 3 spacetime supersymmetry. There are several reasons why the possibility of the membrane having an arbitrary radius is especially interesting:

1) Vacuum solutions with a continuous parameter (in this case the membrane radius) are important for a treatment of semi-classical quantisation, since they can provide information about the structure of the theory at the non-perturbative level. In this respect they resemble the rotating membrane solutions of [20,21,22]. Indeed these solutions are rotating solutions, in the sense that the membrane is rotating around the U(1) fibers in the internal space. This explains how the membrane configuration with non-zero radius can be stable against collapse. Corresponding to each isometry of the d = 11 spacetime, there is a conserved Noether charge Q , given by

$$Q = - \int d\sigma d\rho \left[\sqrt{-\gamma} \; \gamma^{oi} K_M \, \partial_i x^M - \epsilon^{oij} \, \partial_i x^N \, \partial_j x^P (K^M B_{MNP} - \Lambda_{NP}) \right] \quad (1.27)$$

where $\mathcal{L}_K B_{MNP} = \partial_{[M} \Lambda_{NP]}$ and where K^M is the Killing vector associated with the isometry. The de-Sitter energy for a solution with radius r is given by

$$E = 4 \pi r/a^2 \qquad (1.28)$$

and the U(1) charge q is given by

$$q = - 8\pi r/a^2 . \qquad (1.29)$$

The arbitrary-radius solutions admit an unbroken membrane supersymmetry when the radius tends to infinity. As discussed in [23], the criterion for a spacetime supersymmetry to be preserved by the membrane configuration is that the spacetime Killing spinor should be annihilated by (1 - Γ), where Γ is given by equation (1.17). It is a remarkable feature of anti-de Sitter space that its Killing spinors become eigenstates of Γ asymptotically, as one moves out to spatial infinity. After making an 11 = 4 + 7 split, Γ tends to $\gamma \otimes 1$ as r tends to infinity, where $\gamma = \gamma_0 \gamma_1 \gamma_2$ and γ_μ denotes the 4-dimensionsal Dirac matrices, and the D = 11 supersymmetry parameter ε behaves like

$$\epsilon \sim (\epsilon^- r^{-1/2} + \epsilon^+ r^{1/2}) \otimes \eta \qquad (1.30)$$

where $\epsilon^\pm$ are 4-component spinors on the boundary of AdS, satisfying $\gamma \epsilon^\pm = \pm \epsilon^\pm$, and η is a Killing spinor in the internal space [17]. Thus we can obtain a supersymmetric membrane vacuum only when we are allowed to take the radius of the membrane to infinity (the Membrane at the End of the Universe).

3) Fluctuations about the Membrane at the End of the Universe are described by a free field theory. In a semi-classical analysis, one quantises the fluctuations around a classical solution. For the arbitrary-radius solutions in this paper, the fluctuations must be rescaled by appropriate powers of r in order to obtain finite and non-zero kinetic terms as r is taken to infinity. Under these same rescalings, however, the interaction terms are damped down by inverse powers of r, so that in the limit r → ∞ the fluctuation Lagrangian describes a free field theory. This may have important consequences for the renormalisability of the D = 11 supermembrane [24,25].

4) Singletons live on the boundary of anti-de Sitter space [26]. Thus one might expect [23] that the fluctuations described above should fall into the ultra-short singleton supermultiplets of the N = 8 supersymmetry discussed above in 2), since this is the only way in which a free (spin 0, spin 1/2) system can have N = 8 supersymmetry. However, explicit calculation reveals that in contrast [24,25] to the supersym-

metric singleton Lagrangian on $S^1 \times S^2$ that requires bosonic mass terms but no fermionic mass terms [26], neither the fermion nor boson rescaled fluctuations that we discussed in 3) have mass terms. This means that while the fermion fluctuations are singletons the bosons are not, and so supersymmetry is broken. The resolution of the paradox seems to be that at the same time as performing the rescalings of the fluctuations described in 3), one would also need to rescale the supersymmetry parameter ϵ in order to obtain sensible supersymmetry transformation rules. However, this rescaling then invalidates the conclusions about the vacuum supersymmetry discussed in 2), since the rescaled ϵ is no longer annihilated by $(1 - \Gamma)$ as r tends to infinity. In fact the absence of a mass term for the bosons means that the time dependence of their solutions is of the form exp $i a \omega t$ with $\omega = \sqrt{\ell(\ell+1)}$ instead of

$$\omega = \sqrt{\ell(\ell+1) + 1/4} = \ell + 1/2$$

and so given the periodicity condition on t, there are no solutions possible except when $\ell = 0$. Thus the fluctuations around the solutions described in [19] apparently are comprised of free fermion singletons but no non-trivial bosons. Upon quantisation, when the classical solutions are promoted to quantum creation and annihilation operators acting on the vacuum state of the theory, the fermionic singletons will generate massless and massive states. There will be an infinity of massless states, corresponding to the product of two singleton operators. These states, which will all be bosonic, are a subset of those discussed in [27].

Finally, as already remarked in Superworld II [1], the membrane at the end of the universe naturally favours compactifications of eleven-dimensional supergravity to four, rather than any other, dimension.

In [24] we conjectured how after compactification to an $(AdS)_4 \times S^7$ spacetime, the 8 + 8 physical degrees of freedom of the eleven-dimensional supermembrane could be described by a <u>superconformal field theory</u> with OSp(8/4) as the superconformal group. This followed from the above-mentioned interpretation of the 8 spin 0 bosons and the 8 spin 1/2 fermions as singleton [1,23] representations of OSp(8/4) which is an anti-de Sitter group as far as the four-dimensional spacetime is concerned, but which acts as a superconformal group on the boundary of $(AdS)_4$ which is then identified with the three-dimensional worldvolume of the supermembrane. The constraints on the background fields of the other extended objects in the brane-scan are weaker than the corresponding supergravity equations and so the situation here is even less clear than the eleven-dimensional supermembrane. Nevertheless, we also imagined that this superconformal interpretation could apply to all the 12 extended objects on the brane-scan of Fig. 1 by noting that the Freund-Rubin compactification mechanism for which

$$F_{\mu_1\mu_2\cdots\mu_d} \; \alpha \; \epsilon_{\mu_1\mu_2\cdots\mu_d} \tag{1.31}$$

naturally leads to an $(AdS)_{d+1} \times S^{D-d-1}$ spacetime, whose boundary $S^1 \times S^{d-1}$ could be identified with the d-dimensional world-volume of the corresponding super extended object. The same conjecture was made independently by Nicolai, Sezgin and Tanii [25] who also pointed out that the twelve corresponding superconformal groups were precisely those supergroups that admitted (spin 0, spin 1/2) singleton representations! The results are shown in Table 2.

Table 2. Superconformal Groups

d	SPACETIME	SUPERGROUP	N
2	$(AdS)_3 \times S^0, S^1, S^3, S^7$	OSP(N/2)	1,2,4,8
3	$(AdS)_4 \times S^0, S^1, S^3, S^7$	OSP(N/4)	1,2,4,8
4	$(AdS)_5 \times S^1, S^3$	SU(2,2/N)	1,2
5	$(AdS)_6 \times S^3$	F(4)	2
6	$(AdS)_7 \times S^3$	OSP(6,2/N)	2

In (1.31) F = dB is the (d + 1) - form field strength of the bosonic d- form $B_{\mu_1\mu_2\cdots\mu_d}$ which appears in the Wess-Zumino term of the d-dimensional extended object. Thus the AdS spacetime always has one more dimension than the world-volume (four in the case of the eleven-dimensional supermembrane). This is another way of understanding the upper limit d = 6 on the world-volume of a super-extended object: d + 1 = 7 is the upper limit for an anti-de Sitter supergroup. (Note also that d = 6 is also the upper limit for renormalizable (spin 0, spin 1/2) interactions. So the possible resolution of the renormalizability problem for supermembranes suggested in [24] applies to all 12 supersymmetric extended objects).

As discussed in Section (1.3), the case of strings is special because of the ability to have both left and right movers on the world-sheet. So when we write OSp(N/2) as in Table 2, we have in mind a Type

II string with the same supersymmetry for both left and right movers. More generally we could have $OSp(p/2) \times OSp(q/2)$ i.e. p supersymmetries on the right and q on the left. Group theoretically, we are exploiting the fact that the three-dimensional anti-de Sitter group is a product:

$$SO(2,2) \simeq SO(1,2) \times SO(1,2) \ .$$

The singleton interpretation corresponding to these supergroups has been discussed before in the literature by Gunaydin, Nilsson, Sierra and Townsend [28]. In which case the reader may well ask why the corresponding compactifications $(AdS)_3 \times S^7$ and $(AdS)_7 \times S^3$ have not also appeared in the string literature. In fact they have in a 1982 paper by Duff, Townsend and van Nieuwenhuizen [29] who considered Freund-Rubin compactifications of ten-dimensional supergravity. An interesting feature of these solutions was the part played by the dilaton. Its derivative acted as a conformal Killing vector on $(AdS)_3$ or S^3. (These solutions were criticised in the "Ten into four won't go" paper of Freedman, Gibbons and West [30]. As far as I can tell, however, ten into three (or seven) went and is still going!)

(1.5) The signature of spacetime

If our senses are to be trusted, we live in a world with three space and one time dimensions. However, the revival of the Kaluza-Klein idea, brought about by supergravity and superstrings, has warned us that this may be only an illusion. In any case, there is a hope, so far unfulfilled, that the four-dimensional structure that we apparently observe may actually be predicted by a "Theory of Everything". Whatever the outcome, imagining a world with an arbitrary number of space dimensions has certainly taught us a good deal about the properties of our three-space-dimensional world.

In spite of all this activity, and in spite of the popularity of Euclidean formulations of field theory, relatively little effort has been devoted to imagining a world with more than one time dimension. This is no doubt due partially to the psychological difficulties we have in treating space and time on the same footing. As H. G. Wells reminds us in The Time Machine, "There is, however, a tendency to draw an unreal distinction between the former three dimensions and the latter, because it happens that our consciousness moves intermittently in one direction along the latter from the beginning to the end of our lives." There are also more justifiable reasons associated with causality. Nevertheless, one might hope that a Theory of Everything should predict not only the dimensionality of spacetime, but also its signature.

For example, quantum consistency of the superstring requires 10 spacetime dimensions, but not necessarily the usual (9,1) signature. The signature is not completely arbitrary, however, since spacetime supersymmetry allows only (9,1), (5,5) or (1,9). Unfortunately, superstrings have as yet no answer to the question of why our universe appears to be four-dimensional, let alone why it appears to have signature (3,1).

Blencowe and Duff [31] therefore considered a world with an arbitrary number T of time dimensions and an arbitrary number S of space dimensions to see how far classical supermembranes restrict not only S + T but S and T separately. To this end they also allowed an (s,t) signature for the worldvolume of the membrane where s ≤ S and t ≤ T but are otherwise arbitrary. It is not difficult to repeat the analysis of Section (1.3) for arbitrary signatures, and to show that there is once again a matching of the bosonic and fermionic degrees of freedom as a consequence of the κ-symmetry. However severe constraints on possible supermembrane theories will now follow by demanding spacetime supersymmetry.

We restrict ourselves in this section to N = 1 flat superspace, and require θ^α to be a minimal spinor i.e. Majorana, Weyl, Majorana-Weyl etc., whenever it is possible to impose such a condition. We furthermore assume invariance under the generalized super-Poincare group super-IO(S,T), as required by the superspace construction of Section (1.3). Such supermembranes (and their compactifications) are the only ones currently known. Later we shall consider the possibility of other supermembrane theories obtained by requiring spacetime supersymmetry but with a different supergroup. For the moment, however, we shall require super-Poincare which means, in particular, that the anticommutator of two supersymmetry charges Q yields a translation

$$\{\theta, \theta\} \sim P .\qquad (1.32)$$

This is only possible for certain values of S and T when Q is a minimal spinor.

Those values of S and T permitting minimal spinors have been determined by Kugo and Townsend [32]. See also the works of van Nieuwenhuizen [33], Coquereaux [34] and Freund [35]. For the Clifford algebra given by $\Gamma^a\Gamma^b + \Gamma^b\Gamma^a = 2\eta^{ab}$ there exist matrices A and B for which

$$\Gamma_a{}^\dagger = (-1)^T A\Gamma_a A^{-1}, \quad AA^\dagger = 1 \qquad (1.33)$$

$$\Gamma_a = \eta \; B^{-1} \; \Gamma_a^* B, \quad BB^\dagger = 1 \; , \qquad\qquad (1.34)$$
$$B^* B = \epsilon$$

where ϵ and η are given in Table 3.

<u>Table 3. Values of ϵ and η</u>

S-T mod 8	ϵ	η
0,1,2	+1	+1
6,7,8	+1	-1
4,5,6	-1	+1
2,3,4	-1	-1

We can choose a basis such that

$$\Gamma_a^\dagger = \begin{cases} -\Gamma_a & a = 1,\ldots,T \\ +\Gamma_a & a = T+1,\ldots D \end{cases} \qquad\qquad (1.35)$$

and

$$A = \Gamma_1 \Gamma_2 \ldots \Gamma_T \; . \qquad\qquad (1.36)$$

The charge conjugation matrix is defined as $C = \tilde{B}A$. The properties of A and B then imply

$$\tilde\Gamma = (-1)^T \eta \; C\Gamma C^{-1} \qquad\qquad (1.37)$$

$$\tilde C = \epsilon \eta^T (-1)^{T(T+1)/2} C, \quad C^\dagger C = 1 \qquad\qquad (1.38)$$

where the tilda denotes transpose. For D even, we can also define the projection operator

$$P_\pm = \frac{1}{2} \left[1 \pm (-1)^{(S-T)/4} \; \Gamma^{D+1} \right] \qquad\qquad (1.39)$$

where

$$\Gamma^{D+1} = \Gamma^1 \Gamma^2 \ldots \Gamma^D \; . \qquad\qquad (1.40)$$

Using the above properties we find that we can have the minimal spinors given in Table 4.

The next task is to check which of these possibilities admits the super-Poincare algebra. The part of the superalgebra which is the same in each case is

$$[M_{ab}, M_{cd}] = -i \, (\eta_{bc} \, M_{ad} - \eta_{ac} \, M_{bd} - \eta_{bd} \, M_{ac} + \eta_{ad} \, M_{bc})$$

$$[M_{ab}, P_c] = i \, (\eta_{ac} \, P_b - \eta_{bc} \, P_a)$$

$$[P_a, P_b] = 0$$

$$[M_{ab}, Q_\alpha] = - \frac{i}{2} \, (\Gamma_{ab} Q)_\alpha \, . \tag{1.41}$$

We now examine the (Q,Q) anticommutator. Consider first $S - T = 0, 1, 2$ mod 8 for which $Q\alpha$ is Majorana. The only possible form for the anticommutator is

$$\{Q_\alpha, Q_\beta\} = (\Gamma^a C^{-1})_{\alpha\beta} \, P_a \, . \tag{1.42}$$

Since the left hand side is symmetric under interchange of α and β we require

$$(\Gamma_a C^{-1}) = \Gamma_a C^{-1} \tag{1.43}$$

but

$$(\Gamma_a C^{-1}) = \tilde{C}^{-1} \tilde{\Gamma}_a$$

$$= \epsilon \, \eta^T (-1)^{T(T+1)/2} C^{-1} \, \tilde{\Gamma}_a$$

$$= \epsilon \, \eta^{T+1} (-1)^{T(T-1)/2} \Gamma_a C^{-1} \, . \tag{1.44}$$

Now from Table 1, $\epsilon = \eta + 1$ for $S - T = 0,1,2$ mod 8 and hence

$$(\Gamma_a C^{-1}) = (-1)^{T(T-1)/2} \Gamma_a C^{-1} \tag{1.45}$$

which is compatible with (1.43) only if $T = 0,1$ mod 4. Now consider the subcase $S - T = 0$ mod 8, $T = 0, 1$ mod 4. Define

$$Q_{\pm\alpha} = (P_\pm Q)_\alpha \, . \tag{1.46}$$

From (1.42) we have

$$\{Q_{\pm\alpha}, Q_{\pm\beta}\} = P_{\pm\alpha\gamma} \, \{Q_\gamma, Q_\beta\} \, P_{\pm\delta\beta}$$

$$= (P_\pm \Gamma_a \, C^{-1} \tilde{P}_\pm)_{\alpha\beta} P_a \tag{1.47}$$

but

52

$$C^{-1} \tilde{P}_{\pm} = C^{-1}(1 \pm \tilde{\Gamma}_D \tilde{\Gamma}_{D-1} \cdots \tilde{\Gamma}_1)$$

$$= (1 \pm \Gamma_D \Gamma_{D-1} \cdots \Gamma_1)$$

$$= (1 \pm (-1)^{D(D-1)/2} \Gamma^{D+1}) C^{-1}$$

and therefore

$$P_{\pm} \Gamma_a C^{-1} \tilde{P}_{\pm} = P_{\pm} \left[1 \mp (-1)^{T(2T-1)} \Gamma^{D+1} \right] \Gamma_a C^{-1}$$

$$= \begin{cases} 0 & T = 0 \bmod 4 \\ P_{\pm} \Gamma_a C^{-1} & T = 1 \bmod 4 \ . \end{cases} \qquad (1.48)$$

Thus splitting up $\{Q,Q\} = \Gamma^a C^{-1} P_a$ into its chiral parts, we get

$$\{Q_{\pm}, Q_{\pm}\} = P_{\pm} \Gamma^a C^{-1} P_a$$

$$\{Q_{\mp}, Q_{\pm}\} = 0 \qquad (1.49)$$

for $T = 1 \bmod 4$ and

$$\{Q_{\pm}, Q_{\pm}\} = 0$$

$$\{Q_{\pm}, Q_{\mp}\} = P_{\pm} \Gamma^a C^{-1} P_a \qquad (1.50)$$

for $T = 0 \bmod 4$. Thus for $S - T = 0 \bmod 8$, only for $T = 1 \bmod 4$ can we set $Q_- = 0$ say, and obtain $\{Q,Q\} \sim P$ with Q a Majorana-Weyl minimal spinor.

We can proceed in this way to exhaust all the possible values of S and T admitting super-Poincare symmetry. These are summarized in Table 4, where $\bar{Q} = Q^{\dagger} A$.

Combining these results with those of Section 2 allows to draw the S/T plot of Fig. 2 whose points correspond to possible supermembrane theories. Once again we have used the symbols O, H, C and R to denote objects with 8+8, 4+4, 2+2, or 1+1 degrees of freedom, respectively. For pictorial reasons, we call this the "brane-molecule".

Several comments are now in order:

1) In the absence of any physical boundary conditions which treat time differently from space, and which we have not yet imposed, the mathematics will be symmetric under interchange of S and T. This can easily be seen from Fig. 2. For every supermembrane with (S,T) signature, there is another with (T,S). Note the self-conjugate theories that lie on the S = T line which passes through the (5,5) superstring.

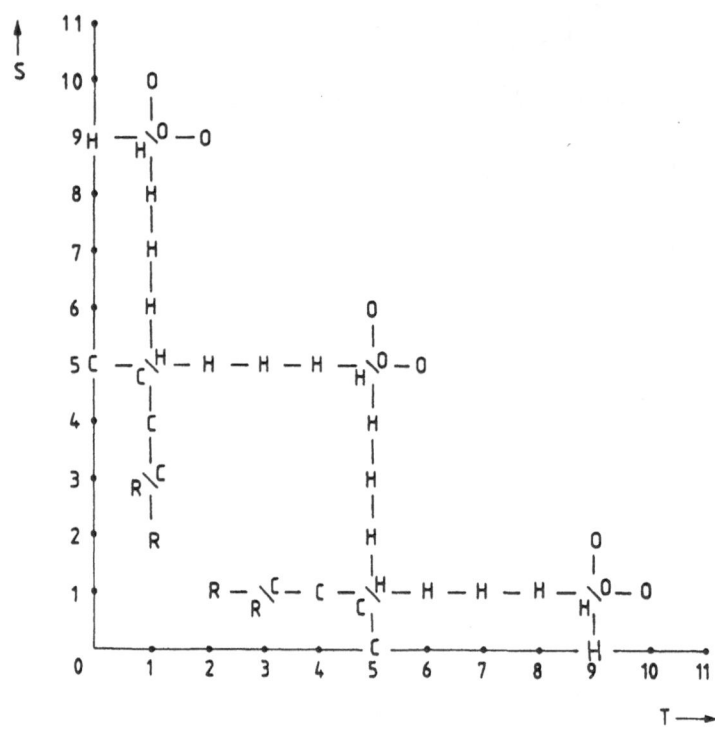

Fig. 2. The brane-molecule, assuming super-Poincaré invariance.

Table 4. Minimal spinors for different values of S - T mod 8 and
         super-Poincare algebras for different values of T mod 4.

S - T mod 8	Minimal spinor type	T mod 4	anticommutator
1,2	Majorana	0,1	$\{Q,Q\} = \Gamma^a C^{-1} P_a$
6,7	pseudo-Majorana	1,2	$\{Q,Q\} = i\Gamma^a C^{-1} P_a$
0	Majorana-Weyl $\}$	1	$\{Q_+,Q_+\} = P_+\Gamma^a C^{-1} P_a$
	pseudo-Maj-Weyl $\}$		$\{Q_+,Q_+\} = i P_+\Gamma^a C^{-1} P_a$
3,5	Dirac	0,1	$\{Q,\bar{Q}\} = \Gamma^a P_a$
		2,3	$\{Q,\bar{Q}\} = i\Gamma^a P_a$
			$\{Q,Q\} = \{\bar{Q},\bar{Q}\} = 0$
4	Weyl	1	$\{Q_+,\bar{Q}_+\} = P_+\Gamma^a P_a$

2) There is, as yet, no restriction on the world-volume signatures beyond the original requirement that s≤S and t≤T.

3) If we were to redraw the D/d brane-scan of Fig. 1 allowing now arbitrary signature, there would be no new points on the plot, but rather the new solutions would be superimposed on the old ones. For example, there would now be six solutions occupying the (d = 3, D = 11) slot instead of one.

4) Perhaps the most interesting aspect of the brane-molecule is the mod 8 periodicity. Suppose there exist signatures (s,t) and (S,T) which satisfy both the requirements of bose-fermi matching and super-Poincare invariance. Now consider (s′,t′) and (S′,T′) for which

$$s' + t' = s + t \tag{1.51}$$

$$S' + T' = S + T . \tag{1.52}$$

As a consequence of the modulo 8 periodicity theorem for real Clifford algebras [20,18], the minimal condition on a spinor is modulo 8 periodic e.g. S - T = 0 mod 8 for Majorana-Weyl. So if, in addition to (1.51) and (1.52) we also have

$$S' - T' = S - T + 8n \quad n\epsilon Z \tag{1.53}$$

then (s′,t′) and (S′,T′) satisfy bose-fermi matching. (1.52) and (1.53) imply

$$S' = S + 4n$$

$$T' = T - 4n . \tag{1.54}$$

Since, from Table 4, the existence of a super-Poincare algebra with minimal spinors is modulo 4 periodic in T, we see from (1.54) that the super-Poincare invariance is also satisfied. Thus given the vertical sequence $(S,T) = (10,1) \rightarrow (2,1)$, modulo 8 periodicity implies the existence of the two other vertical sequences of Fig. 2, namely $(6,5) \rightarrow (0,5)$ and $(2,9) \rightarrow (0.9)$. The three horizontal sequences $(1,10) \rightarrow (1,2)$, $(5,6) \rightarrow (5,0)$ and $(9,2) \rightarrow (9,0)$ are similarily related via modulo 8 periodicity.

Note the special crossover points at $(9,1)$, $(5,5)$ and $(1,9)$ which permit Majorana-Weyl spinors and which correspond to the top horizontal line in the brane-scan of Fig. 1. Similarly Weyl spinors are permitted at the crossover points $(5,1)$ and $(1,5)$ corresponding to the middle horizontal line of Fig. 1. It is curious that the fundamental extended objects at the top of the H and C sequences are chiral, while those at the top of the O and R sequences are not. We shall return to this in Section (2.5).

In the usual signature all extended objects appear to suffer from ghosts because the kinetic term for the X^o coordinate enters with the wrong sign. These are easily removed, however (at least at the classical level) by the presence of diffeomorphisms on the world-volume which allow us to fix a gauge where only positive-norm states propagate e.g. the light cone gauge for strings and its membrane analogues (See Section 2.2). Alternatively we may identify the d world-volume coordinates ξ^i with d of the D space-time coordinates $X^i (i=1,2,3)$ leaving D - d coordinates $X^I (I=1..D-d)$ with the right sign for their kinetic energy [23]. Of course, this only works if we have one world-volume time coordinate τ that allows us to choose a light-cone gauge or else set $\tau = t$.

In the same spirit, we could now require absence of ghosts (or rather absence of classical instabilities since we are still at the classical level) for arbitrary signature by requiring that the "transverse" group $SO(S - s, T - t)$ which governs physical propagation after gauge-fixing, be compact. This requires $T = t$.

It may be argued, of course, that in a world with more than one time dimension, ghosts are the least of your problems. Moreover, in contrast to strings, unitarity on the worldvolume does not necessarily imply unitarity in spacetime (I am grateful for discussions on this point with J. Polchinski). This is because the transverse group no longer coincides with the little group. (For example, the $(2,1)$ object in $(10, 1)$ spacetime and the $(1,2)$ object in $(9,2)$ spacetime both have transverse group $SO(8)$, but the former has little group $SO(9)$ and the latter $SO(8,1)$.) Nevertheless, it is an interesting exercise to see how compactness of the gauge group restricts the possible super-extended-objects. For example, the superstring in $(9,1)$ survives with $SO(8)$, but the superstring in $(5,5)$ with $SO(4,4)$ does not. What about the superstring in $(1,9)$? Here we once again encounter the problem that, in the absence of any physical input, we cannot distinguish (S,T) signature form (T,S). Since positivity of the energy is only a convention in field theory, ghosts can still be avoided by choosing S = s instead of T = t. To avoid this repetition, let us cut the Gordian knot and demand from now on that S ≥ T. The possible ghost-free solutions are those shown in Fig. 3. Note the special case (S,T) = (2,1) which permits not only (s,t) = (1,1) but also (2,0) since a single field with negative energy can, by convention, be ghost-free.

Perhaps the most striking aspect of Fig. 3 is that the majority of super-extended objects do indeed lie on the T = 1 axis.

(1.6) Spinning Membranes

As discussed in Section (1.2), Lindstrom and Rocek [9] having recently constructed a "spinning membrane" by supersymmetrizing the conformally invariant action (1.8). (N. B. Since the conformal action is equivalent to the Dirac-Nambu-Goto or Howe-Tucker-Polyakov actions, it must presumably be possible to supersymmetrize these actions also, in spite of previous claims to the contrary. However, the transformation rules may be horribly complicated and outside the framework of the usual tensor calculus.) This raises several interesting questions:

1) In string theory, we know by going to the lightcone gauge that the NSR spinning string is actually equivalent to the Green-Schwarz superstring. Can one demonstrate, with suitable choice of gauge-fixing, that the spinning membrane and supermembrane are also equivalent? We note also that a conformally invariant version of the supermembrane based on (1.8) has been given by Lindstrom [36].

2) Can the connection between singletons and membranes which seems so suggestive but which has so far escaped rigorous proof, be put on a firmer basis now that superconformal invariance on the worldvolume is now manifest?

3) Can one exploit this conformal invariance, as in NSR string theory, to obtain critical dimensions and equations of motion from vanishing β functions? If so, we learn that <u>flat Minkowski cannot be a solution</u>; because the three dimensional σ- model with vanishing background fields is not a finite theory (Minkowski space is of course a solution of the D=11 supergravity equations that follow from Siegel symmetry. However this is a tree-level result which, as in string theory, is presumably subject to quantum corrections).* This raises once again the intriguing possibility [1] that the theory will be finite only for a very special choice of non-trivial background fields. Thus membranes offer the possibility of avoiding the vacuum degeneracy problems of string theory and, in particular, of answering the question of why we do not live in ten (or eleven) flat dimensions.

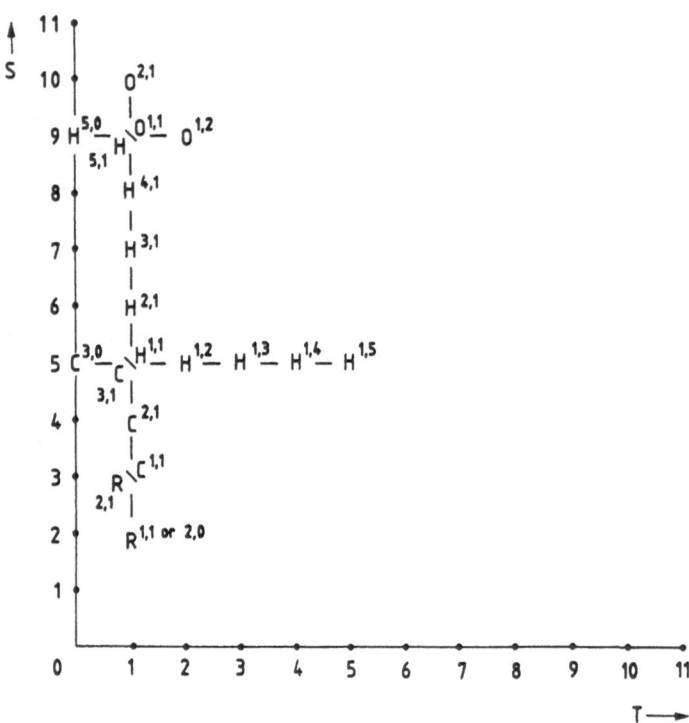

Fig. 3. The brane-molecule, assuming compact transverse group
and $S \geqq T$. The superscripts denote (s,t).

2. QUANTUM SUPERMEMBRANES

(2.1) Consistent quantum gravity

A consistent quantum theory of gravity probably requires the following four ingredients:

1) It must contain a massless spin-two particle in its spectrum (in
order to describe gravity)

2) It must be described by an extended object (in order to avoid
ultraviolet divergences)

3) It must be supersymmetric (in order to avoid other divergences)

4) It must be free from anomalies (in order to guarantee unitarity).

We shall argue that these requirements narrow us down to just two
*Quantum corrections and κ symmetry have recently been treated by
Paccanoni, Pasti and Tonin [50] who show the absence of anomalies in the
case of the D=11 supermembrane.
possibilities: the D=10 superstring or the D=11 supermembrane.

(2.2) The lightcone gauge: area-preserving diffeomorphisms

In string theory, the light cone gauge is convenient for quantization because it allows the elimination of all unphysical degrees of freedom and unitarity is guaranteed. Of course, one loses manifest Lorentz
invariance and one must be careful to check that it is not destroyed by
quantization.

In membrane theory, however, the lightcone gauge does not eliminate
all unphysical degrees of freedom. Let us split

$$X^\mu = \left\{ X^\pm = \frac{1}{\sqrt{2}} (X^0 \pm X^{D-1}); \quad X^I, \ I = 1,\ldots,(D-2) \right\} \tag{2.1}$$

and then set

$$\dot{X}^+ = p^+ \tag{2.2}$$

where the dot denotes differentiation with respect to τ, the time coordi
nate on the worldvolume. One can then solve for X^- leaving the (D-2)
variables X^I. For membranes, however, only (D-d) variables are physical
as explained in Section (1.3). Thus the lightcone gauge must leave a
residual gauge invariance [37]. Let us focus our attention on a d=3
bosonic membrane in flat spacetime. The lightcone action turns out to be

$$S = \frac{1}{2} \int d\tau \int d^2\sigma \left[(D_0 X^I)^2 - \det \partial_a X^I \partial_b X^I \right] \tag{2.3}$$

where

$$D_0 = \partial_0 + u^a(\sigma,\tau)\partial_a; \ \partial_0 = \frac{\partial}{\partial\tau}, \ \partial_a = \frac{\partial}{\partial\sigma^a}, \ a = 1,2 \ . \tag{2.4}$$

D_o is a "covariant time derivative" with "gauge field" u^a satisfying

$$\partial_a u^a = 0 .$$ (2.5)

The action possesses a gauge invariance that ensures that only (D - 3) of the (D - 2) X^I are physical.

For a membrane of spherical topology, the solution to (2.5) is

$$u^a = \epsilon^{ab} \partial_b \omega .$$ (2.6)

If we now introduce the Lie bracket

$$\{f,g\} = \epsilon^{ab} \partial_a f \, \partial_b g$$ (2.7)

the action may be rewritten in polynomial form

$$S = \frac{1}{2} \int d\tau \int d^2\sigma \left[(\partial_o X^I - \{\omega, X^I\})^2 - \frac{1}{2} \{X^I, X^J\}\{X^I, X^J\} \right] .$$ (2.8)

Remarkably, this looks like a (D-1) dimensional Yang-Mills theory dimensionally reduced to one time dimension. To be more precise, consider the correspondence

$$X^I \to A^I$$

$$\omega \to A_o$$

$$\{,\} \to [,]$$

$$\int d^2\sigma \to \text{tr}$$ (2.9)

then (2.8) describes a Yang-Mills theory with infinite dimensional gauge group. This group is, in fact, the subgroup of the worldvolume diffeomorphism group that preserves the Lie bracket (2.7) and is known as the group of area-preserving diffeomorphisms.

For spherical membranes, this group is given [37] by $\lim N \to \infty$ SU(N). The toroidal case is discussed in [38]. Here, as for Riemann surfaces of genus $g \geq 1$, one must take into account that the solution (2.6) is valid only locally.

One can generalize these results to the d = 3 supermembranes [39] in D = 4,5,7 and 11, and one finds super Yang-Mills quantum mechanical models corresponding to the dimensional reduction of super Yang-Mills in D = 3,4,6 and 10, which provides yet another way of understanding the allowed values of D. (One might conjecture a similar relationship between the d > 3 membranes and quantum mechanical models, but this time the gauge symmetry could not be of the Yang-Mills type. It has been suggested [40] that they are given by infinite dimensional non-Abelian antisymmetric tensor gauge theories.)

(2.3) Massless states and anomalies

The connection between supersymmetric Yang-Mills quantum-mechanical models and supermembranes has recently proved useful in the debate [20,7,21,22,40,39] about the spectrum of the supermembrane and the issue of whether or not it contains massless particles. The supersymmetric action looks like

$$S = \frac{1}{2} \int d\tau \ \text{tr} \ \left\{ (D_0 A^I)^2 - \frac{1}{2} [A^I, A^J][A^J, A^I] + i\bar{\lambda}D_0\lambda + i\bar{\lambda}\gamma^I[A^I, \lambda] \right\} \tag{2.10}$$

where the fields

$$(A_0, A^I) \qquad I = 1, 2 \ldots D-1$$

and

$$\lambda^{\alpha} \quad \alpha = 1, 2 \ldots 2(D-1) \tag{2.11}$$

are all in the adjoint representation of $SU(\infty)$. The membrane supercharge and the Hamiltonian are given by

$$Q = \int d^2\sigma \left[P_I \gamma^I + \frac{1}{2} \{X^I, X^J\}\gamma^{IJ} \right]$$

$$H = \int d^2\sigma \left[\frac{1}{2}P_I^2 + \frac{1}{2} h^{ab}\lambda^{\dagger}\gamma_a\gamma\partial_b\lambda \right] \tag{2.12}$$

where

$$\gamma_a = \partial_a X^I \gamma^I$$

and

$$\gamma = \frac{1}{2} \epsilon^{ab}\partial_a X^I \partial_b X^J \gamma^{IJ} \ . \tag{2.13}$$

One sees, classically, that the membrane Hamiltonian is positive semi-definite and that the zero energy configurations are those for which the potential energy

$$h = \det \partial_a X^I \partial_b X^J = \frac{1}{2} \{X^I, X^J\}\{X^I, X^J\} \tag{2.14}$$

vanishes. These are just the "collapsed" membranes with zero area. (Intuitively, one would expect membranes and higher dimensional objects to be different from strings in this respect, because of the ability to deform the object without increasing its volume.) Since the charges Q and H obey

$$\{Q^A, Q^{\dagger}_B\} = 2 \ \delta^A_B H \tag{2.15}$$

and the quantum Hamiltonian is positive semi-definite and the zero-energy states are those annihilated by Q i.e. the supersymmetric states, it

61

seems reasonable to conclude that the supermembrane contains massless states provided that supersymmetry is not broken.

At the level of perturbation theory, Bars, Pope and Sezgin [41] have shown that this expectation is borne out. They conclude that in flat D=11 spacetime, the massless sector of the supermembrane is just D=11 supergravity, just as the massless sector of the superstring is D=10 supergravity. Curiously, they find that although the other super extended objects in the R, C and H sequences also have massless particles, the corresponding supermultiplet is not the supergravity multiplet. An important caveat is that it is assumed throughout that the Clifford vacuum is a singlet. Different results would be obtained with different choices of Clifford vacua. Unfortunately, with our incomplete knowledge of the supermembrane at the non-linear level, we are unable to fix this choice uniquely. Nevertheless, the apparent mismatch between the background fields in the supermembrane action and the massless states for the R, C and H sequences, suggests that these theories are inconsistent. As discussed in Section (2.2), the price to be paid for a unitary gauge is the loss of manifest Lorentz invariance. Sure enough, it has been shown by Bars [43] and Bars and Pope [44] that all the members of the R, C and H sequences suffer from anomalies. Thus, as promised in Section (2.1), the eleven-dimensional supermembrane, alone of all other extended objects, emerges as a possible rival candidate to the ten-dimensional superstring for a consistent quantum theory of gravity.

Does it really pass all the tests?

Doubts were recently raised by de Wit, Nicolai and Hoppe [39] who questioned whether, at the non-perturbative level, the ground state really was supersymmetric, and concluded that the D=11 supermembrane was unlikely to have massless states in its spectrum. Pope and Stelle [42], on the other hand have invoked Witten index arguments to claim that supersymmetry is indeed unbroken, even non-perturbatively. Most workers in the field now see the main threat not being the absence of massless particles, but rather the presence of infinitely many of them! Closely related to this problem is the threat of a continuous, rather than discrete, mass spectrum.

(2.4) Discrete supermembrane spectrum in AdS

The issue of whether or not the supermembrane spectrum is discrete has been the subject of recent debate [42,45,47]. As discussed in the last section one would intuitively expect membranes and higher-dimensional extended objects to be different from strings, because of the ability to deform the object without increasing its volume. Thus, for example, a membrane can grow "hairs", without suffering any increase in its area. In particular there is an infinity of zero-energy classical solutions for which the membrane has zero area. Thus one might expect that the spectrum of the quantised theory would be continuous. Remarkably, it appears that this naive intuition is incorrect in the case of the bosonic membrane. Specifically, it has been shown that the quantum-mechanical spectrum of SU(N) one-dimensional bosonic Yang-Mills models is discrete [46]. And, as we discussed before, it has been shown that these models, in the limit N → ∞ are equivalent to a membrane of spherical topology propagating in Minkowski spacetime, in the light-cone gauge [37]. The reason for the discreteness of the spectrum in this case is that although there are flat directions in the classical potential for the theory, the steepness of the walls grows in such a way that the increasingly-confined wave-function acquires a zero-point energy that increases with distance along the "valley", and thus there is a confining effective potential for the valley mode and hence it has a discrete spectrum [46]. Unfortunately, the presence of fermions in supersymmetric theories destroys this mechanism, since the zero-point energies will now be zero. Thus intuition now suggests that the spectrum for the supermembrane should be continuous. This expectation is supported by a recent explicit calculation for supersymmetric SU(N) one-dimensional Yang-Mills models [45], which, in the N → ∞ limit are equivalent to a spherical supermembrane propagating in Minkowski spacetime, in the light-cone gauge [39].

Following Duff, Pope and Sezgin [47] we now turn our attention to the supermembrane in Anti-de Sitter space. Anti-de Sitter space can be defined as the four-dimensional hyperboloid in R^5 given by the equation

$$\eta_{ab} y^a y^b = - \frac{1}{a^2} \qquad (2.16)$$

where $\eta_{ab} = \text{diag}(-1,1,1,1,-1)$ and a is a constant. Points in AdS may be parametrised by the coordinates (t,ρ,θ,ϕ) where

$$y^0 = a^{-1} \sec \rho \sin at$$

$$y^1 = a^{-1} \tan \rho \sin \theta \cos \phi$$

$$y^2 = a^{-1} \tan \rho \sin \theta \sin \phi$$

$$y^3 = a^{-1} \tan \rho \cos \theta$$

$$y^4 = a^{-1} \sec \rho \cos at . \qquad (2.17)$$

The time coordinate t is therefore periodic, with period $\Delta t = 2\pi/a$. Of course one may choose a q - fold covering of AdS for which $\Delta t = 2\pi q/a$ or even the universal covering space CAdS, for which t takes values on the real line. However, the beauty of the AdS or its finite coverings is that the allowed lowest-energy eigenvalues E_o used in the classification of SO(3,2) representations are necessarily discrete. Since the time dependence of wavefunctions is $e^{iE_o t}$ we must have

$$E_o = \frac{\pi n}{\Delta t} = \frac{na}{2q} \qquad (2.18)$$

where n is an integer. It follows that the E_o values of the massive states in the supermembrane spectrum must satisfy (2.18). In a previous paper [19], it was pointed out that there is another reason why membranes appear to favour certain finite coverings of AdS. It was shown that in the cases of AdS x RP^7, $\widetilde{AdS}$ x S^7 and AdS x N(0,1) backgrounds there exist stable solutions for which the membrane is a sphere of arbitrary radius in the four dimensional spacetime, where $\widetilde{AdS}$ denotes the 2-fold cover of AdS. (Curiously enough, S^7 and RP^7 provide the only Freund-Rubin compactifications of d=11 supergravity for which the Kaluza-Klein massive modes have E_o values compatible with any finite covering of AdS. Thus only in these cases can the Kaluza-Klein modes be part of the super-membrane spectrum in AdS.) Finally, we note that another advantage of AdS is that it is not possible for instabilities to arise if the time coordinate is periodic.

It remains to be seen whether a periodic time is too high a price to pay for having a stable supermembrane vacuum with massless particles and a discrete spectrum.

(2.5) Further possibilities

It is interesting to ask whether we have exhausted all possible theories of extended objects with spacetime supersymmetry and fermionic gauge invariance on the worldvolume. This we claimed to have done in Section (1.5) by demanding super-Poincare invariance but might there exist other Green-Schwarz type actions in which the supergroup is not necessarily super-Poincare? Although we have not yet attempted to construct such actions, one may nevertheless place constraints on the dimensions and signatures for which such theories are possible. We simply impose the bose-fermi matching constraints of Section (1.5) and those in the first two columns of Table 3 but relax the constraints in the second two columns which specifically assumed super-Poincare invariance. The results of Blencowe and Duff [31] are shown in Fig. 4.

Although the possibilities are richer than those of Fig.2, there are still severe constraints. Note in particular that the maximum space-time dimension is now D=12 provided we have signatures (10,2), (6,6) or (2,10). These new cases are particularly interesting since they belong to the ℓ sequence and furthermore admit Majorana-Weyl spinors. In fact,

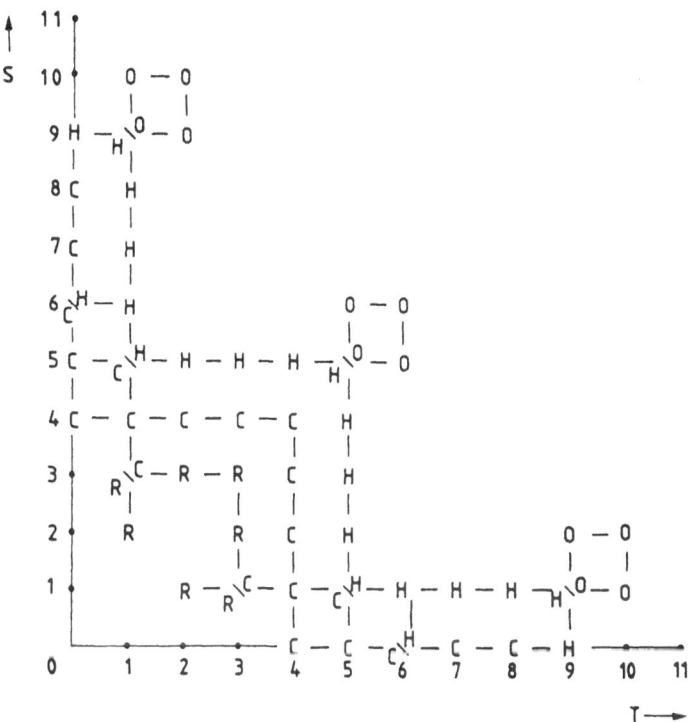

Fig. 4. The brane-molecule, without assuming
super-Poincaré invariance.

Table 5. Do the four forces of nature correspond to the
four division algebras?

Algebra	Symmetry	Subgroup	Chiral?	Force
O	SO(8)	SU(3)	No	Strong
H	SO(4)	SU(2)	Yes	⎧Weak
C	SO(2)	U(1)	Yes	⎨E.M.
R	1	1	No	Gravity

twelve-dimensional supersymmetry algebras have been discussed before in
the supergravity literature [48]. The RHS of the (Q,Q) anticommutator
yields not only a Lorentz generator but also a six index object so it is
certainly not super-Poincare. We conjecture (together with C. Hull and
K. Stelle) that the (2,2) extended object moving in (10,2) space-time may
(if it exists) be related by simultaneous dimensional reduction [17,7] to
the (1,1) Type IIB superstring in (9,1) just as the (2,1) supermembrane
in (10,1) is related to the Type IIA superstring [17]. We are encouraged
in this conjecture by the appearance of Majorana-Weyl spinors and self-
dual tensors in both the twelve-dimensional and Type IIB theories. We
concluded tentatively in Section (2.3), that the octonionic sequence of
Fig. 1 is quantum consistent while the quaternionic, complex and real sequences suffer from anomalies. (If this can be generalized to arbitrary
signature, then only the nine extended objects marked O of Fig. 2 would
survive, and only the three marked O of Fig. 3 would survive if we
further demand T = t and S ≤ T. We note that of these three, two have
T=1 and one has T=2. It would obviously be interesting to see how far
this "reductionist" approach can be carried and whether one may ultimately be able to narrow things down completely to a unique "theory of
everything". An alternative strategy has been suggested by Townsend [49]
in which one keeps all possible extended objects but regards the quaternionic, complex and real sequences not as fundamental but rather as "cosmic p-branes" i.e. as solitonic excitations of some underlying theory.

Finally one might ask what all this has to do with the real world.
As discussed above, one may adopt either a narrow reductionist view or
else take a broader stance. The reductionist, on examining Figs. 2 or 4,
would declare that one point one the graph (e.g. the (1,1) superstring in
(9,1) space-time) is the theory of everything while all the other points
are theories of nothing. Alternatively, one could entertain the idea
that the real world requires more than one, or possibly all, of the
theories permitted by the mathematics. In particular, we recall the observation of [31] discussed in Section (1.5) regarding the appearance of
chiral theories. This suggests something along the lines of Table 5. Do
the four forces in Nature correspond to the four division algebras?

ACKNOWLEDGEMENTS

I am grateful for conversations with Miles Blencowe, Paul Townsend, Ergin Sezgin, Kelly Stelle and Chris Pope.

REFERENCES

1) M. J. Duff, "From supersphagetti to superravioli" in Superworld II, Proceedings of the XVth Erice Summer School on Subnuclear Physics, 1987 (Ed. A. Zichichi, Plenum, 1988).

2) P. A. M. Dirac, Proc. R. Soc. A $\underline{268}$ (1962) 57.

3) Y. Nambu, Lectures at the Copenhagen conference (1970); T. Goto, Prog. Theor. Phys. $\underline{46}$ (1971) 1560.

4) P. S. Howe and R. W. Tucker, J. Phys. A$\underline{10}$ (1987), L155.

5) A. M. Polyakow, Phys. Lett. $\underline{103B}$ (1981) 207, 211.

6) M. J. Duff, S. Deser and C. J. Isham, Nucl. Phys. $\underline{B114}$, (1976) 29.

7) M. J. Duff, T. Inami, C. N. Pope, E. Sezgin, and K. S. Stelle, Nucl. Phys. $\underline{B297}$ (1988) 515.

8) E. Bergshoeff, E. Sezgin and P. K. Townsend, Phys. Lett. $\underline{B209}$ (1988) 451.

9) U. Lindstrom and M. Rocek, Stony Brook preprint ITP-SB-88-61.

10) J. H. Schwarz, in Supersymmetry and its applications, eds. G. W. Gibbons, S. W. Hawking and P. K. Townsend (C.U.P. 1985).

11) J. Hughes, J. Lui and J. Polchinski, Phys. Lett. $\underline{180B}$ (1986) 370.

12) E. Bergshoeff, E. Sezgin and P. K. Townsend, Phys. Lett. $\underline{189B}$ (1987) 75.

13) A. Achucarro, J. M. Evans, P. K. Townsend and D. L. Wiltshire, Phys. Lett. 1988 (1987) 441.

14) G. Sierra, Class. Quantum Grav. $\underline{4}$ (1987) 227.

15) J. M. Evans, Nucl. Phys. $\underline{B298}$ (1988) 92.

16) M. J. Duff, P. S. Howe, T. Inami and K. S. Stelle, Phys. Lett. $\underline{B191}$ (1987) 70.

17) M. J. Duff, B. E. W. Nilson and C. N. Pope, Phys. Rep. $\underline{130}$ (1986) 1.

18) P. G. O. Freund and M. A. Rubin, Phys. Lett. $\underline{97B}$ (1980) 233.

19) E. Bergshoeff, M. J. Duff, C. N. Pope and E. Sezgin. Texas A&M preprint CTP-TAMU-76/88.

20) K. Kikkawa and M. Yamasaki, Prog. Theor. Phys. $\underline{76}$ (1986) 116.

21) L. Mezincescu, R. Nepomechie and P. van Nieuwenhuizen preprint UMTG-139/ITP-SB-84-43.

22) S. Ghandhi and K. S. Stelle, Class. Quantum Grav. 5 (1988) L127.

23) E. Bergshoeff, M. J. Duff, C. N. Pope and E. Sezgin, Phys. Lett. $\underline{B176}$ (1987) 69.

24) M. P. Blencowe and M. J. Duff, Phys. Lett. $\underline{B203}$ (1988) 229.

25) H. Nicolai, E. Sezgin and Y. Tanii, Trieste preprint IC/88.

26) C. Fronsdal, Phys. Rev. D26 (1981) 26.

27) E. Bergshoeff, A. Salam, E. Sezgin and Y. Tanii, Trieste preprint IC/88/5.

28) M. Gunaydin, B. E. W. Nilson, G. Sierra and P. K. Townsend Phys. Lett. 176B (1986) 45.

29) M. J. Duff, P. K. Townsend and P. van Nieuwenhuizen. Phys. Lett. 122B (1983) 232.

30) D. Z. Freedman, G. W. Gibbons and P. C. West, Phys. Lett. 124B (1983) 491.

31) M. P. Blencowe and M. J. Duff, Nucl. Phys. B310 (1988) 387.

32) T. Kugo and P. K. Townsend, Nucl. Phys. B266 (1983) 440.

33) P. van Nieuwenhuizen, in Relativity, Groups and Topology II. Eds. DeWitt and Stora, North-Holland 1983.

34) R. Coquereaux, Phys. Lett. 115B (1982) 389.

35) P. G. O. Freund, Introduction to Supersymmetry (C.U.P. 1986).

36) U. Lindstrom and G. Theodoridis, ITP-SU-88-3.

37) J. Hoppe, Aachen preprint PITHA 86/24.

38) E. Floratos and J. Illiopoulos, Phys. Lett. 201B (1988) 237.

39) B. de Wit, J. Hoppe and H. Nicolai, preprint KA-THEP 6/88.

40) E. Bergshoeff, E. Sezgin and P. K. Townsend, preprint IC/88/186.

41) I. Bars, C. N. Pope and E. Sezgin, Phys. Lett. 1988 (1987) 455.

42) C. N. Pope and K. S. Stelle, preprint Imperial/TH/87-88/27.

43) I. Bars, preprint USC-87/HEPO6.

44) I. Bars and C. N. Pope, Class. Quantum Grav. 5 (1988).

45) B. de Wit, M. Luscher and H. Nicolai, preprint DESY 88-162 and THU-88-43.

46) B. Simon, Ann. Phys. 146 (1983) 209.

47) M. J. Duff, C. N. Pope and E. Sezgin, Texas A&M preprint CTP-TAMU-01/89.

48) J. W. van Holten and A. van Proeyen, J. Phys. A15 (1982) 3763.

49) P. K. Townsend, Phys. Lett 202B (1988) 53.

50) F. Paccanoni, P. Pasti and M. Tonin, Padova preprint DFPD/88/TH/9.

Chairman: M.J. Duff

Scientific Secretaries: L. Cappiello and B. Warr

DISCUSSION I [*]

– Horne:

To calculate tree level scattering in $D = 11$ supermembrane theory, one would presumably want to integrate over all simply connected compact three manifolds. How do we get around the problem that mathematicians haven't proved or disproved the Poincaré conjecture? Presumably, to calculate higher loop effects, one would need to know all three dimensional manifolds. Is it possible to classify all three manifolds?

– Duff:

If the quantization of membranes really does require summing over all Euclidean 3–manifolds in the functional integral, then we do have a serious problem because as yet these have not been classified. Indeed, this is one of the classic unsolved problems of pure mathematics. I can give you two answers. First, if I were Ed Witten, I might turn the problem around and give you a membrane–inspired proof of the Poincaré conjecture (laughter). Secondly however, I am not yet convinced of the validity of the Euclidianisation programme in general. This is a particularly difficult problem for supermembranes because three–dimensional Majorana spinors exist only in signatures (2,1) or (1,2) and not (3,0) or (0,3).

– Polychronakos:

In order to calculate the critical dimensions we used Siegel symmetry. Is this necessary for the consistency of the theory, or is it simply an aesthetic consideration?

– Duff:

The Siegel invariance was indeed used for the critical dimension. I believe most workers consider the symmetry to be crucial for the consistency of the theory, but this has not been proven. If you do not have the symmetry, there is a mis–match in the numbers of fermions and bosons after gauge–fixing, and we lose the world–volume supersymmetry. This makes the analysis much more difficult, and I think it is likely that theory becomes inconsistent. You would certainly lose the massless particles in the spectrum.

---

[*] In written version, both lectures will be combined

– *Pollak:*

You mentioned that by dimensional reduction the type II superstring theory can be obtained from the eleven dimensional supermembrane. Now there are serious doubts that realistic phenomenology can be obtained from the type II superstrings, so what does this imply about the phenomenology of supermembranes?

– *Duff:*

At the beginning I mentioned that the main thrust here is whether or not there exists a consistent quantum theory of gravity based on membranes. I have little to say at this point about "supermembrane–inspired phenomenology". From a phenomenological point of view, it is of course necessary to get chiral fermions in four dimensions with realistic representations.

This is already difficult for type IIB superstring and even more so for type IIA superstrings, which are not chiral in $D = 10$. So chiral fermions are certainly something to worry about if you believe that their origin lies in the $D = 11$ supermembrane. Since you twist my arm to say something about low–energy phenomenology, let me make the following comment: as discussed in my lectures, one may adopt a narrow "reductionist" view or else take a broader stance. The reductionist, upon examining the "brane–molecule" would declare that one point on the graph (e.g. the (1,1) superstring in (9,1) spacetime) is the theory of everything while all other points are theories of nothing. Alternatively, one could entertain the point of view that the real world requires more than one, or possibly all, of the theories permitted by the mathematics. In particular, we recall the observation that the fundamental extended objects at the top of $\mathcal{H}$ and $\mathcal{C}$ sequences are chiral, while those at the top of the $\mathcal{O}$ and $\mathcal{R}$ sequences are not. (One might even tolerate the anomalous $\mathcal{H}$, $\mathcal{C}$ and $\mathcal{R}$ sequences if, following Townsend, one regards them not as fundamental but rather as "cosmic p–branes", i.e. as solitonic excitations of some other underlying object). This suggests something along the following lines:

Algebra	Symmetry	Subgroup	Chiral?	Force
$\mathcal{O}$	SO(8)	SU(3)	No	Strong
$\mathcal{H}$	SO(4)	SU(2)	Yes	Weak
$\mathcal{C}$	SO(2)	U(1)	Yes	E.M.
$\mathcal{R}$	1	1	No	Gravity

Speculating wildly, could the four forces in nature correspond to the four division algebras?

– *Giannakis:*

1) What is the role of diffeomorphisms of the membrane which are not connected to the identity?

2) Is there a symmetry, analogous to conformal symmetry in strings, which prevents us to add more terms with powers of the Reimann and Ricci tensors into the action?

3) Are there both open and closed membranes?

– *Duff:*

1) Let me paraphrase your question into asking what is the analogue of "modular invariance" in membrane theories? This I am sure is very important, but I know of no investigations into this difficult subject.

2) No, I don't believe so. The Siegel symmetry, for example doesn't forbid higher derivative terms in the action as far as we know. However, these terms may affect the question of renormalizability of the d = 3 quantum field theory which is a question I discussed in the lectures.

3) No. I should have stated in the lecture that all super extended objects for $d > 2$ must be <u>closed</u>. The reason is that in going from the closed string to the open string the change in boundary conditions effectively divides the number of supersymmetries by 2, and we go from $N = 2$ to $N = 1$. For $d > 2$ we start from only $N = 1$ supersymmetry, and so we cannot divide by 2 and so they must be closed.

– *Balek:*

Can we use the membrane theory to solve the chirality problem in eleven dimensions, that is the problem of finding a chiral 4 dimensional theory as a compactification from $D = 11$?

– *Duff:*

Let me talk first about chirality in $D = 10$. You remember that we obtained the non–chiral type IIA theory from "simultaneous dimensional reduction", instead of the chiral type IIB. This process was the identification of one of the coordinates on the world–volume with one of those of the spacetime (compactified on a circle). But maybe there is a more complex identification involving, say, twists of some kind, so that we get type IIB.

– *Mannel:*

Are there any restrictions on the possible topologies of closed supermembranes which are to be summed over in the quantum mechanical formulation?

– Duff:

None that I know of. In the supersymmetric case we do require a spin–structure in order to define world–volume fermions.

– Ferrara:

What are the possible gauge groups that can be obtained from the supermembranes? Is there an analogous classification to that of string theory compactified to 4 dimensions?

– Duff:

Well, it seems reasonable that we can, in principle, get the obvious symmetries obtained à la Kaluza–Klein from $D = 11$ supergravity. It is very interesting to find out if there is any mechanism analogous to Frenkel–Kac which could enlarge these symmetries. A hint comes from Bergshoff, Salam Sezgin and Tanii, who compactify the supermembrane onto the seven–sphere. They appear to find 56 gauge bosons rather than the naïve expectation of 28 of $SO(8)$. The mechanism for this, and the general case is as yet unclear.

– Polychronakos:

Can we find the general solution of the membrane equation of motion by, say, a clever choice of gauge, as in string theory?

– Duff:

As far as we know, no gauge choice will linearize the equations of motion. So we accept that we are stuck with intrinsic non–linearity, and such equations are notoriously hard to solve. A recent paper from Bulgaria by Stoyanov claims to have obtained general exact solutions by picking a convenient gauge, but I think this deserves further study.

DISCUSSION II

– Skarke:

When you presented models with more than one time dimension, why did you immediately rule out possibilities with $S < s$ or $T < t$?

– Duff:

I am taking the point of view that the extended object should be immersed in spacetime otherwise there are problems with the definition of the induced metric.

– Cappiello:

Could you say something about the Fradkin–Vasiliev approach to higher spin fields and the associated algebras; and what could be their role in supermembranes?

– Duff:

It is an old problem in quantum field theory to construct an interacting theory for massless particles with spin greater than two. There are "no–go theorems" which rest on reasonable assumptions, that tell us we must stop at spin 2. Fradkin and Vasiliev succeeded in writing down a consistent lagrangian including massless fields of spin greater than two, provided you have the following unusual ingredients:

1) there must be an infinite tower of higher spin fields; and
2) the cosmological constant Λ does not vanish, because the lagrangian is not analytic as Λ goes to zero.

Now, as in general relativity, this lagrangian can be regarded as corresponding to the gauging of a global symmetry. This global symmetry, which they call "super higher spin algebra" is an infinite dimensional algebra which contains the super de Sitter algebra that I mentioned in my talk.

So you may regard this as a pure field–theoretic result, but the existence of an infinite number of particles suggests the connection to an extended object. Indeed, work by Bergshoff, Salam, Sezgin and Tanii shows that the spectrum of states generated by the singleton operators acting on the vacuum is in one to one correspondence with that of Fradkin and Vasiliev. There lies the connection with supermembranes.

– Horne:

It is very interesting that for a specific configuration of the target space, the 3 dimensional nonlinear sigma model is renormalizable. However we do not seem to live in anti–de Sitter space. Have other backgrounds been found that give renormalizable solutions?

– Duff:

In Minkowski space this trick does not work and the sigma model remains non–renormalizable. However, with regard to the phenomenology, we may be able to use the anti–de Sitter space solution, even though it has a non zero cosmological constant if Coleman's recent work is to be believed. Here you nevertheless seem to get zero at the end.

– Warr:

In the case where you can quantise the supermembrane (due to the special choice of background) can you give formulae for the scattering amplitudes analogous to those in string theories?

– Duff:

This is a point where the connection of membranes to the work of Fradkin and Vasiliev may be important, because one could presumably use the explicit form of their interacting lagrangian to compute scattering amplitudes.

– Warr:

In string theories, even when you can easily define the quantisation, due to the world sheet being non–interacting, you still have the possible problem of anomalies. So can anomalies afflict the membrane theory?

– Duff:

The issue of anomalies has been addressed by Salam et al. What they check is the closure of $OSp(8/4)$ algebra, which is the analogue of the Lorentz algebra in string theories, and they find no anomalies.

– Gourdin:

May I ask a non–membrane question?

– Duff:

Yes.

– Gourdin:

Consider the point like description of particles. Within the framework of axiomatic quantum field theory you can derive two important theorems: the spin-statistics connection and the TCP theorem. In the latter case two main ingredients are invariance under the special Lorentz group and locality. I suspect that in string theories and membrane theories you can manage with an extension of Lorentz invariance. But what can we do with locality, and is the TCP theorem derivable in these new frameworks?

74

– Duff:

I agree that the conditions under which the TCP theorem is proved in axiomatic QFT no longer obtain in string or membrane theories, or indeed, in general
relativity. If there were to be a breakdown in TCP, however, one could expect to
see it only at Planckian energies. In the meantime I would follow the maxim of
Sir Denys Wilkinson and not throw away an old trusted idea until one is forced
to.

– Nilles:

You have provided two ways to get a consistent theory of quantum gravity.
One was the $D = 10$ type IIA superstring, and the other was $D = 11$ super
membrane. But what about the heterotic string? One reason why this can exist is
that we have a consistent theory of the purely bosonic string. However, you have
not spoken of purely bosonic membrane, so I now ask if you know this theory to
be inconsistent?

In particular I refer to work by Kikkawa on the bosonic membrane, which
showed that it had some undesireable features, including the lack of a graviton or
indeed any massless states.

– Duff:

I did briefly discuss the bosonic membrane. The reason why it seems to be
uninteresting, as first pointed out by Kikkawa, is that in contrast to the bosonic
string it does not have massless particles in any integer dimension. Thus, as a
candidate for a quantum theory of gravity it is no good, since it does not have a
graviton. This problem is cured by the super–membrane.

Now, I must comment that when I listed the consistent quantum theories of
gravity, I did not only include the type IIA superstring, I included all five consistent
superstrings. In particular I am not excluding the heterotic string. In the lectures
I asked the question of what else can there be apart from all these superstring
theories.

Referring to the point about the bosonic membrane, let me state what Kikkawa's result was, and then we can consider whether his result truly forces us
to abandon the quantisation of the bosonic membrane. He noticed that in string
theories we have the mass–spin relation that gives rise to linear Regge trajectories,
and in order to have a mass–zero spin–two particle you must have a specific value
for the "intercept". It is remarkable that the $D = 26$ bosonic string does reproduce
exactly this intercept, as first calculated by Brink and Nielsen.

So Kikkawa tried to reproduce this for the bosonic membrane. He wrote down
the mass–spin relation, and then looked to see if there were any non zero particles
for integer spin J. He found that the "intercept", as computed from from the

Casimir energy, took some arbitrary irrational value, so that for a mass set equal to zero you do not get any integer spin J in any integer spacetime dimension. It was on this basis that he ruled out the use of a bosonic membrane.

– *Volkas:*

I was wondering about the connection between the very small membranes hypothesied to generalise particles and the very large "membrane at the end of the universe". Is this the same phenomenon as when strings join up into a type of cosmic string?

– *Duff:*

I regard the "membrane at the end of the universe" as defining the vacuum state, and the membrane fluctuations about this vacuum as describing the particles.

EFFECTIVE LAGRANGIANS
FOR SUPERSTRING COMPACTIFICATIONS

S. Ferrara, CERN, Geneva, Switzerland

and

University of California, Los Angeles, USA

ABSTRACT

In this lecture I will discuss some macroscopic (target space–time) properties versus microscopic (world–sheet) properties of four–dimensional superstrings.

The possible occurrence of an enhanced space–time supersymmetry (N = 2 or N=4) in certain internal conformal field theories (when used to "compactify" type II supertrings) allows us to obtain strong constraints on the effective (supergravity) Lagrangians. In particular the Kähler metric for the moduli of Calabi–Yau spaces (which correspond to deformation of the Kähler class) can be determined by symmetry and topological considerations, at least to each finite order in the α' expansion.

The Kähler metric for the moduli which corresponds to deformation of the complex structure is also shown to be compatible with N=2 space–time supersymmetry.

These results imply that the full low–energy superstring Lagrangian for the massless modes can be computed from pure topological data on Calabi–Yau spaces.

1. INTRODUCTION

In the last couple of years, superstring theories have been reformulated by using some general properties of two–dimensional superconformal invariant theories [1].

At present, barring complications due to possible inconsistencies in higher–genus surfaces, a "consistent" string model can be formulated for any space–time dimension $D \leq D_{crit} = 10$, in terms of a corresponding 2d superconformal theory where certain requirements for the absence of local and global anomalies must be implemented [2–4].

In particular, the main ingredients which have been used so far to construct superstring theories are the specific conformal properties of the "internal" degrees of freedom. The conformal structure of the "internal" degrees of freedom is, in fact, related to the possible occurrence of one or more supersymmetries in target space–time [5]. To be more specific, the local anomaly cancellation demands the vanishing of the "central charge" for left– and right–moving conformal fields separately [6].

$$C_L^{TOT} = -26 + 11 + \frac{3}{2}D + C_L^{int} = 0 \tag{1.1}$$

$$C_R^{TOT} = \begin{cases} C_L^{TOT} \text{for } (1,1) \text{ 2d local susy (type II s.strings)} \\ -26 + D + C_R^{int} = 0 \text{ for } (1,0) \text{ 2d local susy (heterotic s.strings)} \end{cases} \tag{1.2}$$

In the 4D, $C_L^{int} = 9$ and $C_R^{int} = 9$ or 22, depending on whether one is in type II strings [7] or heterotic superstrings [3,4].

The structure of the conformal field theory is further determined by the representations of C_L^{int} and C_R^{int} and the modular properties of the conformal fields.

In 4D superstrings, as pointed out by many authors, the existence of one or more space–time supersymmetries requires an enhanced symmetry [5] of the "internal" superconformal theory.

The N = 1 space–time supersymmetry requires at least a (2,0) extended superconformal system [5]; N = 2 supersymmetry requires at least a (4,0) or (2,2) superconformal system, and N = 4 requires a free internal system (six $C = \frac{3}{2}$ superfields) for heterotic strings or a (4,4) system for type II superstrings [5]. The (2,2) and (4,4) systems may have a lower (N = 1 and N = 2, respectively) supersymmetry in the case of heterotic superstrings with (1,0) local world–sheet supersymmetry.

Recently, a large class of (2,2) superconformal systems with (1,0) local supersymmetry has been constructed [8] by using unitary representations [9] of the discrete series of the N = 2 superconformal algebra [10]. There is strong evidence that such systems correspond to exact Calabi–Yau compactifications [11] with N = 1 space–time supersymmetry. In these models, $C_R^{int} = C_L^{int} + C'$, where $C' = 13$ corresponds to conformal free fields and

$$C_L^{int} = 9 = \sum_K C_K^{int}(C_K \text{ discrete unitary series}) \tag{1.3}$$

Toroidal (or orbifold) [12] compactifications correspond to the case of free conformal systems for the internal degrees of freedom, although it has recently been pointed out that some of these compactifications can be formulated on group manifolds [13].

On the contrary, (2,0), (2,2), (4,0), and (4,4) conformal systems correspond to non–trivial σ–models on the world–sheet [14]. The (4,4) system seems to be rather unique and it corresponds to the manifold $K_3 \times T_2$, where K_3 is the (unique) compact Calabi–Yau space in two complex dimensions. For such a system,

$$C^{int} = 6 + 3 \tag{1.4}$$

where 3 is the central charge of a complex free superfield which realizes an n = 2 superconformal algebra, and 6 is the central charge of the representation of the n = 4 superconformal algebra [5].

In the σ–model approach to superstring models, one is led to introduce background fields for the massless degrees of freedom occurring in a given superconformal theory. The equations of motion of these background fields are then determined by implementing conformal invariance (vanishing β functions) in a given background in σ–model perturbation theory [15, 16]. This approach gives, in principle, a rationale for the construction of the effective Lagrangians for the massless particles in terms of general properties of the underlying 2d superconformal theory. In particular, the equations of motion which we get from the σ–model approach should be viewed as an effective dynamics for the light degrees of freedom upon integration over the stringy massive modes.

The lowest–derivative terms for the effective Lagrangian of a given superstring model correspond to a standard 4D supergravity Lagrangian [17]. Higher (loop)–order corrections in the σ–model give higher–derivative and higher–curvature modifications to standard supergravity [18]. These higher–order corrections for N = 1, 4D supergravity have recently been investigated [19].

It is even possible to introduce background gravitational auxiliary fields [20]. In particular, the U(1), U(2), and SU(4) auxiliary fields of N = 1,2 and 4 extended supergravities are related to the world–sheet U(1), U(1) × SU(2) and SU(4) Kac–Moody algebras of the associated extended superconformal systems.

In the present lecture, some recent results on the effective Lagrangians for different superstring models will be reviewed. These results have been derived by using properties of the underlying two–dimensional superconformal theory as well as the strong constraints coming from the requirement of one or more space–time supersymmetries.

A very useful property of two–dimensional field theories, in the case of toroidal compactification, is the existence of hidden (or accidental) symmetries related to two–dimensional duality transformations of scalar field strengths [21, 22].

A particular role in these effective Lagrangians is played by the background scalar fields. Their vacuum expectation values (v.e.v.'s) correspond to possible different conformal field theories. These theories may exist not only for isolated values of the parameters; they may also depend on a continuous set of parameters. In the effective Lagrangian approach, this means that the equations of motion of the scalar fields,

$$\beta_{<\phi>} = \frac{\partial V}{\partial \phi} = 0 \tag{1.5}$$

leave some scalar fields undetermined. These are the so–called flat directions, and they correspond to the moduli space of a conformal field theory [23]. This is the reason why in superstring theories the effective Lagrangians for these massless degrees of freedom have the form of a "no–scale" supergravity model [24].

The structure of this moduli space for different superconformal systems has recently been considered by several authors [21, 22, 25]. Here we will report on the known results and the new results related to (2,2) superconformal systems which can be embedded in type II superstrings.

The review is organized as follows. In Section 2 we will report on the structure of the effective Lagrangians for trivial conformal systems, related to toroidal or orbifold compactifications. This structure can be uniquely determined, at least if one confines the investigation to some sectors of the theory (untwisted sector), by using hidden symmetries of the underlying conformal theory [21, 22]. This approach will be discussed in Section 3. This method gives the same result as that obtained by a naïve truncation of the N = 4 supergravity action or by toroidal compactification of N = 1, D = 10 supergravity [26, 27].

Section 4 will be devoted to the analysis of type II superstrings. In particular, we will be able to improve a recent analysis given by Seiberg [25] for this class of theories. The structure of the moduli space for several (2, 2) superconformal theories will be reported. A striking result of this analysis is the concept of "dual quaternionic manifold" which can be obtained, through a certain operation, from any N = 2 admissible Kähler manifold. Details of this correspondence will be given in Section 9.

In Section 5 it will be shown that the N = 2 couplings coming from type II superstrings are of the standard supergravity form. In fact, they are consistent with the rules of N = 2 superconformal tensor calculus developed in Refs. [28–30]. In Section 6 it will be proven that the coefficients of the cubic holomorphic function which determines the N = 2 Kähler potential for vector multiplets in type II–A supergravity are related to the intersection matrices of Calabi–Yau manifolds in terms of pure topological data which do not require the knowledge of the metric of the Calabi–Yau space.

80

A general formula for vector multiplets in type II–B strings is given in Section 7. These results allow the determination of the complete Kähler metric for the moduli space of Calabi–Yau manifolds.

In Section 8 the implication of these results for the case of heterotic strings is given. Concluding remarks are given in Section 10.

2. EFFECTIVE LAGRANGIANS FOR TOROIDAL OR ORBIFOLD COMPACTIFICATIONS

Effective Lagrangians for (generalized) toroidal compactifications have been studied extensively in the literature.

The first example was provided by Witten [26], by making a suitable truncation of the N = 1, D = 10 supergravity Lagrangian. By retaining only SU(3) singlets in the truncated spectrum, he was able to recover an N = 1, 4D supergravity Lagrangian of a standard form.

Let us recall that a standard N = 1 Lagrangian [31] is specified by a Kähler potential $J(z, z^*)$, a superpotential $g(z)$, and the gauge "modulator" $f_{AB}(z)$. Here z denotes the complex scalar fields occurring in N = 1, 4D Wess–Zumino (chiral) multiplets [they can be interpreted as coordinates of a Kähler (Hodge) manifold]; J is a real function, whilst g and f are analytic functions of the z variables.

Local supersymmetry actually requires that the entire Lagrangian depends only on the combination

$$G \; = \; J \; + \; log|g|^2 \tag{2.1}$$

and on the f function. Note that the Kähler manifold in which the scalars live can have isometries which can be broken by the superpotential term (gauge couplings).

In the example of Ref. [26], the 4D gauge group was $E_6 \times E_{8'}$, a maximal subgroup of $E_8 \times E_{8'}$ commuting with SU(3). The Kähler potential turns out to be

$$J = -log\,(S + S^*) \; - 3\,log\,(T + T^* - 2C_i C^{*i}) \tag{2.2}$$

and it corresponds to the coset space,

$$\frac{SU(1,1)}{U(1)} \; \times \; \frac{SU(1, n+1)}{SU(n+1) \times U(1)}$$

The g and f functions are

$$g(C) \; = \; d_{abc} C^a C^b C^c, \; f_{AB} \; = \; \delta_{AB} S \tag{2.3}$$

In terms of 10D degrees of freedom, the S and T fields (neutral under E_6) come from the $g_{i\bar{j}}, b_{\mu\nu}, b_{i\bar{j}}$ components of the metric and antisymmetric tensor and from the dilation ϕ. The C fields, transforming in the **27** of E_6, come from the SU(3)–invariant components of the 10D gauge field A_j^A, where SU(3) is the diagonal

SU(3) of two SU(3)'s acting respectively on external indices [SU(3) $\subset$ O(6) $\subset$ O(1,9)] and the gauge indices [SU(3) $\subset$ E_8, commuting with E_6]. This method was then applied to more general theories.

The second example [26] is the derivation of the effective Lagrangian for the untwisted sector of the orbifold $T_6/\mathcal{Z}_3$. In this case, one uses the same N = 1, D = 10 Lagrangian as a starting point, but one retains $\mathcal{Z}_3$ rather than SU(3) singlets. As a consequence, one gets, from the gravitational sector, 1 + 9 Wess–Zumino multiplets. The scalar degrees of freedom come from $b_{\mu\nu}, \phi, g_{i\bar{j}}, b_{i\bar{j}}$, where now $g_{i\bar{j}}$ and $b_{i\bar{j}}$ are arbitrary Hermitian (3 × 3) matrices. From the gauge sector one gets nine E_6 families rather than one. Also, in the orbifold limit [12], the gauge group is enhanced to SU(3) $\times E_6$.

The Kähler potential is in this case

$$J(z, z^*) \; = \; -lg(S + S^*) - lg \; det(T_{i\bar{j}} + T_{i\bar{j}} - 2C_i^{ma}C_{\bar{j}}^{*\bar{m}\bar{a}}) \qquad (2.4)$$

where

$$T_{i\bar{j}} \; = \; g_{i\bar{j}}e^{-\phi} + C_i^{ma}C_{\bar{j}}^{*\bar{m}\bar{a}} - i\sqrt{2}B_{i\bar{j}}$$

It corresponds to the coset space

$$\frac{SU(3, 3(1 + n))}{SU(3) \times SU(3(1 + n)) \times U(1)}$$

The superpotential and f function are given respectively by

$$g(C) \; = \; d_{abc}\varepsilon_{emn}\varepsilon^{ijk}C_i^{la}C_j^{mb}C_k^{nc} \qquad (2.5)$$

$$f_{AB}(S) \; = \; \delta_{AB}S \qquad (2.5b)$$

The previous method can be used, in principle, for all types of orbifolds, with different $\mathcal{Z}_N$ symmetries. The result is summarized in Table IV.

To obtain the effective Lagrangian for the massless states of the untwisted sector, one can use either a truncation from D = 10, N = 1 supergravity or, equivalently, a truncation of the D = 4, N = 4 supergravity which is obtained from D = 10, N = 1 supergravity [32]. In the second case, the $\mathcal{Z}_N$ group is embedded in the SU(3) subgroup of the SU(4) group which rotates the four 4D gravitinos.

The same strategy can now be extended to general toroidal compactifications with a maximal gauge group of rank 22 by using the fact that the effective Lagrangian for N = 4, 4D supergravity coupled to matter is completely fixed by local supersymmetry once the gauge group G is known.

In particular the 6n scalar fields of the n = dim G vector multiplets can be viewed as coordinates of the manifold [33].

$$\frac{SO(6,n)}{SO(6) \times SO(n)} \tag{2.6}$$

The isometries of this manifold are broken by the G–gauge interactions. However, if we confine ourselves to the Cartan subalgebra of G, the N = 4 scalar potential vanishes identically, since it is proportional to the structure constants of G. This means that SO(6, 22) is an exact symmetry of the effective Lagrangian [21].

Generalized toroidal compactifications with a lower number of space–time supersymmetry can be obtained by making suitable $\mathcal{Z}_2$ projections of the N = 4 theory [4] ($\mathcal{Z}_2$ twists). By making a single $\mathcal{Z}_2$ projection, N = 4 can be reduced to N = 2. By making a further $\mathcal{Z}_2'$ projection, one can get an N = 1, 4D theory [4].

The effective Lagrangians for the untwisted sector of these theories can be obtained with the same technique [27]. One starts with an N = 4 Lagrangian and then considers suitable $\mathcal{Z}_n$ symmetries in the SU(4) group which rotates the four gravitinos. In this case, the $\mathcal{Z}_n$ groups are $\mathcal{Z}_2$ and $\mathcal{Z}_2 \times \mathcal{Z}_2'$, generated by the following SU(4) elements:

$$\alpha = \begin{pmatrix} 1 & & & \\ & 1 & & \\ & & -1 & \\ & & & -1 \end{pmatrix}, \ \beta = \begin{pmatrix} 1 & & & \\ & -1 & & \\ & & -1 & \\ & & & 1 \end{pmatrix} \tag{2.7}$$

$\mathcal{Z}_2$ and $\mathcal{Z}_2'$ leave (separately two gravitino invariants, but $\mathcal{Z}_2 \times \mathcal{Z}_2'$ leaves only one. In this way the original manifold given by formula (2.6) reduces to [27]

$$\frac{SO(6,n)}{SO(6) \times SO(n)} \xrightarrow[\text{under} \ \mathcal{Z}_2]{} \frac{SO(2,n_1)}{SO(2) \times SO(n_1)} \times \frac{SO(4,n_2)}{SO(4) \times SO(n_2)} \tag{2.8}$$

$$\frac{SO(6,n)}{SO(6) \times SO(n)} \xrightarrow[\text{under} \ \mathcal{Z}_2 \times \mathcal{Z}_2']{} \prod_{i=1}^{3} \frac{SO(2,n_i)}{SO(2) \times SO(n_i)} \tag{2.9}$$

Formula (2.8) corresponds to an N = 2 superstring theory with the Kähler manifold K given by

$$K_{N=2} = \frac{SU(1,1)}{U(1)} \times \frac{SO(2,n_1)}{SO(2) \times SO(n_1)} \tag{2.10}$$

and the quaternionic manifold Q given by

$$Q_{N=2} = \frac{SO(4,n_2)}{SO(4) \times SO(n_2)} \tag{2.11}$$

Formula (2.9) gives rise to an N = 1 theory with the Kähler manifold given by the product of four Kähler manifolds:

$$K_{N=1} = \frac{SU(1,1)}{U(1)} \times \prod_{i=1}^{3} \frac{SO(2,n_i)}{SO(2) \times SO(n_i)} \tag{2.12}$$

Up to now we have applied this strategy to type I (heterotic) superstrings. However, the same analysis can be extended to type II strings with (1,1) local world–sheet supersymmetry [8].

In this case the manifold structure is more restricted since toroidal or orbifold compactifications of these theories can be viewed as descending from N = 2, D = 10 non–chiral and chiral supergravities [34], or equivalently from N = 8, D = 4 supergravity [35]. The coset space of the untwisted sector of type II strings will always be a particular truncation of the N = 8, 4D manifold [35] $E_{7,7}/SU(8)$. This is analogous to the fact that in type I superstrings the manifolds in which the (untwisted) scalars live are all submanifolds of [SU(1,1)/U(1)] × [(SO(6,n)/SO(6) × SO(n)].

Type I (heterotic) and type II superstrings have some common sectors, in which the scalar fields live in the same manifold. This situation occurs for theories which correspond to the same (2,2) superconformal system. Indeed, a (2,2) superconformal system can equally well be used for (1,0) (heterotic) or (1,1) (type II) world–sheet supersymmetry. From the point of view of the truncation of supergravity Lagrangians, this phenomenon can be understood from the fact that the [SU(1,1)/U(1)] × [SO(6,n)/SO(n)] manifolds, for n ≤ 6, are submanifolds of the $E_{7,7}/SU(8)$ manifold. However, the result is more general [25] in the sense that (2,2) superconformal systems can be severely constrained even in cases in which the effective theory (for the massless excitations) cannot be obtained from a naïve truncation of the N = 2, D = 10 (or n = 8, D = 4) Lagrangians.

3. HIDDEN NON–COMPACT SYMMETRIES IN SUPERSTRING THEORIES

In this section we would like to make a string argument for the occurrence of the non–compact symmetries described in the previous section.

For scalar fields with vanishing scalar potential, these symmetries are exact invariances of the low–energy effective supergravity Lagrangian. This applies, for instance, to the S and T fields of orbifold compactifications. More generally, scalar background fields with vanishing scalar potential (flat directions) correspond to the so–called moduli space of the underlying superconformal field theory. These fields parametrize the connected component of the space of continuous parameters for which a conformal field theory exists.

In all the examples described in Section 2, for the untwisted sector of generalized toroidal (orbifold) compactifications, the scalar background fields parametrize manifolds of the type SO(p,q)/SO(p) × SO(q) or of the type SU(p,q)/SU(p) × SU(q) × U(1).

A natural question is, do these non–compact symmetries correspond to an exact property of the underlying superconformal theory, or are they just an artefact of the low–energy approximation?

It has recently been shown [21] that the microscopic origin of these non–compact symmetries is quite similar to the way in which non–compact (sometimes called hidden) symmetries originate in 4D N–extended supergravities, i.e. through "duality" transformations [36]. The only difference is that in 2d, the vector field strength and its dual (which are two–forms) are replaced by a scalar field strength and its dual (one–forms). Now we see that the crucial difference is that, because commuting objects (two–forms) are replaced by anticommuting ones (one–forms), the maximal duality group of n–scalar field strengths is SO(n,n), rather than Sp(2n) as in the four–dimensional case. The 4D analysis was given some years ago by Gaillard and Zumino [37].

In 2d, the field strength of a scalar field

$$F_\mu^a = \partial_\mu \phi^a \tag{3.1}$$

satisfies the Bianchi identity

$$\partial \widetilde{F}^{a\mu} = 0 \ , \ \widetilde{F}^{a\mu} = \varepsilon^{\mu\nu} \partial_\nu \phi^a \tag{3.2}$$

If the Lagrangian L depends on ϕ^a only through their field strengths F_μ^a, the equations of motion read

$$\partial_\mu \widetilde{G}_a^\mu = 0 \tag{3.3}$$

$$\widetilde{G}_a^\mu = \frac{\delta \mathcal{L}}{\delta F_\mu^a} \tag{3.4}$$

Equations (3.2) and (3.3) are invariant under any linear transformation among the $F's$ and $G's$. However, we are interested in the transformations that are compatible with Eq. (3.4). In general, this will require that also other fields χ^i (typically scalars) transform under those duality transformations: $\delta \chi^i = \xi^i(\chi)$.

In superconformal field theories corresponding to toroidal compactification, the scalar that describe internal degrees of freedom appear only through derivatives, and their kinetic term is multiplied by X^μ–dependent background fields. So in the application of two–dimensional duality to superstrings the one–forms are related to derivatives of the "internal" degrees of freedom, whilst the remaining fields χ^i, transforming non–linearly under the duality group, will be identified with the (space–time) scalar background fields.

Following the analysis of Ref. [37], the generators X of the duality group

$$\delta_X \begin{pmatrix} F \\ G \end{pmatrix} = X \begin{pmatrix} F \\ G \end{pmatrix} \tag{3.5}$$

and

$$\delta_X \chi^i = \xi_X^i(\chi) \tag{3.6}$$

will satisfy the following condition:

$$X^T \Omega = -\Omega X \tag{3.7}$$

with

$$\Omega = \begin{pmatrix} 0 & 1 \\ 1 & 0 \end{pmatrix} \tag{3.8}$$

In 2d Ω is symmetric whilst in 4D it is antisymmetric; this is related to the fact that in 2d, $F_1 \cdot \tilde{F}_2 = -\tilde{F}_1 \cdot F_2$, whilst in 4D, $F_1 \cdot \tilde{F}_2 = \tilde{F}_1 \cdot F_2$.

The (real) matrix X satisfying Eq. (3.7) has the general form

$$X = \begin{pmatrix} A & B \\ C & -A^T \end{pmatrix}, \ B = -B^T, \ C = -C^T \tag{3.9}$$

The $X's$ define an SO(n,n) algebra [21].

The general form of a Lagrangian $\mathcal{L}$, leading to invariant equations of motion, is

$$\mathcal{L} = \frac{1}{2} F\tilde{G} + \mathcal{L}_{inv} \tag{3.10}$$

Note that $\mathcal{L}$ itself is not invariant since

$$\delta_X \mathcal{L} = \frac{1}{2} \left(FC\tilde{F} - GB\tilde{G} \right) \tag{3.11}$$

If the vectors $(H_\mu^a, K_{\mu a})$ transform under Eq. (3.9) like $(F_\mu^a, G_{\mu a})$, we can have an interaction

$$\mathcal{L} = \frac{1}{2} F\tilde{G} + \frac{1}{2} \left(K\tilde{F} + H\tilde{G} \right) + \mathcal{L}_{inv} \tag{3.12}$$

By varying Eq. (3.12), we get

$$\tilde{G}_a = \frac{\delta \mathcal{L}}{\delta F^a} = \frac{1}{2} \left(\tilde{G}_a - \tilde{K}_a \right) + \frac{1}{2} \frac{\delta \tilde{G}_b}{\delta F^a} \left(F^b + H^b \right) \tag{3.13}$$

that is,

$$\tilde{G}_a + \tilde{K}_a = \frac{\delta \tilde{G}_b}{\delta F^a} \left(F^b + H^b \right) \tag{3.14}$$

By differentiating Eq. (3.14) once more, we find that the matrix $\delta \tilde{G}_b / \delta F^a$, which is independent of F^a, is symmetric. Therefore we get

$$\tilde{G}_a + \tilde{K}_a = [g_{ab}(\chi) + b_{ab}(\chi)^*] \left(F^b + H^b \right) \tag{3.15}$$

for some matrices g and b ($g = g^T, b = -b^T$). Equation (3.15) can be written in an SO(n,n) covariant way. Let us define

$$\hat{\mathcal{F}} = \begin{pmatrix} \mathcal{F}^a & + & H^a \\ G_b & + & K_b \end{pmatrix} \tag{3.16}$$

Then Eq. (3.15) is equivalent to

$$\nu \widehat{\mathcal{F}} = \Omega \nu \widetilde{\widehat{\mathcal{F}}} \tag{3.17}$$

when the matrix ν is an SO(n,n) group element:

$$\nu^T \Omega \nu = \Omega \tag{3.18}$$

Equation (3.17) is invariant under SO(n, n) global transformations $\nu \to \nu g^{-1}$, $\mathcal{F} \to g\mathcal{F}$, and under SO(n) local transformations $\nu \to h\nu$, since $h\Omega = \Omega h$.

Comparing Eqs. (3.17) and (3.15) we see that the matrix

$$T_{ab} = g_{ab} + b_{ab} \tag{3.19}$$

parametrizes the coset SO(n,n)/SO(n) $\times$ SO(n).

This result can be extended to chiral bosons. If we have p left–movers and q right–movers [38], then the self–couplings of the chiral bosons compatible with duality parametrize the coset [21]

$$SO(p,q)/SO(p) \times SO(q) \tag{3.20}$$

The coset representatives $A_{ia}(X^\mu)$ are in this case the background scalars corresponding to the Abelian bosonization of the system of six (1,0) leftons coupled, through these gauge scalars, to forty-four right-moving fermions. The conformal Lagrangian is invariant under SO(6,22) rotations. In the case of constant background fields, we get free theories whose vacua are then parametrized by coordinates of the coset SO(6,22)/SO(6) $\times$ SO(22). Indeed, in this case we recover the Narain [4] construction of new 4D string theories.

If these considerations are extended to scalars which are not in the Cartan subalgebra of SO(44), it is found that these symmetries are broken by gauge couplings.

By using functional methods, it can be proved that the above symmetries are quantum symmetries of the conformal field theory, but they are not expected , in general, to be symmetries when higher–genus surfaces are considered [21].

This implies that they are exact symmetries to all orders in α', including higher–curvature terms, but they may be spoiled by string loops.

Moreover, these considerations can be extended to superconformal theories with background scalars which may be restricted by GSO projections [4]. In the latter case, the relevant duality group is a subgroup G' of SO(n,m) commuting with the GSO projections. In this way, we get, for instance [27] (untwisted sector):

$$G' = SO(2, n_1) \times SO(4, n_2) \tag{3.21}$$

for an N = 2 heterotic superstring

$$G' = \prod_{i=1}^{3} SO(2, n_i) \qquad (3.22)$$

for an N = 1 heterotic superstring, and

$$SU(3, n) \qquad (3.23)$$

for a T_6/Z_3 orbifold.

These results coincide with the macroscopic analysis performed in Section 2.

At this point we notice that the above considerations do not apply to a non–trivial σ–model for the internal degrees of freedom. In superstrings with N = 1 or N = 2 space–time supersymmetry, non–trivial σ–models correspond to representations of n = 2 and n = 4 superconformal algebras on the world–sheet.

However, we will see that for most of these models, a complete determination of (some sectors of) the effective theory is possible.

4. TYPE II SUPERSTRINGS AND THE MUDULI SPACE OF (2,2) SUPERCONFORMAL SYSTEMS

In this section we will extend our analysis of effective Lagrangians to superstring theories which correspond to (2,2) superconformal systems. We make contact with an approach of Seiberg [25], who elucidates the connection of different superstring models based on the same superconformal system for the internal degrees of freedom. In particular:

i) we establish the general (perturbative) structure of some sectors of the moduli space of a (2,2) superconformal system;

ii) we give the precise relation between vector multiplets and hypermultiplets in type IIA and IIB superstrings and introduce the notion of "dual" quaternionic manifold;

iii) we show that the universal hypermultiplet, containing the dilaton and the antisymmetric tensor has no trilinear coupling to two vector field strengths in type II couplings, as required by N = 2 supergravity tensor calculus [28].

Type II superstrings, as recently emphasized by Seiberg [25], are a suitable tool with which to study general properties of superconformal systems with (2,2) or (4,4) global world–sheet supersymmetry.

In fact, from the general arguments of Section 1, such superconformal systems can be used as backgrounds of type II rather than type I (heterotic) closed superstrings.

Under these circumstances we can use the more powerful constraints of the enhanced N = 2 and N = 4 local supersymmetries of the effective Lagrangians to study the general structure of the background fields.

In particular, the moduli space of such superconformal theories is the same whether we are dealing with type II or type I superstrings.

In the case of (4,4) superconformal theories, the analysis of Ref. [25] is complete. In this case $C_{LR}^{int} = 6 + 3$, where 6 comes from the (4,4) system and 3 is the central charge contribution of a complex free superfield.

In D = 4, the moduli space of such theories is SO(6,22)/SO(6) × SO(22) [(the (4,4) system time T^2)], and the type II string based on this system has N = 4 supersymmetry. In this case, the uniqueness of the manifold of the moduli space can be understood either in terms of the uniqueness of matter interactions in N = 4 supergravity, or mathematically as being due to the uniqueness of K_3 as the only compact hyper–Kähler manifold. Note that N = 4 supergravity does not give only the coset space but also its explicit metric (with a definite value of the curvature). It is amusing to observe, as pointed out in Ref. [25], that the maximal rank 22 of the N = 4 superstring theories can be understood if one compactifies the (4,4) system in a D = 6, N = 2 chiral (2,0) theory, simply by the mechanism of anomaly cancellation.

Since the N = 4 theory is unique, whether we compactify the heterotic string on a torus or a type IIA or type IIB on a (4,4) system, one gets in all cases the same structure for the moduli space. Note that in D = 4 space–time dimensions, the IIA and IIB effective theories with N = 4 space–time supersymmetry just coincide.

The analysis becomes more complicated if we consider (2,2) superconformal field theories. In this case the analysis becomes more involved because, even if we use these theories as backgrounds of type II superstrings, they correspond to theories with N = 2 space–time supersymmetry. The (2,2) superconformal field theories correspond to non–trivial Kählerian σ–models. The two space–time supersymmetries come one from the left–movers and one from the right–movers.

This has to be contrasted with N = 2 space–time supersymmetry in heterotic strings in which both supersymmetries may come from the left–movers [(4,0) superconformal system].

In type II strings, the number of moduli is specified by two integers, n_v and n_s, which are related to harmonic forms (1,1) and (2,1) of an abstract σ–model on a three–dimensional complex manifold (with vanishing first Chern class).

If we denote by n_ν the number of N = 2 vector multiplets and by n_s the number of hypermultiplets in type IIA and type IIB superstrings, then

$$n_v^A = h_{1,1} \tag{4.1}$$

$$n_s^A = h_{2,1} + 1 \tag{4.2}$$

whilst the role of $h_{1,1}$ and $h_{2,1}$ is interchanged, exchanging IIA $\rightarrow$ IIB, i.e.

$$n_v^B = h_{2,1} = n_s^A - 1 \tag{4.3}$$

$$n_s^B = h_{1,1} + 1 = n_v^A + 1 \tag{4.4}$$

From these formulae we can see that there is a universal hypermultiplet, in type II strings, which corresponds to the identity operator of the superconformal field theory ($h_{1,1} = h_{2,1} = 0$).

This hypermultiplet contains, as bosons, the dilaton ϕ, the antisymmetric tensor $b_{\mu\nu}$ and two other (Ramond–Ramond) scalars A, B[25].

In heterotic strings corresponding to (2,2) superconformal systems, with N = 1 space–time supersymmetry, the $h_{1,1}$ and $h_{2,1}$ forms both correspond to Wess–Zumino multiplets. From a recent analysis [39], it turns out that the Kähler manifold, in this case, is the product manifold of the vector multiplets of type IIA and IIB superstrings compactified on the same (2,2) superconformal system.

In the present review I will report on the main results of the investigation of the moduli space of (2,2) superconformal systems and will only sketch some of the proofs. Details can be found in Ref. [39].

The main properties of type II superstring compactifications are the following:
a) The low–energy theory is described by a standard N = 2 supergravity with no unusual couplings.
b) The universal sector (see above) is described by a standard coupling of a hypermultiplet to N = 2 supergravity, where the scalars of the universal hypermultiplet parametrize the quaternionic coset SU(2,1)/U(1). This coset is in general a submanifold of the quaternionic space Q in which the hypermultiplets live.
c) In type IIA superstrings (under a mild assumption) the perturbative Kähler potential of the vector multiplets is of the general form

$$J = -logY$$

with *)

$$Y = -\frac{i}{2} d_{IJK}(z - z^*)^I (z - z^*)^J (z - z^*)^K, I, J, K = 1 \ldots h_{1,1} \tag{4.5}$$

This is an admissible N = 2 Kähler potential [30], generated by a scalar analytic function

$$f(z) = i\, d_{IJK} z^I z^J z^K \tag{4.6}$$

---

*) This result is valid in finite order σ–model perturbation theory but it is spoiled by instanton effects.

90

through the following operation:

$$Y = f + f^* - \frac{1}{2} (f_{,I} - f_{,I}^*) (z_I - z_I^*) \qquad (4.7)$$

The numerical coefficients d_{IJK} are related to a cubic norm of a generalized Jordan algebra [40]. Equation (4.6) rules out minimal $N = 2$ vector couplings in type IIA superstrings. The same result is not true for type IIB superstrings [39].

d) For type II superstrings, not all quaternionic manifolds can appear for the hypermultiplet σ–model. Only a special class, which we will call dual quaternionic manifolds, can occur; these manifolds are completely determined by a holomorphic function related to the $N = 2$ admissible Kähler metric of the vector multiplet manifold of the type II chirality–reversed theory $(A \rightleftarrows B)$.

There is an explicit operation which gives the metric of the $n_S^{A,B}$ hypermultiplets in terms of two holomorphic functions which determine the Kähler metric of the $n_V^{A,B} = n_S^{A,B} - 1$ vector multiplets. Since for type IIA superstrings the Kähler metric is known up to the d_{ABC} constants (which can be computed in any given superstring model, in terms of a three–point vertex correlator), it follows that the hypermultiplet quaternionic spaces for type IIB superstrings are known.

Surprisingly, the only symmetric quaternionic spaces which are not dual are the "minimal" quaternionic spaces (as described in the $N = 2$ tensor calculus [28]), based on the manifolds $Sp(2,n)/Sp(2) \times Sp(n)$.

In the present section we confine our discussion to points (a), (b), and (c). Point (d) will be discussed in the last section.

Statement (a) means that in type II superstrings there do not appear couplings which are not explained by the $N = 2$ supergravity tensor calculus of Refs. [28]–[30].

In fact, there are different ways of proving that in $D = 4$ there are no $bF_{\mu\nu}\widetilde{F}_{\mu\nu}$ (or $\phi F_{\mu\nu}^2$) couplings, where $b_{\mu\nu}$ is an antisymmetric tensor field, $F_{\mu\nu}$ is a (Ramond–Ramond) vector, and ϕ is the dilaton. The first way is by using a pure 4D superstring calculation which shows that the $\phi F_{\mu\nu} F_{\rho\sigma}$ correlator is proportional to

$$< \phi F_{\mu\nu} F_{\rho\sigma} > \propto Tr(\gamma^\lambda \sigma_{\mu\nu} \gamma_\lambda \sigma_{\rho\sigma}) = 0 \qquad (4.8)$$

The second way is to make a toroidal compactification or orbifold compactification of the $N = 2$, $D = 10$ (type IIa or type IIB) supergravity Lagrangian. This makes sense since the universal hypermultiplet does not depend on the internal degrees of freedom.

By making the Witten SU(3) truncation [26] of type IIA supergravity, one is left with the universal hypermultiplet, the graviphoton, and a vector multiplet with bosonic degrees of freedom given by

$$A_\mu, A_{\mu i j}, g_{i \bar j}, b_{i \bar j}, b_{\mu\nu}, \phi, A_{ijk} \tag{4.9}$$

where A_μ comes from the N = 2, D = 10 vector, and $A_{\mu i j}$ comes from the N = 2, D = 10 three–form $A_{\mu\nu\rho}$. An explicit calculation [40] shows that no couplings like (4.8) exist. Another argument is an anomaly argument [39]. The $bF_{\mu\nu}\widetilde F_{\mu\nu}$ coupling (b is the axion) would imply an anomalous transformation for $b_{\mu\nu}$ and the explicit appearance of a Chern–Simons coupling. However, Ramond vertices appear only through field strengths, so such coupling is impossible. In other words, the argument for the absence of an axion coupling follows the same argument as the one which forbids gauge couplings for (R–R) vectors. The last argument uses the duality symmetry of the equations of motions of the Cremmer–Julia theory [35], of which the universal sector can be obtained through a suitable SU(3)–invariant projection.

The explicit N = 2 theory which corresponds to an SU(3) invariant truncation of N = 2, D = 10 type IIA supergravity, with six scalars given by $g_{i \bar i}, b_{i \bar i}, A_{ijk}, \phi$, and $b_{\mu\nu}$, gives a unique σ–model based on the coset space

$$\frac{SU(1,1)}{U(1)} \times \frac{SU(2,1)}{SU(2) \times U(1)} \tag{4.20}$$

If we interchange type IIA with type IIB theory, the same truncation gives rise to two hypermultiplets: one is the universal one, and the other corresponds to the $h_{1,1} = 1$ form of the type IIA theory.

In terms of the type IIB supergravity, the eight scalar states are now given by

$$g_{i \bar i}, b_{i \bar i}^c, b_{\mu\nu}^c, A_{\mu\nu i \bar i}, \phi^c \tag{4.11}$$

where $b_{\mu\nu}^c$ is the complex 10D antisymmetric tensor, and $A_{\mu\nu\rho\lambda}$ is the four–form with self–dual field strength [34].

The coset space in which these hypermultiplets live is [39]

$$G_{2,2}/SO(4) \tag{4.12}$$

where $G_{2,2}$ is a non–compact form of the exceptional group G_2.

Note that the coset spaces given by formulae (4.10) and (4.12) are uniquely obtained from a truncation of 4D, N = 8 supergravity with the following embedding of $SO(6) \approx SU(4)$ in SU(8):

Type IIA supergravity $8 \rightarrow 4 + \bar 4$,
Type IIB supergravity $8 \rightarrow 4 + 4$.

The two coset spaces emerge from an SU(3) invariant truncation of the original [35] coset space $E_{7,7}/SU(8)$.

The same strategy can be used for more general compactifications. In all orbifold compactifications of type II superstrings the untwisted sector can be deduced from a $\mathcal{Z}_N$ invariant truncation of the D = 10, N = 2 supergravity Lagrangians. Therefore, at least for the untwisted scalar fields, i.e. the moduli space of orbifolds, we know all the Kähler and quaternionic manifolds of vector and hypermultiplets.

The list of the $\mathcal{Z}_N$ symmetric orbifolds and the related moduli space (for untwisted moduli) are reported in Tables I, II and III.

Note that these manifolds are uniquely determined in terms of two holomorphic functions of $h_{1,1}$ and $h_{2,1}$ variables, respectively. If we also consider the $h_{2,1}$ vector multiplets of type IIB superstrings, then the product space of these N = 2 Kähler manifolds coincides with the Kähler space of the moduli of heterotic superstrings [22, 27] (with N = 1 space–time supersymmetry) compactified on the same (2,2) superconformal system. The only difference is that now the universal multiplet is a Wess–Zumino multiplet which parametrizes an SU(1,1)/U(1) manifold (of given curvature).

To give a further example, let us consider a type II superstring compactified on the orbifold $T_6/\mathcal{Z}_3$ which was already considered in Sections 2 and 3. In the untwisted sector,

$$h^0_{1,1} = 9 \quad , \quad h^0_{2,1} = 0 \tag{4.13}$$

Therefore for a type IIA theory we have the universal hypermultiplet and nine vector multiplets which parametrize the coset space [39]:

$$\frac{SU(2,1)}{SU(2) \times U(1)} \times \frac{SU(3,3)}{SU(3) \times SU(3) \times U(1)} \tag{4.14}$$

These states come from the supergravity fields

$$g_{i\bar{j}}, b_{i\bar{j}}, b_{\mu\nu}, \phi, A_{ijk} \tag{4.15}$$

If we interchange types IIA $\rightleftarrows$ IIB we get a quaternionic manifold of dimension $4(9+1) = 40$ (no N = 2 vector multiplets are present), which is [39]

$$E_{6(2)}/SU(2) \times SU(6) \tag{4.16}$$

This manifold is uniquely determined from a $\mathcal{Z}_3$ invariant projection of the original $E_{7,7}/SU(8)$ manifold of the N = 8, 4D supergravity theory.

The two manifolds shown above are examples of dual quaternionic manifolds.

We would like to emphasize that in order to show that the coset manifolds, discussed until now, correspond to standard N = 2 coupling, we have to use duality properties of the 4D N = 2 supergravity theory [30, 31]. This duality property

Table I – Symmetric $\mathcal{Z}_N$ Coxeter Orbifolds

Case	Lattice	$\mathcal{Z}_N$	(a,b,c,)	N	χ	$h_{1,1}$	$h_{2,1}$	$h_{1,1}^{(0)}$	$h_{2,1}^{(0)}$
1	$[SU(3)]^3$	$\mathcal{Z}_3$	1/3(1,1, -2)	27	72	36	0	9	0
2	$SU(3) \times (G_2)^2$	$\mathcal{Z}_6$	1/6(1,1, -2)	3	48	29	5	5	0
3	$[SU(4)]^2$	$\mathcal{Z}_4$	1/4(1,1, -2)	16	48	31	7	5	1
4	$SU(3) \times SO(8)$ $SU(2) \times SU(6)$	$\mathcal{Z}_6$	1/6(1,2, -3)	12	48	35	11	3	1
5	$SO(9) \times SO(4)$ $SO(8) \times SO(4)$	$\mathcal{Z}_8$	1/8(1,2, -3)	8	48	28	4	3	1
6	$SO(4) \times F_4$	$\mathcal{Z}_{12}$	1/12(1,4, -5)	4	48	31	7	3	1
7	$SO(5) \times SO(9)$ $SO(5) \times SO(8)$	$\mathcal{Z}_8$	1/8(1,3, -4)	4	48	27	3	3	0
8	$SU(3) \times SO(8)$ $SU(3) \times F_4$ E_6	$\mathcal{Z}_{12}$	1/12(1,4, -5)	3	48	27	3	3	0
9	$SU(7)$	$\mathcal{Z}_7$	1/7(1,2, -3)	7	48	24	0	3	0

Table II – Scalar σ–models for the UNTWISTED sector of type IIA string compactified on the orbifolds of Table I

Case	Vector multiplet manifold	$\dim_{\mathcal{C}}$	Jordan Algebra	Hypermultiplet manifold	dim.
1	$U(3,3)/[U(3) \times U(3)]$	9	$J_3^{\mathcal{C}}$	$SU(2,1)/U(2)$	4
2	$\dfrac{SU(1,1)}{U(1)} \times \dfrac{SO(2,4)}{SO(2) \times SO(4)}$	5	$\mathcal{R} + Q(4)$	$SU(2,1)/U(2)$	4
3	$\dfrac{SU(1,1)}{U(1)} \times \dfrac{SO(2,4)}{SO(2) \times SO(4)}$	5	$\mathcal{R} + Q(4)$	$U(2,2)/[U(2) \times U(2)]$	8
4,5,6	$[SU(1,1)/U(1)]^3$	3	$\mathcal{R} + \mathcal{R} + \mathcal{R}$	$U(2,2)/[U(2) \times U(2)]$	8
7,8,9	$[SU(1,1)/U(1)]^3$	3	$\mathcal{R} + \mathcal{R} + \mathcal{R}$	$SU(2,1)/U(2)$	4

Table III – Same as Table II, for type IIB

Case	Vector manifold	$\dim_{\mathcal{C}}$	Hypermultiplet manifold	dim.	Jordan algebra
1	-	0	$E_{6(+2)}/[SU(2) \times SU(6)]$	40	$J_3^{\mathcal{C}}$
2	-	0	$SO(4,6)/[SO(4) \times SO(6)]$	24	$\mathcal{R} + Q(4)$
3	$SU(1,1)/U(1)$	1	$SO(4,6)/[SO(4) \times SO(6)]$	24	$\mathcal{R} + Q(4)$
4,5,6	$SU(1,1)/U(1)$	1	$SO(4,4)/[SO(4) \times SO(4)]$	16	$\mathcal{R} + \mathcal{R} + \mathcal{R}$
7,8,9	-	0	$SO(4,4)/[SO(4) \times SO(4)]$	16	$\mathcal{R} + \mathcal{R} + \mathcal{R}$

relates the non–compact symmetries of the scalar σ–models to the scalar–vector–vector couplings of the theory.

We conclude this section by discussing point (c).

In the case of type IIA superstrings, the vector multiplets are associated with harmonic (1,1) forms

$$\omega_{ij}^A(x^k, \bar{x}^\ell) \quad A = 1, \ldots h_{1,1}$$

The scalar background fields $b^A(X^\mu)$, associated with these forms, enter in the Lagrangian of the 2d σ–model through a contribution to the torsion two–form B_{ij}

$$i \int_{S^2} \sum_A b^A(x^\mu) \omega_{ij}^A(x^k, \bar{x}^\ell) \varepsilon^{\alpha\beta} \partial_\alpha x^i \partial_\beta \bar{x}^j \tag{4.17}$$

Now, neglecting world–sheet instantons, we have a Peccei–Quinn symmetry as in Ref. [26]. In fact under a constant shift

$$\delta b^A(x) \;=\; C^A \tag{4.18}$$

the torsion three–form H = dB is unchanged. Therefore, Eq. (4.18) is an exact invariance of the σ–model equations of motion to all orders in perturbation theory. As a consequence, the effective Lagrangian should be invariant under formula (4.18). Since b^A is related to the real part of the complex scalar field associated with an N = 2 vector multiplet, the N = 2 Kähler potential should remain invariant under $Re z^A \rightarrow c^A$. In N = 2 tensor calculus (of which we assume the validity) this implies that [30, 31].

$$\delta f(z) \;=\; i P_2(z) \tag{4.19}$$

where P_2 is a polynominal, with real coefficients, of degree 2 in the z fields. The most general function which has this property is

$$f(z) \;=\; i d_{ABC} z^A z^B z^C \;+\; i P_2(z) \;+\; \lambda \tag{4.20}$$

(where λ is a real constant). In all the models that we have constructed $\lambda = 0$, so we get the Kähler manifold as in Eq. (4.5). It can be shown that the scalar manifolds obtained by the reduction of N = 2, D = 10 type IIA supergravity are all of this type. In the simplest example [given by Eq. (4.10)] with one multiplet, $f(z) = iz^3$.

5. STANDARD SUPERGRAVITY FORM FOR TYPE-II EFFECTIVE LAGRANGIANS IN D=4 DIMENSIONS

In this section we will show, in terms of the original 10D fields, the appropriate combinations that give rise to a standard supergravity form in four dimensions. The starting point is to consider N = 2, 10D supergravity, rather than N = 8 supergravity in D = 4. The reason for this is that in the Cremmer–Julia theory [35], all vectors have the same parity [as required by SO(8) symmetry] while *) the reduction from D = 10, has two vectors of opposite parities, in agreement with an N = 2 supergravity coupling of a matter vector multiplet specified by a holomorphic function which is a cubic monomial [39]. †)

Type IIA 10D supergravity can be obtained by dimensional reduction from D = 11 to D = 10 of the maximal supergravity of Cremmer, Julia and Scherk [34].

The 11D supergravity (bosonic) fields are the vierbein $\hat{e}^{\hat{a}}_{\hat{\mu}}$ and the three–form $\hat{A}_{\hat{\mu}\hat{\nu}\hat{\rho}}$. Dimensional reduction to D = 10 is obtained by fixing a triangular gauge for $\hat{e}^{\hat{a}}_{\hat{\mu}}$

$$\hat{e}^{\hat{a}}_{\hat{\mu}} = \begin{pmatrix} e^a_\mu & e^{11}_\mu \\ 0 & e^{11}_{11} \end{pmatrix} \quad \hat{a},\hat{\mu} = 1\ldots 11, a,\mu = 1\ldots 10 \tag{5.1}$$

with

$$e^{11}_{11} = \phi \quad , \quad e^{11}_\mu = \phi Z_\mu \tag{5.2}$$

Furthermore, from the three–form we define the two 10D gauge fields ‡

$$B_{\mu\nu} = \hat{A}_{\mu\nu 11} \quad , \quad H_{\mu\nu\rho} = 3\partial_{[\mu} B_{\nu\rho]} \tag{5.3}$$

$$A_{\mu\nu\rho} = \hat{A}_{\mu\nu\rho} - 3Z_{[\mu} B_{\nu\rho]} \tag{5.4}$$

with $F_{a_1 a_2 a_3 a_4} = \hat{F}_{a_1 a_2 a_3 a_4} - 6Z_{[a_1 a_2} B_{a_3 a_4]}$. It is trivial to show that $B_{\mu\nu}, Z_\mu$, and $A_{\mu\nu\rho}$ transform as 10D, N = 1 supergravity, ϕ and $B_{\mu\nu}$ belong to the gravity multiplet, whilst $\mathcal{Z}_\mu$ and $A_{\mu\nu\rho}$ belong to the multiplet of the second gravitino.

The D = 11 (bosonic part of the) Lagrangian is [34]

$$\mathcal{L} = -\frac{1}{2}\hat{e}\hat{R} - \frac{1}{48}\hat{e}(\hat{F}_{\hat{\mu}_1\ldots\hat{\mu}_4})^2 + \frac{\sqrt{2}}{(12)^4}\varepsilon^{\hat{\mu}_1\ldots\hat{\mu}_{11}}\hat{F}_{\hat{\mu}_1\ldots\hat{\mu}_4}\hat{F}_{\hat{\mu}_5\ldots\hat{\mu}_8}\hat{A}_{\hat{\mu}_9\hat{\mu}_{10}\hat{\mu}_{11}} \tag{5.5}$$

Using the definition given for Eqs. (5.1) to (5.4), we get for the terms in Eq. (5.5) the following Lagrangian in D = 10:

$$-\frac{1}{2}\hat{e}\hat{R} \to -\frac{1}{2}\phi eR - \frac{1}{8}e\phi^3 Z^2_{\mu\nu} \tag{5.6}$$

$$-\frac{1}{48}\hat{e}(F_{\hat{\mu}_1\ldots\hat{\mu}_4})^2 \to -\frac{1}{48}\phi e\{(F_{\mu_1\ldots\mu_4} + 6Z_{[\mu_1\mu_2} B_{\mu-3\mu_4]})^2 + 4\phi^{-2}(H_{\mu_1\mu_2\mu_3})^2\} \tag{5.7}$$

---

*) The two formulations are related by a duality transformation [35] which is not needed here.

†) Here we will work out only the bosonic sector of the Lagrangian, the rest being determined by supersymmetry.

‡ Under the B gauge transformation A is not inert, i.e. $\delta B = d\eta, \delta A = Z \wedge d\eta$.

$$\frac{\sqrt{2}}{(12)^4} \, \varepsilon^{\hat{\mu}_1 \ldots \hat{\mu}_{11}} \, \widehat{F}_{\hat{\mu}_1 \ldots \hat{\mu}_4} \widehat{F}_{\hat{\mu}_5 \hat{\mu}_8} \widehat{A}_{\hat{\mu}_9 \ldots \hat{\mu}_{11}} \rightarrow \frac{\sqrt{2}}{(48)^2} \, \varepsilon^{\mu_1 \ldots \mu_{10}} \{ (F_{\mu_1 \ldots \mu_4} +$$

$$+6 Z_{\mu_1 \mu_2} B_{\mu_3 \mu_4}) F_{\mu_5 \ldots \mu_8} B_{\mu_9 \mu_{10}} + 12 Z_{\mu_1 \mu_2} Z_{\mu_3 \mu_4} B_{\mu_5 \mu_6} B_{\mu_7 \mu_8} B_{\mu_9 \mu_{10}} \} \qquad (5.8)$$

In order to have a canonically normalized Einstein action, we perform a Weyl rescaling by using the general formula, valid in any dimension D:

$$-\sqrt{g} \Omega^{D-2} R \rightarrow -\sqrt{g} R - \sqrt{g}(D-1)(D-2)(\partial_\mu \, log\Omega)^2$$

$$g_{\mu\nu} \rightarrow \Omega^{-2} g_{\mu\nu} \quad (e^a_\mu \rightarrow \Omega^{-1} e^a_\mu) \qquad (5.9)$$

For D = 10, then, $\Omega = \phi^{\frac{1}{8}}$, and we get from formulae (5.6) and (5.7),

$$-\frac{1}{2} \hat{e} \widehat{R} \rightarrow e \left(-\frac{1}{2} R - \frac{9}{16} (\partial_\mu log \, \phi)^2 - \frac{1}{8} \phi^{\frac{9}{4}} Z^2_{\mu\nu} \right) \qquad (5.10)$$

$$-\frac{1}{48} \hat{e} (\widehat{F}_{\hat{\mu}\hat{\nu}\hat{\rho}\hat{\sigma}})^2 \rightarrow -\frac{1}{48} e \phi^{\frac{3}{4}} (F_{\mu\nu\rho\sigma} + 6 Z_{[\mu\nu} B_{\rho\sigma]})^2 - \frac{1}{12} e \phi^{-\frac{3}{2}} (H_{\mu\nu\rho})^2 \quad (5.11)$$

the topological term being unaffected by the Weyl rescaling. Formulae (5.8), (5.10), and (5.11) give the bosonic sector of type IIA supergravity in D = 10 dimensions.

To perform the dimensional reduction to D = 4, following ref. [26], we keep only the SU(3) singlets in the internal indices:

$$g_{\mu\nu} \rightarrow g_{\mu\nu} \quad , \quad g_{i\bar{j}} = e^\sigma \delta_{i\bar{j}}$$

$$B_{\mu\nu} \rightarrow B_{\mu\nu} \quad , \quad B_{i\bar{j}} = i a \delta_{i\bar{j}} \qquad (5.12)$$

$$Z_\mu \rightarrow Z_\mu \quad , \quad \phi \rightarrow \phi$$

$$A_{\mu i \bar{j}} = i A_\mu \delta_{i\bar{j}} \quad , \quad A_{ijk} = C \varepsilon_{ijk} \qquad (5.13)$$

The new terms, with respect to the N = 1, D = 10 supergravity, are those coming from the Z_μ vector field and the three-from $A_{\mu\nu\rho}$. In D = 4 they reduce to

$$-\frac{1}{8} e \phi^{\frac{9}{4}} Z^2_{\mu\nu} \rightarrow -\frac{1}{8} e \phi^{\frac{9}{4}} e^{3\sigma} Z^2_{\mu\nu} \qquad (5.14)$$

which is Weyl–invariant,

$$-\frac{1}{48} e \phi^{\frac{3}{4}} (F_{\mu\nu\rho\sigma} + 6 Z_{[\mu\nu} B_{\rho\sigma]})^2 \rightarrow$$

$$- \frac{3}{4} \, e \phi^{\frac{3}{4}} \, e^{\sigma} (F_{\mu\nu} + a Z_{\mu\nu})^2 \; - e \sigma^{\frac{3}{4}} e^{-3\sigma} |\partial_\mu C|^2 \; - \; \frac{1}{48} \, e \sigma^{\frac{3}{4}} e^{9\sigma} (F_{\mu\nu\rho\sigma} + 6 Z_{[\mu\nu} B_{\rho\sigma]})^2$$

$$(5.15)$$

after a Weyl rescaling $e \to \Omega^{-1} e$, with $\Omega = e^{3\sigma/3}$, which brings the Einstein term to the canonical normalization - 1/2 e R.

Finally, the topological term (which is Weyl–invariant) reduces to

$$- \frac{i}{\sqrt{26}} \, \varepsilon^{\mu\nu\rho\sigma} \overline{C} \overset{\leftrightarrow}{\partial}_\mu C H_{\nu\rho\sigma} + \frac{1}{2\sqrt{2}} \, \varepsilon^{\mu\nu\rho\sigma} (a^3 Z_{\mu\nu} Z_{\rho\sigma} + 3 a^2 Z_{\mu\nu} F_{\rho\sigma} + 3 a F_{\mu\nu} F_{\rho\sigma})$$

$$(5.16)$$

Formulae (5.14), (5.15), and (5.16) must be added to the part of the N = 1, D = 10 bosonic reduction (worked out in Ref. [26]), which is

$$- \frac{1}{2} \, eR - 3e(\partial_\mu \sigma)^2 \; - \; \frac{9}{16} \, e(\partial \, log\phi)^2 \; - \; \frac{3}{2} e\phi^{-\frac{3}{2}} e^{-2\sigma} (\partial_\mu a)^2$$

$$- \frac{1}{12} \, e\sigma^{-\frac{3}{2}} e^{6\sigma} (H_{\mu\nu\rho})^2 \qquad (5.17)$$

with

$$H_{\mu\nu\rho} = 3 f_{\mu\nu\rho} [\text{ref. 10}], \quad B_{\mu\nu} = a_{\mu\nu} \; [\text{ref. 10}]$$

If we now perform a duality transformation on the $B_{\mu\nu}$ field, this amounts to adding a Lagrange multiplier

$$\frac{1}{3} \, \phi^{-\frac{3}{2}} e^{6\sigma} H_{\mu\nu\rho} = e\varepsilon_{\mu\nu\rho\lambda} \left(\partial^\lambda D + \frac{i}{3\sqrt{2}} \, \overline{C} \overset{\leftrightarrow}{\partial}^\lambda C \right) \qquad (5.18)$$

and to integrating over $H^{\mu\nu\rho}$ treated as an auxiliary field. Owing to the topological term [Eq. 5.16)], this gives an interference between the D and C fields.

If we now define the complex field

$$\begin{aligned} S &= \phi^{-\frac{3}{4}} e^{3\sigma} + i3\sqrt{2} D + \overline{C}C, & A_\mu^0 &= \sqrt{2} Z_\mu \\ iT &= \phi^{\frac{3}{4}} e^{\sigma} - i\sqrt{2}a & A_\mu^1 &= 2 A_\mu, \tfrac{x^1}{x^0} = T \end{aligned} \qquad (5.19)$$

we get, for the scalar kinetic terms, a Kähler manifold with the Kähler potential given by

$$K = -3 log \, i(T - \overline{T}) \; - \; log(S + \overline{S} - 2\overline{C}C) \qquad (5.20)$$

In agreement with Ref. [39] this is a product of an N = 2 Kähler manifold [18] SU(1,1)/U(1) and a quaternionic manifold SU(2,1)/SU(2) × U(1) which corresponds to the "universal sector" of any (2,2) conformal system in type II superstrings. The vector kinetic terms depend only on the T field, and they are in a standard supergravity form if, in the language of the tensor calculus [28–30], one defines the holomorphic F function to be

$$F(T) = iT^3 = i \left(\frac{X^1}{X^0} \right)^2 = (X^0)^{-2} f(X^1, X^0) \qquad (5.21)$$

In terms of Eq. (5.21), the normalization of the vector kinetic term and of the topological $F\tilde{F}$ terms is given by the real and the imaginary part of the kinetic matrix [28] (I,J = 0,1):

$$\mathcal{N}_{IJ} = \frac{1}{4} \overline{F}_{IJ} - \frac{(NX)_I (NX)_J}{(X^L N_{LM} X^M)} \qquad (5.22)$$

with

$$F_{IJ} = \partial_I \partial_J F , \quad N_{IJ} = \frac{1}{4}(F_{IJ} + \overline{F}_{IJ})$$

The entries 0,1 refer to the Z_μ and A_μ fields, respectively. The three entries of the $\mathcal{N}$ matrix are given by

$$
\begin{aligned}
\mathcal{N}_{00} &= \tfrac{i}{32} (T - \overline{T})^3 - \tfrac{3i}{32} (T - \overline{T})(T + \overline{T})^2 - \tfrac{i}{16} (T + \overline{T})^3 \\
\mathcal{N}_{01} &= \tfrac{3i}{16} (T - \overline{T})(T + \overline{T}) + \tfrac{3i}{16} (T + \overline{T})^2 \\
\mathcal{N}_{11} &= -\tfrac{3i}{8} (T - \overline{T}) - \tfrac{-3i}{4} (T + \overline{T})
\end{aligned}
\qquad (5.23)
$$

By using the formula

$$\frac{1}{4} F_{\mu\nu}^{+I} \mathcal{N}_{IJ} F_{\mu\nu}^{+J} + h.c. (F_{\mu\nu}^+ = \text{self–dual combination})$$

one easily reproduces the terms given in formulae (5.15) and (5.16). We note, in particular, the Peccei–Quinn symmetry [39],

$$a \to a + c \quad , \quad A_\mu \to A_\mu - cZ_\mu \qquad (5.24)$$

under which the Lagrangian is invariant, up to a total derivative induced by the topological term in Eq. (5.16). Formulae (5.24) tell us that the Peccei–Quinn symmetry is a rotation between the graviphoton and the matter vector. In the language of the N = 2 superconformal tensor calculus, it is a rotation between a matter vector multiplet and the compensating vector multiplet.

In conclusion, we see that no coupling is present between the four–dimensional (dilaton) S and the vector fields. Therefore no unusual couplings are generated in N = 2 supergravity theories, even when the two supersymmetries come one from the left–movers and one from the right–movers.

From the general analysis given in Ref. [39] it is rather obvious that these results are model–independent and are properties of the effective Lagrangian of type II superstrings compactified on an arbitrary (2,2) superconformal system.

6. KÄHLER MANIFOLD FOR (1,1) MODULI FROM INTERSECTION MATRICES

In the previous sections we have shown that the Kähler potential associated with complex fields in K_A, i.e. those massless modes corresponding to deformations of the Kähler class, has the general perturbative form

$$K_A = - \ln Y_A \tag{6.1}$$

with

$$Y_A = - \frac{i}{2} d_{abc}(z^a - z^{xa})(z^b - z^{xb})(z^c - z^{xc}) \tag{6.2}$$

In Ref. [39] it was also shown that an explicit formula for the coefficients d_{abc} in Eq. (4.5) can be obtained in terms of the three–point functions of the underlying superconformal field theory.

Here we want to observe further that the relevant three–point functions are just the same as those defining the Yukawa coupling [41] [for the families in heterotic strings corresponding to $h_{(1,1)}$ massless modes], hence they are just proportional to the intersection matrix.

$$g_{abc} = \int_{K_3} \omega_a \wedge \omega_b \wedge \omega_c \tag{6.3}$$

where $\{[\omega_a]\}$ is a (real) basis for $H^{1,1}(K)$.

The simplest way to realize that Eq. (6.3) is related (modulo an irrelevant normalization) to the coefficients d_{abc} is to look at the effective couplings in the field theory limit [41].

In the 4D Lagrangian there is a term proportional to [28]

$$Rez^a F^b_{\mu\nu} F^c_{\rho\sigma} \varepsilon^{\mu\nu\rho\sigma} \tag{6.4}$$

whose coefficient is precisely d_{abc}.

In the case of compactification on a smooth Calabi–Yau manifold the coefficients for this coupling can be obtained by Kaluza–Klein techniques. The couplings we are looking for, containing an $\varepsilon_{\mu\nu\rho\sigma}$ symbol, should come from a term [42]

$$\int_{M_4 \times K_3} B \wedge F \wedge F \tag{6.5}$$

of the 10D Lagrangian. Here F = dA is the field strength of the three–form A, and B is a two–form of 10D type IIA supergravity.

Then, in terms of the (1,1) harmonic functions ω^a, and the space–time (4D) scalars $\phi^a = Rez^a$ and vectors A^a_μ, the relevant coupling reads:

$$\int_{M_4} d^4x \phi^a F_{\mu\nu}^b F_{\rho\sigma}^c \varepsilon^{\mu\nu\rho\sigma} \int_{K_3} \omega_a \wedge \omega_b \wedge \omega_c \qquad (6.6)$$

Comparing this equation with the coefficients d_{abc}, we get the identification [29]

$$d_{abc} = g_{abc} \qquad (6.7)$$

In Ref. [41] a kinetic matrix for the space–time scalars corresponding to (1,1) form was given. We will now compare our results with the one given in Ref. [41].

From the way in which Peccei–Quinn symmetry acts on the space–time fields, we get the following identification for the "torsion" and Kähler classes:

$$[\tau] = 2Re\, z^a [\omega_a] \qquad (6.8)$$

$$[\omega] = -2Im\, z^a [\omega_a] = Y^a [\omega_a] \qquad (6.9)$$

from which we infer for the Kähler class J,

$$J = [\omega_0] = \langle Y^a \rangle [\omega_a] \qquad (6.10)$$

The above identification is made possible by the fact that in N = 2 supegravity the Kähler manifold has a preferred set of coordinates [28] (in the N = 1 formulation this structure is completely hidden).

From the relation between the d_{abc} coefficients and the intersection numbers g_{abc} [Eqs. (6.6) and (6.7)], we get the following identification between Ref. [41] and our results [42]:

$$g(J, J, J) \rightarrow d_{abc} Y^a Y^b Y^c = d_{yyy} \qquad (6.11)$$

$$g(J, J, F_a) \rightarrow d_{abc} Y^b Y^c = (d_{yy})_a \qquad (6.12)$$

$$g(J, F_a, F_b) \rightarrow d_{abc} Y^c = (d_y)_{ab} \qquad (6.13)$$

The Kähler metric corresponding to the Kähler potential in Eq. (6.1) is [30]

$$G_{ab} = 9 \frac{(d_{yy})_a (d_{yy})_b}{(d_{yyy})^2} - 6 \frac{(d_y)_{ab}}{d_{yyy}} \qquad (6.14)$$

which differs from the normalization given in Ref. [41] by an extra volume factor

$$24[d_{yyy}]^{-1} = \frac{1}{2} \, [\text{vol of } K_3]^{-1} \qquad (6.15)$$

This extra factor is due to the Weyl rescaling, which is needed in order to get the Einstein term in canonical form. Without this conformal factor the metric would not be Kählerian.

The fact that the Kähler metric obtained from our $N = 2$ technique is the Kähler metric, conformal with the one of Ref [41], is a further confirmation of our correct space–time identification [Eq. (6.10)] of the Kähler class.

Our results can be checked in explicit examples, and they allow us to compute the non–linear σ–model of the space–time scalar fields in terms of pure topological data of an abstract (2,2) superconformal theory.

As an example we consider the orbifold $T_6/\mathcal{Z}_3$ [12]. This manifold is a singular limit of a Calabi–Yau manifold. It is realized by a six–dimensional torus, identifying points under the $\mathcal{Z}_3$ action and then blowing up 27 singular points. The Hodge numbers for $T_6/\mathcal{Z}_3$ are $h_{(1,1)} = 36$ and $h_{(2,1)} = 0$.

Nine $(1,1)$ forms come from radial deformations, the so–called untwisted sector and 27 $(1,1)$ forms come from the blowing up modes.

Since the blowing up modes have non–zero intersection which is completely diagonal [41], the associated holomorphic function for the 36–dimensional complex manifold of the moduli is [21, 27, 39 42, 43, 44]

$$F(T,B) = det(\lambda \cdot T) + a \sum_{i=1}^{27} (B_i)^3$$

where $\lambda \cdot T$ is a 3×3 matrix, and λ are the nine Gell–Mann matrices. For $B_i = 0$ the holomorphic function $F(T)$ gives rise to the Kähler metric of the symmetric Kähler space $\frac{U(3,3)}{U(3) \times U(3)}$. This result has been obtained in the literature with several methods [21, 27, 39, 42, 43, 44]. Among the $\mathcal{Z}_N$ orbifolds corresponding to (2,2) superconformal systems this is the symmetric space for the untwisted moduli with a larger hisotropy group. This is merely due to the fact that $\mathcal{Z}_3$ is the center of $SU(3)$.

For $B_i \neq 0$ the Kähler metric corresponds to a non homogeneous space. Note that the corresponding holomorphic function was already considered in the literature as an example of Kähler ($N = 2$) manifolds giving rise to a flat potential. In the orbifold limit $< B >= 0$ and the metric becomes singular in the twisted directions, so the twisted modes do not propagate. This is in agreement with the fact that in the point–field theory limit the twisted modes are confined at the singular points of the orbifold. The kinetic terms for the twisted states come from σ model non–perturbative corrections.

Another interesting example is the untwisted sector of the T_6/Z_4 orbifold with Hodge numbers $h_{(1,1)} = 5$ and $h_{(2,1)} = 1$. In this case the two holomorphic functions for the (1,1) and (2,1) moduli are respectively [39, 44]

$$f_A(Z_1, Z_2, Z_3, Z_4, Z_5) = iZ_1(Z_2^2 - Z_3^2 - Z_4^2 - Z_5^2)$$
$$B_B(\eta) = 1 - \phi^2$$

These functions give rise to the following Kähler manifolds [39, 44]

$$k_A = \frac{SU(1,1)}{U(1)} \times \frac{SO(2,4)}{SO(2) \times SO(4)} \; , \; k_B = \frac{SU(1,1)}{U(1)}$$

We observe that the F function for the (2,1) modulus is quadratic rather than cubic. This is consistent with the fact that this function is not protected by a Peccei–Quinn symmetry [39].

7. N=2 KÄHLER MANIFOLD FOR (2,1) MODULI

Recently it has been shown the connection of the holomorphic F_B function of the (2,1) moduli with the Calabi–Yau geometry [43, 45,46].

Let us consider deformations of the complex structure, namely (2,1) forms denoted by

$$G_a = G_{ak\lambda\overline{\mu}}dx^k \wedge dx^\lambda \wedge d\overline{x}^\mu \qquad \alpha = 1, \ldots h_{(2,1)} \qquad (7.1)$$

From Ref. [41] it follows that the metric for the (2,1) moduli in target space is

$$G_{a\overline{b}} = -\int_{K_3} G_a \wedge G_b / \int_{K_3} \Omega \wedge \overline{\Omega} \qquad (7.2)$$

where Ω is the (3,0) holomorphic three form on the Calabi–Yau three–fold. A usual formula gives the variation of the three form Ω under the variation of the complex structure

$$\frac{\partial \Omega}{\partial \phi_a} = k_a \Omega + G_a \qquad (7.3)$$

A little algebra reveals that

$$G_{a\overline{b}} = -\frac{\partial}{\partial \phi_a \partial \overline{\phi}_b} log \int_{K_3} \Omega \wedge \overline{\Omega} \qquad (7.4)$$

which shows that the moduli space of (2,1) form is indeed Kähler. This is what is required by N = 1 space–time supersymmetry.

However we need more since N = 2 space–time supersymmetry would allow the determination of $Y = i \int_{K3} \Omega \wedge \overline{\Omega}$ in terms of a single holomorphic function of local coordinates on the moduli space. This is achieved, following a recent work of Candelas, Green and Hubsch [47], by describing the space of complex structures by studying the periods of holomorphic three–forms.

By defining the A_a, B_b cycles $a, b = 1 \ldots h_{(2,1)} + 1$ and the cohomology basis α_a, β^b so that

$$\int_{A^a} \alpha_a = \int_{K_3} \alpha_a \wedge \beta^b = \delta_a^b$$

$$\int_{B^a} \beta^b = \int_{K_3} \beta^b \wedge \alpha_a = -\delta_a^b \tag{7.5}$$

the complex moduli can be defined as

$$Z^a = \int_{A^\alpha} \Omega = \int_{K_3} \Omega \wedge \beta^b \tag{7.6}$$

Since a rescaling of $Z^a \to \lambda Z^a$ just rescales Ω, Z^a can be regarded as local projective coordinates for the moduli space, with Ω homogeneous of degree 1 in Z. The holomorphic three–form can be expanded in terms of the basis

$$\Omega = Z^a \alpha_a + i F_a(Z) \beta^a \tag{7.7}$$

since

$$\int_{K_3} \Omega \wedge \frac{\partial \Omega}{\partial Z^a} = 0 \tag{7.8}$$

it follows that

$$2 F_a = \frac{\partial}{\partial Z^a} (Z^c F_c) \tag{7.9}$$

from which it follows that locally F_a is the gradient of a holomorphic function of degree 2:

$$F_a(Z) = \partial/\partial Z^a F(Z), \quad F(\lambda Z) = \lambda^2 F(Z) \tag{7.10}$$

It also follows that

$$i \int \Omega \wedge \overline{\Omega} = Z^a \overline{F}_a + \overline{Z}^a F_a = \overline{Z}^a (F_{ab} + \overline{F}_{ab}) Z^b \tag{7.11}$$

If we go to inhomogeneous coordinates

$$\phi^a = Z^a/Z^0 , \quad F(\phi^a) = Z_0^{-2}F(Z)$$

then we obtain

$$Y = F + \overline{F} - \frac{1}{2}(F_a - \overline{F}_a)(\phi^a - \overline{\phi}^a) \tag{7.12}$$

which gives the standard form of the Kähler potential for N = 2 supergravity.

It also follows that

$$\int \Omega \wedge \Omega = \int \Omega \wedge \frac{\partial \Omega}{\partial Z^a} = \int \Omega \wedge \frac{\partial \Omega}{\partial Z^a \partial Z^b} = 0$$

and

$$\int \Omega \wedge \frac{\partial^n \Omega}{\partial Z^{a_1} \dots \partial Z^{a_n}} = i(2-n)\frac{\partial^n F}{\partial Z^{a_1} \dots \partial Z^{a_n}} , \quad n \geq 3$$

These integrals determine F up to quadratic function $n_{ab}Z^a Z^b$ ($a, b = 1 \dots h_{(2,1)} + 1$). The third derivative of F is related to the Yukawa couplings in heterotic superstrings [41, 46, 47]. In the orbifold example of the previous section, with a quadratic F, the Yukawa coupling for the (2,1) untwisted family therefore vanishes. This fact was noted in Ref. [48] where the effective Lagrangian for the untwisted sector of the $T_6/\mathcal{Z}_4$ orbifold, including the charged fields, was given.

8. CALABI–YAU COMPACTIFICATIONS FOR HETEROTIC STRINGS

In the previous sections we have seen that the metric on the moduli space is parametrized in terms of two holomorphic functions $F_A(\phi_A), F_B(\phi_B)$ where $A = 1 \dots h_{(1,1)}, B = 1, \dots h_{(2,1)}$. This follows from N = 2 space–time supersymmetry. The function $F_A(\phi_A)$, at least in the weak coupling regime of string theories, may be interpreted as a complexification of the volume of the Calabi–Yau space [42, 43] (see eq. (6.15)) while the gradient of the F_B function is related to the deformation of the complex structure. In homogeneous coordinates $Z^B(B = 1 \dots h_{(2,1)} + 1)$

$$Z^A = \int \Omega \wedge \beta^A \tag{8.1}$$

$$-iF_a(Z) = \int \Omega \wedge \alpha_A \tag{8.2}$$

It is of great physical interest to consider the case of heterotic compactifications on Calabi–Yau manifolds [11], since this situation leads to N = 1 space–time supersymmetry with chiral families with gauge group $G \subseteq E_6$.

A natural question which arises is whether we can compute the Kähler potential for the charged fields C in terms of the metric of the moduli. This should be possible since, in Ref. [41], it was shown that the two metrics are in fact related, at least in the vacuum configurations with unbroken gauge symmetry $< C >= 0$. The charged fields are in the (27) of E_6 for (2,1) families and in $(\overline{27})$ for E_6 antifamilies.

It is easy to compute the scalar manifold including the family fields in some limiting situations of Calabi–Yau manifolds, i.e. the $T_6/\mathcal{Z}_N$ orbifolds [12].

If we confine our discussion to the untwisted sector of such orbifolds the low energy effective Lagrangian can be extracted by performing a $\mathcal{Z}_N$ projection of the ten–dimensional action [48] or by computing directly the Polyakov–Zamolodchikov metric [49]. The list of the scalar manifolds is given in Table IV.

This technique can be used in all toroidal type compactifications and gives a correct result provided the truncation is consistent, i.e., provided the integration over the heavy Kaluza–Klein modes does not affect the low energy Lagrangian.

However if we work in a four–dimensional superstring model all the massive modes are taken into account by the σ–model effective action for the background fields. The most general form of such an action is dictated by conformal invariance. So if the σ–model on a Calabi–Yau manifold is conformal invariant the correct integration over the massive modes should be encoded in the equations of motion for the background fields in σ–model perturbation theory. These equations are nothing but the statement of conformal invariance. It turns out that, while the metric for the moduli fields is exact in σ–model perturbation theory, the metric for the matter fields is expected to suffer string corrections since the charged fields are not protected by hidden (or accidental) symmetries [21, 22].

To explore the geometry of the (1,1) and (2,1) moduli we will consider here some examples.

Before doing that let us recall once more that the scalar background scalar fields in type II supergravity come from

$$g_{i\bar{j}}, b_{i\bar{j}}, g_{ij}, A_{ij\bar{k}}, \phi, A_{ijk}, b_{\mu\nu}$$

in the type IIA and from

$$g_{i\bar{j}}, b^c_{i\bar{j}}, A_{\mu v i \bar{j}}, g_{ij}, \phi^c, b^c_{\mu\nu}$$

in type IIB superstrings.

The vector multiplets and hypermultiplets correspond to (1,1) and (2,1) forms in type IIA superstrings while their role is reversed in type IIB [25, 39].

This is best seen by noticing that the vector fields (beyond the graviphoton) are (1,1) forms in type IIA and (2,1) forms in type IIB:

$$\text{matter vectors: } A_{\mu i \bar{j}} \text{ (type IIA), } A_{\mu i j \bar{k}} \text{ (type IIB)}$$

Case	$\mathcal{K} = \mathcal{K}^C \times \mathcal{K}^{NC} \times SU(1,1)/U(1) \quad (n_i = i)$
1	$\dfrac{SU[3,3(1+n_{27})]}{SU(3)\times SU[3(1+n_{27})]\times U(1)} \;\times\; \dfrac{SU(1,1)}{U(1)}$
2	$\dfrac{SU[2,2(1+n_{27})]}{SU(2)\times SU[2(1+n_{27})]\times U(1)} \;\times\; \dfrac{SU(1,1+n_{27})}{SU(1+n_{27})\times U(1)} \;\times\; \dfrac{SU(1,1)}{U(1)}$
3	$\dfrac{SU[2,2(1+n_{27})]}{SU(2)\times SU[2(1+n_{27})]\times U(1)} \;\times\; \dfrac{SO[2,2(1+n_{27})]}{SO(2)\times SO[2(1+n_{27})]} \;\times\; \dfrac{SU(1,1)}{U(1)}$
4,5,6	$\left[\dfrac{SU(1,1+n_{27})}{SU(1+n_{27})\times U(1)}\right]^2 \;\times\; \dfrac{SO[2,2(1+n_{27})]}{SO(2)\times SO[2(1+n_{27})]} \;\times\; \dfrac{SU(1,1)}{U(1)}$
7,8,9	$\left[\dfrac{SU(1,1+n_{27})}{SU(1+n_{27}\times U(1))}\right]^3 \;\times\; \dfrac{SU(1,1)}{U(1)}$
$\mathcal{Z}_N \subset SU(2)$	$\mathcal{M} = \mathcal{K} \times Q \quad (n_i = i)$
$\mathcal{Z}_2$	$\dfrac{SO(2,2+n_{133}+n_3+n_{248})}{SO(2)\times SO(2+n_{133}+n_3+n_{248})}) \;\times\; \dfrac{SU(1,1)}{U(1)} \;\times\; \dfrac{SU(4,4+2n_{56})}{SO(4)\times SO(4+2n_{56})}$
$\mathcal{Z}_{3'}, \mathcal{Z}_4, \mathcal{Z}_{6''}$	$\dfrac{SO(2,2+n_{133}+n_1+n_{248})}{SO(2)\times SO(2+n_{133}+n_1+n_{248})} \;\times\; \dfrac{SU(2,2+n_{56})}{SU(2)\times SU(2+n_{56})\times U(1)}$

graviphoton: A_μ (type IIA), $A_\mu \varepsilon_{ijk}$ (type IIB)

We consider the examples of the $T_6/\mathcal{Z}_3$ and $T_6/\mathcal{Z}_4$ orbifolds where the $\mathcal{Z}_3, \mathcal{Z}_4$ groups are generated by the SU(3)–elements $e^{i\pi/3}1$, $\begin{pmatrix} i & & \\ & +i & \\ & & -1 \end{pmatrix}$.

The untwisted sector is given, for the $\mathcal{Z}_3$ orbifold, by:

$$g_{i\bar{j}}, b_{i\bar{j}} \quad i, \bar{j} = 1, .3$$

and for the $\mathcal{Z}_4$ orbifold by:

$$g_{i\bar{j}}, b_{i\bar{j}}, g_{3\bar{3}}, b_{3\bar{3}}, g_{33} \quad (i, i = 1, 2)$$

So there are nine (1,1) forms on $T_6/\mathcal{Z}_3$ and no (2,1) forms while we have five (1,1) forms and one (2,1) form on $T_6/\mathcal{Z}_4$.

The Kähler manifold for these models were computed in Ref. [27, 47] and it turned out to be

$$\frac{SU(3,3)}{SU(3) \times SU(3) \times U(1)} \text{ and } \frac{SU(2,2)}{SU(2) \times S(2) \times U(1)} \times \frac{SU(1,1)}{U(1)}$$

for the (1,1) forms and SU(1,1)/U(1) for the (2,1) form.

Note that all these manifolds are compatible with N = 2 supergravity. The manifold for (2,1) forms leads to a cubic holomorphic functions $F_A(\phi)$ while the manifold for the (1,1) forms corresponds to a quadratic holomorphic function $F_B(\phi)$:

$$F_A(\phi^a) = id_{abc}\phi^a\phi^b\phi^c, F_B(\phi) = 1 - \phi^2$$

In heterotic string the Kähler manifolds of moduli is given by

$$K^{heterotic} = K_A \times K_B \times \frac{SU(1,1)}{U(1)}$$

where we also included the dilaton and axion modes $\phi, b_{\mu\nu}$.

If we now switch on the family background fields a non–trivial interference between K_A and K_B occurs when $h_{1,1}h_{2,1} \neq 0$, i.e. when there are paired family–antifamily sectors. On $T_6/\mathcal{Z}_3$ the manifold is factorized since $h_{2,1} = 0$. It was given in eq. (2.4). For the $\mathcal{Z}_4$ case when the five family and one antifamily are introduced, one get the manifold [48, 49]

$$(K_A \times K_B)^{heterotic} = \frac{SU(2, 2(1+27))}{SU(2) \times SU(2(1+27))} \times \frac{SO(2, 2(1+27))}{SO(2) \times SO(2(1+27))}$$

The first manifold has (complex) dimension $n_c \times 27 = |h_{2,1} - h_{1,1}|27$ while the second has dimension $(2h_{2,1} \times 27)$. This shows that when family replica are introduced, the factorization of the Kähler manifold between (1,1) and (2,1) forms does not occur any longer. Rather the manifold is a product of manifolds whose dimension is related to the net chirality content $|h_{21} - h_{11}|$ and to the paired family replica min $[h_{(1,1)}, h_{(2,1)}]$. If this product structure is not an accident this would imply a further constraint on the metric of the moduli space.

It is obvious that when family replicas are included, the scalar background field metric is only compatible with N = 1 (rather than N = 2) space–time supersymmetry, i.e. it must be Kähler. However, since the metric for the families is induced from the metric of the moduli [41] it should be possible to write this metric in terms of the two holomorphic functions F_A and F_B.

9. DUAL QUATERNIONIC MANIFOLDS

Dual quaternionic manifolds are manifolds which can be obtained by 3D duality from N = 2, 4D admissible Kähler manifolds.

As is well known from the work of Günaydin, Sierra and Townsend [50], if we dimensionally reduce a 4D N = 2 supergravity theory with n (Abelian) vector multiplets to D = 3, we obtain an N = 4 supergravity theory with (n + 1) hypermultiplets which parametrize a quaternionic manifold. However, the contrary is not true, i.e. not all quaternionic manifolds are "dual".

The amazing fact about type II superstrings is that the hypermultiplets of type IIB(A) theories are precisely the "dual quaternionic manifolds" of the vector multiplets of type IIA(B) theories. Here we will not give the proof – it amounts to showing some equivalence between vertices in type IIA and IIB superstring dimensionally reduced to D = 3 space–time dimension [39]. The above arguments give, in principle, a way of constructing the dual quaternionic manifolds from any given N = 2 vector–multiplet Kähler manifold. In the case of manifolds which come from a truncation of N = 8 supergravity, the classification was already given by Günaydin, Sierra and Townsend [50]. Here we will just show a table (Table V) of the dual quaternionic manifolds of all N = 2 symmetric spaces, as given by Cremmer and van Proeyen [51].

It is remarkable that the only missing sequence of Q spaces is the minimal coupling based on the $Sp(2, n)/Sp(2) \times Sp(n)$ Q–manifolds. If we look at the Kähler manifolds in Table V, we may remark that the space (1) for n = 1 appears in orbifold compactifications of type IIB theories. The spaces in (2), (3) and (4)

Table V – Dual quaternionic spaces for symmetric Kähler spaces
(of the restricted type allowed in N = 2 supergravity)

Kähler space	$\dim_{\mathcal{C}}$	Quaternionic space
$\dfrac{U(1,n)}{U(n)\times U(1)}$ (minimal coupling)	n	$\dfrac{U(2,n+1)}{U(2)\times U(n+1)}$
$\dfrac{SU(1,1)}{U(1)} \times \dfrac{SO(n-1,2)}{SO(n-1)\times SO(2)}$	$n \geq 2$	$\dfrac{SO(n+1,4)}{SO(n+1)\times SO(4)}$
$\dfrac{SU(1,1)}{U(1)}$	1	$\dfrac{G_{2(+2)}}{SO(4)}$
$\dfrac{U(3,3)}{U(3)\times U(3)}$	9	$\dfrac{E_{6(+2)}}{SU(6)\times SU(2)}$
$\dfrac{Sp(6,\mathcal{R})}{U(3)}$	6	$\dfrac{F_{4(+4)}}{USp(6)\times SU(2)}$
$\dfrac{SO^*(12)}{U(6)}$	15	$\dfrac{E_{7(-5)}}{SO(12)\times SU(2)}$
$\dfrac{E_{7(-26)}}{E_6\times SO(2)}$	27	$\dfrac{E_{8(-24)}}{E_7\times SU(2)}$

are all of the type allowed by the type IIA theories. Note that spaces (1) (for $n = 1$) and (3) differ for the value of the scalar curvature ($R = -2$ and $- 2/3$, respectively). The spaces (4), (5) and (6) can be obtained by a truncation of the $N = 8$ supergravity. The latter example is the exceptional case which is not contained in $N = 2$, $D = 10$ superstrings.

We conclude this section by giving the notion of a dual hyper–Kähler manifold in the case of $N = 2$ global supersymmetry.

Here, a drastic simplification in explicit formulae exists owing to the fact that hyper–Kähler manifolds are also Kähler manifolds. An admissible $N = 2$ Kähler manifold for $N = 2$ (global) vector multiplets potential has a potential of the form

$$K = f_{,a}\overline{z}^a + \overline{f}_{,a}z^a \tag{9.1}$$

where $f(z)$ is a holomorphic function of the z fields. Equation (9.1) gives a real symmetric Kähler metric

$$K_{ab} = f_{,ab} + \overline{f}_{,ab} \tag{9.2}$$

We can associate a new Kähler potential of 2n variables with the Kähler potential given by Eq. (9.1), as follows:

$$H(z, \overline{z}, w, \overline{w}) = K + K_{ab}^{-1}(w^a + \overline{w}^a)(w^b + \overline{w}^b) \tag{9.3}$$

It can be shown [39] that H defines an hyper–Kähler metric. Equation (9.3) gives a hyper–Kähler potential H of 2n complex variables z, w in terms of a holomorphic function f of n complex variables.

At this point a comment is in order. The dual map in local supersymmetry is not just obtained by generalizing Eq. (9.3) to curved space. Indeed while the duality (in global symmetry) relates a Kähler manifold of complex dimension n to a hyper–Kähler manifold of real dimension 4n, in local supersymmetry the same map associates a Kähler manifold (of complex dimmension n) to a quaternionic manifold of real dimension 4(n + 1). This is due to the occurrence, in local supersymmetry, of the universal hypermultiplet containing the dilaton and antisymmetric tensor. For n = 0 this hypermultiplet lies in the manifold $SU(2,1)/SU(2) \times U(1)$ (see Eq. 4.10). It would be interesting to find the superspace generalization of the $N = 2$ mapping between Kähler and quaternionic manifolds. The harmonic superspace seems to be the best suited approach for attacking this problem [52].

10. CONCLUDING REMARKS

We have presented general features on the effective supergravity Lagrangian coming from superstring compactifications on Calabi–Yau manifolds or more generally for internal 2d–superconformal field theories preserving one or more space–time supersymmetries.

For (2,2) compactifications the metric of the moduli space can be determined by two holomorphic functions and therefore one may envisage new non renormalization theorems, similar to those for the superpotential.

If we denote by R the (1,1) moduli and by G the (2,1) moduli, these results imply that the superpotential for $\overline{27}$ families can only get non–pertubative corrections and cannot depend on the complex structure, while the superpotential for the 27 families is unrenormalized even at the non–perturbative level. These results are in full agreement with recent string calculations [53, 54, 55].

Besides the interest in superstring compactification, the investigation of the Kaluza–Klein decomposition of 10D type II supergravity yields the construction of quaternionic spaces "dual" to (N = 2) Kähler spaces. These spaces have the property of that the quaternionic metric locally depends on a holomorphic function of the "dual" Kähler space [39].

At the string level it would be interesting to connect the constraints given by N = 2 space–time supersymmetry on the Kähler potential of the moduli to intrinsic properties of the two–dimensional conformal field theory. It is known that the kinetic energy term for the background fields in the effective four–dimensional Lagrangian

$$\int d^4 x \, G_{\alpha\beta} \partial_\mu \phi^\alpha \partial_\mu \phi^\beta$$

is related to the Poliakov–Zamolodchikov metric [23]

$$G_{\alpha\beta} \;=\; < V_\alpha(0) V_\beta(1) >$$

Since $G_{\alpha\bar\beta} = \partial_\alpha \partial_{\bar\beta} K$, by integration one could get the Kähler potential and then the $F_{A(B)}$ holomorphic functions. However this is an indirect determination.

In a recent series of paper [56] a new approach to (2,2) superconformal field theories has been pursued, in which the 2d Lagrangian is written as a 2d N = 2 Super Landau–Ginsburg Lagrangian with a D and F term

$$\mathcal{L}_{2d,N=2} \;=\; \int d^4\theta d^2 z \, K(z,\overline{z},\theta,\overline{\theta}) \;+\; Re \int d^2\theta d^2 z \; W(\theta,z)$$

It is likely that the two holomorphic functions which appear in the Calabi–Yau moduli metric are related to the D and F terms which depend on holomorphic background fields corresponding to (1,1) and (2,1) forms respectively.

REFERENCES

[1] A.A. Belavin, A.M. Polyakov and A.B. Zamolodchikov, Nucl. Phys. **B241** (1984) 333.
D. Friedan, E. Martinec and S. Shenker, Nucl. Phys. **B271** (1986) 93.

[2] M. Green and J.H. Schwarz, Phys. Lett. **149B** (1984) 117.

[3] D.J. Gross, J. Harvey, E. Martinec and R. Rohm, Phys. Rev. Lett. **54** (1985) 502; Nucl. Phys. B256 (1985) 253 and **B267** (1985) 75.

[4] K.S. Narain, Phys. Lett. **169B** (1986) 61.
K.S. Narain, M.H. Sarmadi and E. Witten, Nucl. Phys. **B279** (1987) 369.
I. Antoniadis, C. Bachas and C. Kounnas, Nucl. Phys. **B289** (1987) 87.
N. Lerche, D. Lust and A.N. Schellekens, Nucl. Phys. **B287** (1987) 477.
M. Kawai, D.C. Lewellen and S.H. Tye, Phys. Rev. Lett. **57** (1986) 1832; Nucl. Phys. **B288** (1987) 1.
M. Mueller and E. Witten, Phys. Lett. **182B** (1986) 28.

[5] W. Boucher, D. Friedman and A. Kent, Phys. Lett. **172B** (1986) 316.
A. Sen, Nucl. Phys. **B278** (1986) 289; Nucl. Phys. **B284** (1987) 423.
L. Dixon, D. Friedan, E. Martinec and S.H. Shenker, Nucl. Phys. **B282** (1987) 13.
T. Banks, L. Dixon, D. Friedan and E. Martinec, Nucl. Phys. **B299** (1988) 613.
T. Banks and L. Dixon, Nucl. Phys. **B307** (1988) 93.
A. Schwimmer and N. Seiberg, Phys. Lett. **184B** (1987) 191.
M. Dine, N. Seiberg, X.G. Wen and E. Witten, Nucl. Phys. **B278** (1986) 769; Nucl. Phys. **B289** (1987) 319.

[6] A.M. Polyakov, Phys. Lett. **103B** (1981) 207 and 211.

[7] I. Antoniadis, C. Bachas, C. Kounnas and P. Windey, Phys. Lett. **171B** (1986) 51.
L. Dixon, V. Kaplunovsky and C. Vafa, Nucl. Phys. **B294** (1987) 43.

[8] D. Gepner, Nucl. Phys. **296B** (1988) 757; Phys. Lett. **199B** (1987) 380.

[9] P. Di Vecchia, J. Petersen and H. Zheng, Phys. Lett. **162B** (1985) 327 and **174B** (1986) 280.
P. Di Vecchia, J. Petersen amd M. Yu, Phys. Lett. **172B** (1986) 211.
W. Boucher, D. Friedan and A. Kent, Phys. Lett. **172B** (1986) 316.

[10] M. Ademollo, L. Brink, A. d'Adda, R. Auria, E. Napolitano, S. Sciuto, E. Del Giudice, P. Di Vecchia, S. Ferrara, F. Gliozzi, R. Musto and R. Pettorino, Phys. Lett. **62B** (1986) 105; M. Ademollo, L. Brink, A. d'Adda, R. d'Auria, E. Napolitano, S. Sciuto, E. Del Giudice, P. Di Vecchia, S. Ferrara, F. Gliozzi, R. Musto, R. Pettorino and J.H. Schwarz, Nucl. Phys. **B111** (1976) 77.

[11] P. Candelas, G.T. Horowitz, A. Strominger and E. Witten, Nucl. Phys. **B258** (1985) 46.

[12] L. Dixon, J. Harvey, C. Vafa and E. Witten, Nucl. Phys. **B261** (1985) 678 and **B274** (1986) 285.
      L.E. Ibañez, H.P. Nilles and F. Quevedo, Phys. Lett. **187B** (1987) 25 and **192B** (1987) 332.
      A. Font, L. Ibañez, H.P. Nilles and F. Quevedo, CERN preprint TH.4969/88 (1988).

[13] F. Gliozzi and P. Fré, Turin preprint DFTT 13/88 (1988).

[14] C. Hull and E. Witten, Phys. Lett. **160B** (1985) 398.
      L. Alvarez–Gaumé and P. Ginsparg, Commun. Math. Phys. **102** (1985) 311.
      C. Hull, Nucl. Phys. **B260** (1985) 182.
      T. Banks and N. Seiberg, Nucl. Phys. **B273** (1986) 157.

[15] E.S. Fradkin and A.A. Tseytlin, Nucl. Phys. **B261** (1985) 1; Phys. Lett. **158B** (1985) 131 and **160B** (1985) 169.

[16] C.G. Callan, D. Friedan, E. Martinec and M. Peny, Nucl. Phys. **B262** (1985) 593.

[17] For a review on supergravity see, for example, P. van Nieuwenhuizen, Phys. Rep. **C68** (1981), 189.
      S. Ferrara, *in* Supersymmetry (North–Holland Pub. Co., Amsterdam, and World Scientific, Singapore 1987), Vol. 2.

[18] A. Sen, Phys. Rev. Lett **55** (1985) 1846; Phys. Lett. **166B** (1986) 300.
      M.T. Grisaru, A.E.M. van de Ven and D. Zanon, Phys. Lett. **173B** (1986) 423.

[19] S. Cecotti, S. Ferrara, L. Girardello, A. Pasquinucci and M. Porrati, Phys. Rev. **D33** (1986) 2504 and Int. J. Mod. Phys. **A3** (7) (1988) 1675.
      G. Girardi and R. Grimm, Nucl. Phys. **B292** (1987) 181.
      P. Binétruy, G. Girardi, R. Grimm and M. Müller, Phys. Lett. **195B** (1987) 389.

[20] S. Cecotti, S. Ferrara and L. Girardello, Phys. Lett. **198B** (1987) 336.

S. Cecotti, S. Ferrara and L. Girardello, Phys. Lett. **206B** (1988) 451.

B. Ovrut, Phys. Lett **205B** (1988); W. Siegel, Phys. Lett. **211B** (1988) 55.

S.J. Gates, P. Majumdar, R. Oerter and A. van de Ven, Univ. of Maryland Report No. UMDEPP88–209 (1988).

[21] S. Cecotti, S. Ferrara and L. Girardello, Nucl. Phys. **B308** (1988) 436.

[22] M. Cvetič, J. Louis and B.A. Ovrut, Phys. Lett. **206B** (1988) 227.

[23] A.M. Polyakov, Address to the International Mathematics Union, Berkeley (1986).
A.B. Zamolodchikov, Sov. Phys.–JEPT Lett. **43** (1986) 731.

[24] E. Cremmer, S. Ferrara, C. Kounnas and D.V. Nanopoulos, Phys. Lett. **133B** (1983) 61.
J. Ellis, A.B. Lahanas, D.V. Nanopoulos and K. Tamvakis, Phys. Lett. **134B** (1984) 429.
J. Ellis, C. Kounnas and D.V. Nanopoulos, Nucl. Phys. **B247** (1984) 373.

R. Barbieri, E. Cremmer and S. Ferrara, Phys. Lett. **163B** (1985) 143.

[25] N. Seiberg, Nucl. Phys. **B303** (1988) 286.

[26] E. Witten, Phys. Lett. **155B** (1985) 151.

[27] S. Ferrara, C. Kounnas and M. Porrati, Phys. Lett. **B181** (1986) 263.

S. Ferrara, L. Girardello, C. Kounnas and M. Porrati, Phys. Lett. **192B** (1987) 368 and **194B** (1987) 358.
I. Antoniadis, J. Ellis, E. Floratos, D.V. Nanopoulos and T. Tamaras, Phys. Lett. **191B** (1987) 96.

[28] B. de Wit, P.G. Lauwers, R. Philippe, S.Q. Su and A. Van Proeyen, Phys. Lett. **134B** (1984) 37.
B. de Wit and A. Van Proeyen, Nucl. Phys. **B245** (1984) 89.

J.–P. Derendinger, S. Ferrara, A. Masiero and A. Van Proeyen, Nucl. Phys. **140B** (1984) 307.
B. de Wit, P.G. Lauwers and A. Van Proeyen, Nucl. Phys. **B255** (1985) 569.

[29] J. Bagger and E. Witten, Nucl. Phys. **B222** (1983) 1.

[30] E. Cremmer, C. Kounnas, A. Van Proeyen, J.–P. Derendinger, S. Ferrara, B. de Wit and L. Girardello, Nucl. Phys. **B250** (1985) 385.

[31] E. Cremmer, B. Julia, J. Scherk, S. Ferrara, L. Girardello and P. Van Nieu-
wenheizen, Nucl. Phys. **B147** (1979) 105.
E. Cremmer, S. Ferrara, L. Girardello and A. Van Proeyen, Nucl. Phys.
B212 (1983) 413.
E. Witten and J. Bagger, Phys. Lett. **115B** (1982) 202.
J. Bagger, Nucl. Phys. **B211** (1983) 302.

[32] E. Bergshoeff, I.G. Koh and E. Sezgin, Phys. Lett. **155B** (1985) 71.
M. de Roo and P. Wagemans, Nucl. Phys. **B262** (1985) 644.

[33] J.–P. Derendiger and S. Ferrara, *in* Supersymmetry and Supergravity '84,
eds. B. de Wit et al., (World Scientific, Singapore, 1984), p. 159.

[34] E. Cremmer, B. Julia and J. Scherk, Phys. Lett. **76B** (1978) 409.
M.B. Green and J.H. Schwarz, Phys. Lett. **122B** (1983) 143.
J.H. Schwarz and P.C. West, Phys. Lett. **126B** (1983) 301.
J.H. Schwarz, Nucl. Phys. **B226** (1983) 269.
P.S. Howe and P.C. West, Nucl. Phys. **B238** (1984) 181.
L. Castellani, Nucl. Phys. **B294** (1987) 877.

[35] E. Cremmer and B. Julia, Nucl. Phys. **B159** (1979) 141.

[36] S. Ferrara, J. Scherk and B. Zumino, Nucl. Phys. **B121** (1977) 393.

[37] M.K. Gaillard and B. Zumino, Nucl. Phys. **B193** (1981) 221.

[38] S.J. Gates, *in* Superstrings, Cosmology and Composite Structure, eds. S.J.
Gates and R.N. Mohapatra (World Scientific, Singapore, 1987), p. 223.
R.N. Mohapatra (World Scientific, Singapore, 1987), p. 223.
W. Siegel, Nucl. Phys. **B238** (1984) 207.
R. Brooks, S. Gates and F. Muhammad, Nucl. Phys. **B268** (1986) 599.
J.M.F. Labastida and M. Pernici, Nucl. Phys. **B297** (1987) 557.

[39] S. Cecotti, S. Ferrara and L. Girardello, CERN preprint TH.5080/88,
UCLA/88/TEP/21 (1988), to appear in Int'l Journal of Modern Physics.

[40] S. Ferrara and S. Sabbarwal, CERN TH–5219/88–UCLA/88/TEP/15, to
appear in "Classical and Quantum Gravity".

[41] A. Strominger, Phys. Rev. Lett. **55** (1985) 2547.
A. Strominger and E. Witten, Comm. Math. Phys. **101** (1985) 341.

[42] S. Cecotti, S. Ferrara and L. Girardello, Phys. Lett. **213B** (1988) 443.

[43] P. Candelas, P.S. Green and T. Hubsch, "Connected Calabi–Yau compact-ifications", Proceedings Maryland Superstring 88. Workshop (1988); and Texas preprint UTTG–24–88 (1988).

[44] M. Cvetič, J. Louis and B.A. Ovrut, Phys. Lett. **B206** (1988) 229; M. Cvetič, University of Pennsylvania preprint UPR–T–350, Proceeding of the String Workshop, Maryland (1988); M. Cvetič, B.A. Ovrut and J. Louis, University of Pennsylvania preprint UPR–0380T (1989).

[45] S. Ferrara and A. Strominger, CERN–TH.5291/89, UCLA/89/TEP/6 (1989).

[46] P. Candelas, Nucl. Phys. **B298** (1988) 458.

[47] M. Dine, P. Huet and N. Seiberg, Princeton preprint IASSNS–HEP–88/54; CCNY–HEP–88/20.

[48] S. Ferrara and M. Porrati, Phys. Lett. **216** (1989) 289.

[49] M. Cvetič, J. Molera and B.A. Ovrut, University of Pennsylvania preprint UPR–0376T (1989).

[50] M. Günaydin, G. Sierra and P.K. Townsend, Phy. Lett. **133B** (1983) 72.

[51] E. Cremmer and A. Van Proeyen, Classical and Quantum Gravity **2** (1985) 445.

[52] J. Bagger, A. Galperin, E. Ivanov, V. Ogievetsky, Nucl. Phys. **B303** (1988) 522.

[53] M. Cvetič, Phys. Rev. Lett. **59** (1987) 1795.

[54] J. Distler and B. Green, Nucl. Phys. **B309** (1988) 295.

[55] M. Dine and G. Lee, Phys. Lett. **203B** (1988) 371.

[56] D.A. Kastor E.J. Martinec and S.H. Shenker, Enrico Fermi Institute preprint EFI–88–31.
C. Vafa and N. Warner, Harvard preprint HUTP–88/A047 (1988)
E. Martinec, Enrico Fermi Institute preprint EFI–88–76 (1988).
J.I. Latorre and C.A. Lütken, Nordita preprint 88/42P (1988).

DISCUSSION

– Sennan:

As you told us in your lecture, there seems to be a deep connection between supersymmetry on the world sheet and space–time supersymmetry. What is the underline principle of this connection?

– Ferrara:

The basic fact of space–time supersymmetry is that the supersymmetry algebra has a kind of rotating symmetry. That is in general if you have extended supersymmetries the supersymmetry algebra has a U(N) symmetry corresponding to a rotation of the supercharges. For instance, in N=1 you have a U(1) charge which is the rotation symmetry. For 2 charges the rotation symmetry is a non–abelian U(2) symmetry, and for 4 charges there is a SU(4) non–abelian symmetry These symmetries you have to get in some sense at the level of the world sheet if you want to realize string theories that have space–time SUSY. In order to have N=1 space–time SUSY you must have at the level of the world sheet a symmetry which is an analog of the rotational symmetry of the space–time supersymmetry, in other words a U(1) symmetry. Now, this U(1) symmetry at the level of the world sheet is nothing but the internal U(1) symmetry of the N=2 Kac–Moody algebra which is generated by the N = 2 superconformal algebra. If you have an N = 2 superconformal algebra then there is a U(2) which rotates the charges of this algebra. Then this U(2) will be connected to the U(2) of the space–time supersymmetry

– Zichichi:

As you know I have always been reluctant to consider naturally more than one supersymmetry. Can you defend having more than one supersymmetry N=2,3 etc.?

– Ferrara:

From a mathematical point of view you can have up to four space–time supersymmetries in 4 dimensions (in heterotic strings). Mathematically these are completely natural, the point is that the only space–time SUSY that is consistent with realistic models for particle physics is N = 1 because this is the only one that tolerates chiral fermions. So the reason why we are interested in N =1 in four dimensions is because we want a realistic spectrum for spin–1/2 fermions.

– Zichichi:

What I find unnatural is the immense number of possibilities which are continuously considered. Would it not be better to make a reasonable choice of what is true?

– Ferrara:

This is the problem of strings. In principle we have an enormous set of vacua corresponding to different models in string theory. The number of models increases when you decrease the dimension of space–time. If you are in the critical dimension $D = 10$ the number of models is very limited. For heterotic strings we essentially have only two consistent models, one with gauge group $E_8 \times E_8$ and one with gauge group $SO(32)$. As soon as we go to lower dimensions and reach the dimension 4, which is the dimension we are interested in, the number of mathematically consistent models is enormous. This is related to the fact that there are an enormous number of compactifications in going from 10 to 4 dimensions. As I said in 10 dimensions we essentially have a very few gauge groups. When we go to 4 dimensions there is an enormous variety of theories with all sorts of gauge groups. One restriction on the gauge group is that the rank of the gauge group in 4 dimensions has to be less than or equal to 22. And then you have the possibility of having $N = 1, N = 2, N = 4$ space–time SUSY (for heterotic strings) in 4 dimensions. Why nature finds out a particular vacuum state with $N = 1$ softly broken, in order to have a realistic model, is a problem whose a solution (if any) is very far from being reached.

– Zichichi:

My point is that you should choose what could be considered a really natural choice. $N = 1$ should by faith be the only choice. All the other theories should be abandoned because $N > 1$ is just nonsense for a physical taste.

– Ferrara:

We want to understand why Physics chooses $N = 1$ softly broken. We are not making a postulate.

– Zichichi:

Please use some intuition! Nobody will take you seriously if you take more than one space–time SUSY.

– Green, comment:

Could I just make a comment. In Ferrara's talk part of the point of what he said was that in order to get restrictions on the $N = 1$ theories you find that by studying the $N = 2$ theories, there are common features and that the features of

$N = 2$ theories put constraints on the physically more interesting $N = 1$ theories. So that is one argument for studying them.

– *Ferrara:*

The point is that you start with $N = 1$ in 10 dimensions. But you cannot live in 10 dimensional space–time, you have to go to 4 dimensional space–time. When you go to 4 dimensions you have to curl up these extra 6 dimensions. In the curling of these 6 dimensions you gain a certain arbitraryness in your model because a large variety of compactifications are in principle allowed in 4 dimensions. So even if you confine within $N = 1$ space–time SUSY in 4 dimensions you will loose this uniqueness of 10 dimensional theory. There are plenty of models which have N =1 space–time SUSY in 4 dimensions. So the problem of strings is what is the criterion one uses to select among these models even with $N = 1$ space–time SUSY. This is probably something which goes beyond perturbation theory and there will be important nonperturabtive effects which would select the true vacuum. The problem is that in 4 dimensions, $N = 1$ theories are not unique. As M. Green was pointing out, however, in string theory there is this fact that the $N = 1$ theories do not forget the fact that they are coming from 10 dimensions. And in 10 dimensions you have an enlarged SUSY. So $N = 1$ theories in 4 dimensions are constrained in string theory by the fact that these theories can be promoted to higher space–time SUSY.

– *Zichichi:*

$N = 1$ is only some choice.

– *Ferrara:*

$N = 1$ is restricting your class of models, but as I told you, it is not sufficient to select a unique model.

– *Volkas:*

Is the Peccei–Quinn symmetry you use related to the usual strong CP problem or is it another anomalous global symmetry?

– *Ferrara:*

It may be related to the strong CP problem if you have a realistic model. What I call the usual Peccei–Quinn symmetry is the fact that these are in these theories some elementary scalar fields which I call Q which have a coupling of the form QFF^* where F is the strength of some vector field. Fields which appear in this way give you an axion coupling in the theory, so I call Peccei–Quinn symmetry whenever you have some field coupled in this way. In string models you have many fields which have such a kind of coupling. The number of fields which have such a kind of coupling depend on your particular model. For instance, in the T_6/Z_3 orbifold there are for instance 8 massless fields which have such kind of coupling. So in perturbation theory there are in principle 8 axions which would couple in this way to gauge vectors. However many of this Peccei–Quinn symmetries can be spoiled by non–perturbative effects like world–sheet instantons.

– *Warr:*

  1) Are the Yukawa couplings that you compute in CFT depend with respect to some renormalization scale?

– *Ferrara:*

Yukawa couplings in string theories are related to same topological quantities called intersection matrices. However the absolute normalisation of these couplings depend on the normalisation of the kinetic terms. The normalisations are also topological but in principle they suffer radiative corrections if one has only N +1 space–time SUSY. So, in the point–field limit, these normalisations must be considered as given at the Planck scale and then renormalisation group arguments should determine the couplings at low energies.

– *Colas:*

Nanopolous recently claimed that $D = 4$ superstring allowed very precise predictions, as for example the top mass being ~ 80 GeV. Can you comment on this?

– *Ferrara:*

I cannot really comment on this question because as I told you in order to make contact between string theories and the Standard Model you have to work up a low energy superstring model of the kind that R. Barbieri discussed this morning. Now, when you want to work up a realistic model, you have to introduce the breaking of SUSY. Now, the breaking of SUSY, as far as I know, in all these models is done, by hand. There is no way of obtaining from string theories these soft–breaking terms that R. Barbieri were discussing this morning and which are crucial in order to discuss particle physics. In order to discuss particle physics you have to introduce this SUSY breaking sector and there is no way in string theory to discuss this sector without making ad hoc assumptions. So as far as I can tell, all these predictions which are advocated in these models make crucial assunptions which go beyond the string theory.

– *Giannakis:*

  1) If you consider the space of string theories there are string theories which are continuously connected. Are there supersymmetric string theories which are continuously connected with non–supersymmetric theories?

2) Calabi–Yau manifolds are not exact solutions of string equations. How can you associate a string theory with a C–Y manifold?

Ferrara:

1) There are certainly supersymmetric vacua which are continuously connected. They correspond to "flat directions" or moduli of superconformal systems. However there are arguments that non–supersymmetric vacua cannot be continuously connected (to supersymmetric ones), unlike in point–field supergravities. This is due to a requirement, present in string theories, that broken space–time supersymmetry should be consistent with local world–sheet supesymmetry. This requirement implies that the gravitino mass is quantized in terms of the radius of compactification.

2) From an abstract point of view, Calabi–Yau manifolds should correspond to exact (2,2) superconformal field theories for the internal degrees of freedom on the world–sheet.

TOWARDS A STANDARD STRING MODEL

Hans Peter Nilles
Physik Department, Technische Universität München
and
Max-Planck-Institut für Physik und Astrophysik, München

INTRODUCTION

String theories offer the exciting prospect of a consistent, finite description of gravitation in the framework of quantum mechanics[1]. Since these theories naturally also contain nonabelian gauge interactions, they have been proposed as candidate theories for a unification of all fundamental forces. It is our task to investigate such a possible connection between string theories and the physical world. We should state at the beginning that such questions cannot be answered definitely at the moment. The main reason for this is our poor understanding of the dynamics of string theory. Nonetheless we can ask these questions and see how far it might be possible to embed the standard $SU(3) \times SU(2) \times U(1)$ model of strong and electroweak interactions in such a string theory. Once this is achieved we could then obtain hints to go beyond the standard model and solve some of the problems attached to it. The purpose of these lectures is a review of the progress made in this direction. Before we go into details let us, however, discuss those questions that could (and should) be asked when adopting this framework.

The first question concerns the dimensionality of space-time. While in a field theoretic description one would postpone such a question, this cannot be done here because of the appearence of critical dimensions predicted by string theories. For superstrings one obtains d=10 while, at least macroscopically, we observe d=4 and this needs an explanation, requiring a spontaneous compactification of six spatial dimensions. At the moment we do not understand the reason for this to happen, but string theories force us to think about such a possibility. Such a situation is rather common in the discussion of string theory. It reflects the fact that at the classical level the ground state of these theories is highly degenerate and effects that might lift these degeneracies are not yet well understood. At the present stage we have to investigate these degenerate ground states and see whether there are candidate models in d=4 that ressemble the physical world.

The next important question concerns the chirality of the fermionic representations. In the standard model we require this chirality in order to understand why quark and lepton masses are small and it also explains the presence of parity violation in weak interactions. String theory allows chiral fermions but does not predict them (at least at the level of our present understanding of string theories). A chiral model in d=4 requires a breakdown of the gauge group in the process of compactification[2].

The next question is related to the presence of supersymmetry. A consistent d=10 string theory seems to require supersymmetry and at the moment this seems also to be true for d=4 models. There chiral representations can only be obtained for $N \leq 1$. We eventually have to understand how supersymmetry breaks down. Here we shall exclusively consider $N = 1$ supersymmetry for the pragmatic reason, that they are much simpler than nonsupersymmetric models.

Next there is the question about the gauge group G in d=4. One should first require only that G contains $SU(3) \times SU(2) \times U(1)$. If we consider the $E_8 \times E_8$ heterotic string one can obtain e.g. E_6 in a simple way[2] and this would be a good choice for a grand unified group. One then has to investigate more closely whether strings lead to grand unification or additional gauge groups (like U(1)'s) beyond the standard ones. Here it is very important to discuss methods to lower the rank of the gauge group. In passing we should stress that the requirements of chirality and G containing $SU(3) \times SU(2) \times U(1)$ selects already uniquely the theory based on $E_8 \times E_8$. Theories of type II and those with gauge group O(32) seem not to be able to accomodate the physics of the standard model. This is quite encouraging. A unique model is selected and this shows that even at this level strings have some predictive power.

The next question concerns the structure of families of quarks and leptons. In the naïve approach they appear quite natural in the **27** of E_6 which decomposes into **16+10+1** of O(10). In general one finds a right-handed neutrino in addition to a standard family of quarks and leptons. We shall later discuss the spectrum of matter representations in more detail.

Another question concerns the number of families. The repetition of families has been a deep mystery in particle physics. In string theory this number is tied to the geometrical properties of space-time. Although we do not yet understand what this number is, we can now very well understand that such a repetition occurs. One might even speculate that the number 3 comes from the compactification of 3 complex dimensions.

Next we might address the question about the Higgs-sector of the theory. How many Higgs-doublets do we have? Are there more exotic possibilities like, e.g. SU(2)-triplets? What about the stability of the proton, the masses of the neutrinos or the existence of flavour changing neutral currents? None of these questions can be answered definitely at the momemt and a careful investigation of the models is needed. This includes the computation of Yukawa-couplings to understand fermion masses, mixing angles and the presence of CP-violation. We shall see that these tests are quite restrictive and eliminate many candidate models. But, amazingly, there are models that may pass all these tests. More constraints have to be satisfied, once we try to incorporate the breakdown of $SU(2) \times U(1)$ and its relation to the breakdown of supersymmetry. Here we have some understanding in the framework of the field theory of the massless states but our understanding at the level of string theory is incomplete. Eventually such a relation should give us a key to understand the ratio of the scale of $SU(2) \times U(1)$ breakdown M_W and the Planck mass M_P. Once we understand the breakdown of supersymmetry we should also be able to discuss the way in which the classical degeneracy of the ground states is lifted. But we are not yet at this point.

Two different approaches have been used for the investigation of string theories:

on one hand considerations based on the low energy field theory and on the other hand explicit constructions of string theories in four dimensions. At present both approaches are needed and should be pursued although finally we would like to understand everything in explicit string language. We simply cannot isolate real stringy effects if we do not look at string theory itself. In these lectures we shall therefore concentrate on the construction of string models and only use field theoretic argumentation where it is absolutely needed.

SOME ENCOURAGING FACTS

Before we start a discussion of the technicalities of the explicit construction of string theories, however, let us summarize some qualitative facts that could make us believe that strings have something to do with the physical world.

As we have mentioned already, string theory provides (at least in the $E_8 \times E_8$ version) naturally E_6 as a grand unified gauge group[2]. This group contains SO(10) and SU(5) and we obtain the families in the $\bar{\mathbf{5}}+\mathbf{10}$ structure of SU(5). Since the number of these families is given by the geometry of extended space-time we should therefore not be surprised that families could come in several copies. We do not yet understand why this number should be three but it should no longer be considered as a mystery that this number is larger than one.

The natural appearance of the structure of families in realistic string models leads to a prediction for the $SU(3) \times SU(2) \times U(1)$ representations. *There is simply no place for exotic representations.* Such models contain as chiral superfields only $\mathbf{3}$, $\bar{\mathbf{3}}$, $\mathbf{1}$ of SU(3) and doublets and singlets of SU(2). The observation of sextets or even more exotic representations of SU(3) would simply rule out the picture of quarks being the ground states of fundamental strings. The same is true for SU(2)-triplets, a quite encouraging fact concerning the ρ-parameter in weak $SU(2) \times U(1)$. These remarks are valid for chiral superfields; apart from those we could, of course, have gauginos in the adjoint representation of $SU(3) \times SU(2) \times U(1)$. So we see, that string theory naturally explains the structure of the representations as observed in the real world; a quite remarkable property.

Strings require supersymmetry, at least at the Planck scale, and it seems quite possible that such a symmetry could survive down to low energies and explain why M_W is so small compared to M_P. Such an argumentation has been used already before the consideration of string models and a standard SUSY-model had been constructed. This model necessarily contained gravity and consisted of two sectors: an observable and a hidden sector, coupled only gravitationally with each other[3]. The known particles and their superpartners were supposed to reside in the former and the rôle of the hidden sector was to supply a spontaneous breakdown of supersymmetry. In the original proposal of this model[4] this breakdown of supergravity proceeded through gaugino condensation[4,5] leading to a gravitino mass

$$m_{3/2} \sim \frac{\Lambda^3}{M_P^2} \tag{1}$$

where Λ is the renormalization group invariant scale of the hidden sector gauge group. Thus the weak scale $M_W \sim m_{3/2}$ could be understaood in a similar way as $\Lambda_{QCD} \sim m_{proton}$ through a slow variation of a coupling constant. Also other ways of breakdown of supersymmetry had been considered in the literature[6], but they usually required the introduction of a small scale by hand. In any case the structure of such a model seemed somewhat artificial due to the required existence of the hidden sector.

In string theories, however, such a hidden sector appears naturally. This is most pronounced in the $E_8 \times E_8$ model. The second E_8 provides the candidate hidden sector and also gaugino condensation appears in a natural way[7]; thus explaining the magnitude of the weak scale through nonpertubative effects. A closer inspection of the effective potential of the resulting low energy theory reveals additional interesting facts. It shows that with this mechanism we have a breakdown of supersymmetry without the introduction of a cosmological constant at the classical level[8]. There is actually a dynamical mechanism to cancel the cosmological constant: the dilaton slides (i.e. adjusts it vacuum expectation value) to cancel it. This is very similar to the solution of the strong CP-problem with an invisible axion. Here we observe some sort of axion-dilaton unification since both particles reside in the same supermultiplet. It remains to be seen whether such a mechanism leads to measurable effects like a new fifth force.

Another intriguing fact observed in the construction of realistic string models concerns the number of Higgs-doublets. In the simple models they appear in multiples of three[9]. This Three-Higgs-Rule (T$_{\rm H}$R) appears because of the six (i.e. three complex) compactified dimensions as we shall later discuss in detail. This rule, however, is not universal and it still has to be clarified how far it might be true in those models that are phenomenologically interesting.

In addition, string theories might give us hints for new physics beyond the standard model. These include the possible presence of new gauge interactions, neutrino masses and mixings, axions, leptoquarks and signatures of proton decay, new ways to approach grand unification and the presence of new Yukawa couplings. Although such things are not necessarily predicted in string models, they might appear in many of them, especially in those that might contain the standard model as a low energy approximation.

Finally we should, of course, search for really stringy effects, i.e. effects that would directly show that the underlying structure is string and not field theory. One might also, e.g., try to isolate those effects that are needed in d=10 but not in d=4, like the existence of the Green-Schwarz counterterms[10]. Another approach would be the considerations of finite terms in pertubation theory and see how they compare to a string regularization. The final goal would be to find effects in the low energy regime that cannot be understood in the framework of the effective field theory but find a natural explanation in string theory. Recently some progress has been made in this direction and we shall discuss them in the second lecture.

TORUS COMPACTIFICATION

As a starting point we consider the $E_8 \times E_8$ heterotic string[11] in d=10. The left-movers contribution is given by

$$\frac{1}{2}m_L^2 = N_L - 1 + \frac{p^2}{2} \tag{2}$$

where N_L is the number of excited oscillators and p is a vector in the $E_8 \times E_8$ root lattice; the E_8 root lattice being defined through the vectors

$$(n_1, \ldots, n_8) \text{ and } (n_1 + \frac{1}{2}, \ldots, n_8 + \frac{1}{2}) \tag{3}$$

with $n_i \in Z$ and $\sum n_i$ even. For the right-movers we have

$$\frac{1}{4}m_R^2 = N_R - \frac{1}{2} \tag{4}$$

in the Neveu-Schwarz sector with half integer N_R and

$$\frac{1}{4}m_R^2 = N_R \tag{5}$$

in the Ramond sector (N_R integer). The consistency condition for closed strings

$$m_L^2 = m_R^2 \tag{6}$$

removes the potential tachyon. We are interested in string theories in d=4 and the simplest way to start the discussion is the consideration of torus compactification of the extra dimensions. For simplicity let us first consider a 1-torus, i.e. a circle. Strings are closed if

$$X(\sigma = \pi) = X(\sigma = 0) + 2\pi R L \tag{7}$$

where R is the radius of the circle and L is an integer: the winding number. Strings where (7) is valid with $L \neq 0$ are called winding states; they are closed if we consider the theory on the circle but they would not correspond to closed string states in flat space. Momenta conjugate to the compact coordinate are quantized

$$p = \frac{M}{R} \tag{8}$$

where M is an integer. Let us now consider the motion of a closed string on the circle (for simplicity we consider here the bosonic string):

$$\begin{aligned} X_L &= x_0 + \left(\frac{M}{R} + 2LR\right)(\tau + \sigma) + \ldots \\ X_R &= x_0 + \left(\frac{M}{R} - 2LR\right)(\tau - \sigma) + \ldots \end{aligned} \tag{9}$$

giving rise to the following mass relations

$$\begin{aligned} \frac{1}{4}m_L^2 &= N_L - 1 + \frac{1}{8}\left(\frac{M}{R} + 2LR\right)^2 \\ \frac{1}{4}m_R^2 &= N_R - 1 + \frac{1}{8}\left(\frac{M}{R} - 2LR\right)^2 \end{aligned} \tag{10}$$

which leads to

$$\frac{1}{4}m^2 = N_L + N_R - 2 + \frac{M^2}{4R^2} + L^2 R^2 \tag{11}$$

subject to the constraint

$$N_L - N_R = ML \tag{12}$$

and the spectrum shows a duality between R and 1/2R. Let us now investigate the spectrum in detail. We have a tachyon for $N_L = N_R = 0$. Massless states can be obtained for $N_L = N_R = 1$ and $M, L = 0$ in the usual way and they correspond to the lower-dimensional metric, Kaluza-Klein gauge bosons and scalars just as in the case of field theory

$$g^{MN} = \begin{pmatrix} g^{\mu\nu} & g^{\mu 25} \\ g^{25\nu} & g^{25,25} \end{pmatrix}. \tag{13}$$

Here we have a circle and the gauge group will be $U(1)^2$; one $U(1)$ from the left-handed oscillator in the compact dimension and the other from the right-handed one.

But now in the case of strings there are potentially more massless states: consider, e.g., the case $N_L = 1$, $N_R = 0$ and $ML = 1$. If $R^2 = 1/2$, according to (11), this gives rise to massless states. In our example we obtain two new gauge bosons from the winding states $L = \pm 1$ and the U(1) gauge group is enlarged to SU(2). The same happens for $N_R = 1$, $N_L = 0$. In general one obtains a gauge group $G_L \times G_R$ where the rank of $G_{L,R}$ is given by the number of compact dimensions. For arbitrary radius (and values of other background fields) one would obtain $G_{L,R} = U(1)^n$; only for special values of the radii such groups could be enhanced. Note that the string theory is consistent for any value of R. It is only at the critical points that new massless gauge bosons appear in the spectrum. Changing R from such a critical value amounts to a spontaneous breakdown of the enhanced symmetry accompanied by a Higgs-effect.

For the heterotic string we have

$$
\frac{1}{4}m_L^2 = N_L - 1 + \frac{L^2}{2}
$$
$$
\frac{1}{4}m_R^2 = N_R + \frac{\tilde{L}^2}{2}
$$

(14)

with $(L, \tilde{L})$ a vector in a (16+p,p)-dimensional lattice. The constraint that $L^2 - \tilde{L}^2$ should be even implies a Lorentzian signature for such a lattice. Modular invariance requires the lattice to be selfdual. Such lattices with signature (m,n) exist only if $m = n \bmod 8$. In a given dimension they are unique up to Lorentz boosts in the coset

$$
\frac{SO(16 + p, p)}{SO(16 + p) \times SO(p)}
$$

(15)

thus giving a p(p+16)-parameter family of solutions[12]. These parameters are in one to one correspondence to background fields[13] for the metric g_{ij} (p(p+1)/2), the antisymmetric tensor b_{ij} (p(p-1)/2) and the gauge background fields A_i^I (Wilson lines) around the noncontractible loops (16p). As a result we have for d=4 a 132-parameter family of consistent string vacua with various possibilities for enhanced gauge groups of rank 22 from the left-handers. Notice, that this enhancement only takes place for special values of the background fields.

All these models have one property in common: N=4 supersymmetry. The gravitino as an 8_S of the transverse Lorentz group in d=10 splits into four gravitini (2_S of O(2) in d=4) because of the periodic boundary conditions on the torus (the holonomy group being trivial). N=4 supersymmetry does not allow chiral fermion representations and is therefore not suitable for phenomenological considerations. We have to break supersymmetry down to $N \leq 1$ through a nontrivial holonomy group i.e. twisted boundary conditions[14]. Before we do this in the next section, let us for future reference decompose an N=4 vector

$$
\begin{array}{ccccc}
\text{helicity} = 1 & \frac{1}{2} & 0 & \frac{-1}{2} & -1 \\
\text{multiplicity} = 1 & 4 & 6 & 4 & 1
\end{array}
$$

(16)

in N=1 language where this becomes one N=1 vector multiplet and *three* N=1 chiral superfields. Even in the framework of broken supersymmetry these multiplicities persist in many models and they are at the origin of the Three-Higgs-Rule. Sometimes they are also responsible for the presence of three families.

TWISTED TORI

We want to investigate models with chiral fermions and have therefore to consider the breakdown of supersymmetry in the process of compactification. The simplest way to do this is a twist of the boundary conditions. This can, of course, be done in various ways: twisting bosons[15,16] and/or fermions[17] and in general we could also treat left and right handed modes in a different way. Eventually all these different proceedures should be understood as a kind of generalized compactification[18,19]. We shall here start with the discussion of the simplest method, the orbifold construction[15], and generalize to more complicated cases as we go on. For illustrational purposes consider a simple example in two dimensions. Take a torus $T^2 = R^2/\Lambda$, where Λ is a lattice spanned by two orthonormal basis vectors e_1 and e_2 (see Fig.1). We now want to twist the boundary conditions by a Z^2 symmetry operation, i.e. we identify points through reflections at the origin. The resulting orbifold is

$$O^2 = T^2/P \simeq R^2/S \tag{17}$$

where P is called the point group (here a group with two elements: the identity element and a 180° rotation) and S is called the space group, consisting of P and the shifts in the lattice Λ. Let us first identify those points which are invariant under S. There are four of them, denoted by A, B,C and D in Fig.1. The squares (a) and (a'), as well as (b) and (b') are identified through the symmetry operation; thus (a) and (b) describe the O^2-orbifold. It is a "Ravioli"-like object obtained by folding along DB, flipping (b) over (a) and gluing at the edges.

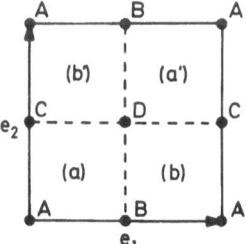

Fig.1. The O^2 orbifold

Having explained the geometry of the orbifold, we now have to understand the motion of closed strings on such an object. As shown in Fig.2, they fall into different sectors. First we have those strings already closed in flat space R^2. Then those which are closed on T^2 but not on R^2. These are a subset of the states obtained in torus compactification and constitute the so-called *untwisted sector* (the latter correspond

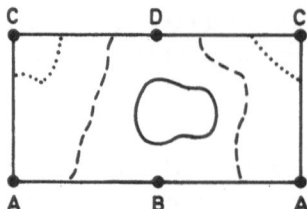

Fig.2. Closed strings on the O^2-orbifold

to the winding states). These states, of course, have to be invariant under the action of the point group. For the winding states, e.g. this implies a superposition of states with positive and negative winding numbers.

But there exist now new closed string states on O^2 which are not closed on T^2. They constitute the so-called *twisted sectors*. Again one has to project on the states invariant under P. The twisted sector states satisfy

$$X(\sigma = \pi) = P[X(\sigma = 0)] + 2\pi l \tag{18}$$

where l is a lattice vector and P is a nontrivial element in the point group. Evaluating (18) explicitly we obtain the following results. The center of mass of a string in a twisted sector is located at one of the fixed points of P, i.e. the number of twisted sectors equals the number of fixed points. Furthermore, if we consider a twist of order N, we find fractional oscillator modes of oscillator number k/N where k is an integer[15].

These twisted sectors have to be included in any model based on twisted tori. This can be seen when we consider the interactions as displayed in Fig.3. Here two strings from twisted sectors join to give a winding state in the untwisted sector. The fact that states in the twisted sectors cannot be disregarded is also a consequence of modular invariance. As a matter of fact, this invariance can be used to give a recipe for the construction of the Hilbert space for strings on orbifolds. Unfortunately we do not have the time here to discuss the constraints from modular invariance in detail and can just quote the results[15].

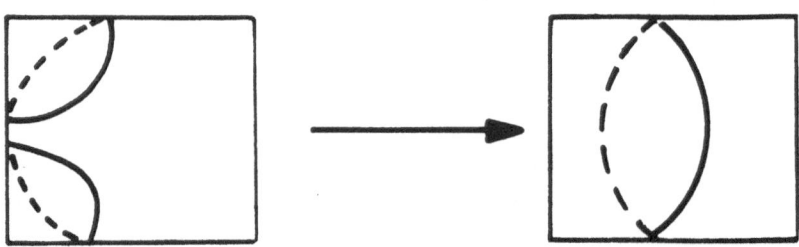

Fig.3. Twisted sector strings join to form a winding state

The recipe is to start from the Hilbert space for strings on a torus and project on all the states invariant under the twist group P (this projection is nothing else than a generalized GSO-projection), and this part of the spectrum corresponds to the untwisted sector. We next have to include all those states which are closed on the orbifold but not on the torus. Again we have to project on those states invariant under the twist. These are now called the twisted sectors and they are characterized by the fixed points of the transformation as well as the elements of P.

Let us now look more closely at these twisted sector states. Their center of mass position is at a fixed point. The massless states correspond to ground states of the string. Many of them do not have oscillator excitations and geometrically one would then view such a ground state as a string collapsed at the fixed point. Reconsidering Fig.3 we would then think that the two twisted sector states could not interact, as long as we adopt this field theoretic approximation. To join and close, these strings have to virtually extend to reach each other and a price has to be paid in such a process: such interactions are exponentially suppressed[20,21], proportional to $\exp(-d^2/\alpha')$, where d is the distance between the fixed points and α' is the inverse string tension. In terms of a σ-model interpretation such an interaction can only appear as a result of world sheet nonpertubative effects. Such couplings will become very useful in a discussion of fermion mass matrices and we shall discuss them explicitly at a later stage.

We have already mentioned in the introduction that a chiral model needs gauge symmetry breakdown in the process of compactification and we shall now discuss such a mechanism in the framework of twisted tori. The key idea is to embed the space-time twist P into the gauge group. We then have a simultaneous action of $P \in O(6)$ on six compactified dimensions and $G(P)$ in $E_8 \times E_8$ with only those states surviving that are invariant under the diagonal subgroup of P and G. The resulting unbroken gauge group is that subgroup of $E_8 \times E_8$ that commutes with G. There are restrictions for such an embedding coming from the group law and also from modular invariance which we do not want to discuss here in detail. They are the analogue of the relations between $\text{Tr}R^2$ and $\text{Tr}F^2$ in the case of compactification on smooth manifolds[2]. Let us discuss one specific example in detail.

THE Z-ORBIFOLD[15]

We consider a six-dimensional torus defined through the product of three SU(3) root lattices (Fig.4), and the twist P is chosen to be a simultaneous 120° rotation of these three lattices: thus $P = Z^3$. On an individual SU(3) root lattice this leaves three fixed points defining the weight lattice of SU(3) and leading to $3^3 = 27$ fixed points in the six-dimensional case.

In order to break the gauge group, one accompanies this space-time twist by a gauge transformation in $E_8 \times E_8$, i.e. we define a homomorphism of $P = Z^3$ in $G \subset E_8^2$. There are many ways to do this; at this moment we want to represent the gauge transformation by a shift v in the E_8 lattice, i.e. $p \to p + v$, where p is a vector of the E_8 lattice as defined in (3). There are, of course, constraints on v. First of all: P is a twist of order 3 which implies that $3v$ should be an E_8 lattice vector. Moreover, restrictions from modular invariance leave only few possible choices. For a detailed discussion of the restrictions of modular invariance see Ref.15. We choose here $v = \frac{1}{3}(1,1,2,0,\ldots,0)$, which is called the standard embedding (for reasons that will become clear later). The Hilbert-space of the physical states will contain the states invariant under the combined action of P and G. Going to a complex notation we can represent the eigenvalues of these transformations by third roots of unity $\alpha = \exp(2\pi i/3)$. For the massless states in the untwisted sector we

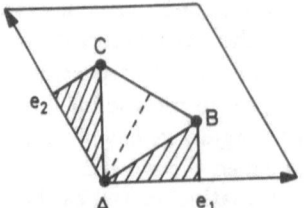

Fig.4. The Z-orbifold

have transformation properties as shown in Fig.5. Upon multiplication of left and right states this gives the usual supergravity multiplet, gauge supermultiplets of $E_6 \times SU(3) \times E_8$ and chiral superfields in the representations $3 \times (3, 27) + 9 \times (1, 1)$ of $SU(3) \times E_6$.

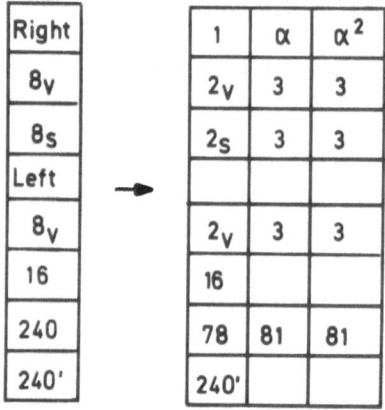

Right		1	α	α^2
8_V		2_V	3	3
8_S		2_S	3	3
Left				
8_V		2_V	3	3
16		16		
240		78	81	81
240'		240'		

Fig.5. Transformation properties of the massless states

The twisted sector comes in 27 copies, each copy with identical spectrum. On the right hand side the sectors twisted by 120° and 240° combine to one chiral superfield for $m = 0$. For the left hand side we have six twisted oscillators: three with $N_L = Z + 1/3$ and three with $Z + 2/3$ and a mass formula

$$\frac{1}{4}m_L^2 = N_L - \frac{2}{3} + \frac{(p+v)^2}{2} \tag{19}$$

Observe that $(p + v)^2 \geq 2/3$ has a minimal length. There are two possibilities to obtain massless states: $N_L = 1/3$ with $(p+v)^2 = 2/3$ and $N_L = 0$ with $(p+v)^2 = 4/3$.

The first one leads to $3 \times (\overline{3}, 1)$ of $SU(3) \times E_6$ while the second one gives $(1, 27)$ and they all come in 27 copies because of the presence of 27 fixed points. In total we thus have 36 families of E_6 in case of the standard embedding. The restrictions from modular invariance can be fulfilled in three additional cases for shifts leading to the gauge groups $(E_6 \times SU(3))^2$, $E_7 \times O(14) \times U(1)^2$ or $SU(9) \times O(14) \times U(1)$.

Although this is not a really realistic model let us look at the massles modes in more detail, since the model shows all the typical characteristics of string ground states. In the untwisted sector we first have $2_V \times 2_V$ with four degrees of freedom. The traceless symmetric part is the graviton, the trace corresponds to the dilaton and the antisymmetric part has all the properties of an axion. It is important to note that such a multiplet exists in any theory of closed strings. The states in $2_V \times 2_S$ give the supersymmetric completion of this gravitational multiplet. Furthermore we have 9 gauge singlet chiral superfields corresponding to the "moduli" (radii and angles) of the underlying torus and 9 families of E_6 ($3 \times (3, 27)$) as "charged matter" multiplets in the untwisted sector. The multiplicity 3 comes from the fact that we have compactified three complex (six real) dimensions. Observe that only the smallest nontrivial representations of $SU(3)$ and E_6 can be obtained as matter fields and this is quite a general property of string ground states. It comes from the fact that the vacuum energy is -1 in the untwisted sector and $-2/3$ in the twisted sectors. For the massless states this requires $p^2 \leq 2$ and does not allow the presence of *exotic* states in representations with a large eigenvalue of the quadratic Casimir-operator. This fact can be regarded as a prediction of these string models. For models with gauge group $SU(3) \times SU(2) \times U(1)$ it implies that we only have 3, $\overline{3}$, 1 of $SU(3)$ and only singlets or doublets of $SU(2)$. It is quite a remarkable property of these models that they predict the structure of the massless spectrum exactly as we observe it.

In the twisted sectors we have a **27** of E_6 at each of the 27 fixed points. This geometrical classification of the states will be very useful for the understanding of the pattern of trilinear Yukawa-couplings. We shall come back to these rules when we discuss the masses of quarks and leptons.

In addition we find states $3 \times (\overline{3}, 1)$ at each fixed point. These states are the so-called blowing-up modes. It can be shown that a nontrivial vev of these fields leads to new consistent string vacua in which $E_6 \times SU(3)$ is broken to E_6. Geometrically this can be understood as a reparation of the orbifold singularities, i.e. a removal of the corners at the fixed points followed by gluing in a smooth disk. In this way one obtains a string model on a Calabi-Yau space, at least pertubatively in the vev's of the blowing-up modes. The corresponding Calabi-Yau space has 36 parameters that can be arbitrarily varied: 9 come from the moduli of the underlying torus and the remaining 27 from the vev's of the blowing-up modes. As soon as the singularities are repaired it is no longer possible to unambiguously identify states belonging to just one of the fixed points and mixings become possible: we can tear the string away from the fixed point. In the language of the low energy field theory this will happen through nonrenormalizable Yukawa couplings into which one inserts the vev's of the blowing-up modes. We shall discuss the importance of higher order Yukawa couplings later in the framework of degenerate orbifolds.

TOWARDS THREE FAMILIES

Let us now discuss the question concerning the number of families of quarks and leptons. The Z_3-model with standard embedding leads to 36 families and this number is embarrassingly large. What about other orbifolds? Since we are interested in models with $N = 1$ supersymmetry we know them all since the allowed twists can be classified through the discrete subgroups of $SU(3)$ (as a subgroup of the maximal holonomy group O(6)). In each case one has several possibilities for the embedding of the twist into the gauge group, and all these models have been worked out[15]. The result concerning the number of chiral families is quite unsatisfactory. No model with three families had been found.

At this point it is important to remember the situation of string models in the case of torus compactification. There it had been observed that, apart from the metric g_{ij}, one also had the freedom of nontrivial background values for the antisymmetric tensor b_{ij} as well as the gauge background fields (Wilson lines) A_μ^I around noncontractible loops. It is natural to ask the question about the rôle of these fields in the framework of generalized orbifold compactification[18]. As a matter of fact, it turns out that a consideration of A_μ^I around a loop

$$\int_i A_\mu^I dx^\mu = 2\pi A_\mu^I e_i^\mu = 2\pi a_i^I \tag{20}$$

is very important, both from the viewpoint of the unbroken gauge group[15] as well as the question concerning the number of families[18]. Let us now try to understand this in detail.

The appearance of gauge background fields can be explained quite easily if one remembers the construction of the Hilbert space for strings on orbifolds. It is nothing else than the embedding of the lattice vectors e_i, defining the underlying torus, into the gauge group; very similar to the embedding of the rotation θ into the gauge group. All this can be summarized as a homomorphism from the space group $S = (P, \Lambda)$ into the gauge group. Let us here first consider the so-called *abelian embedding* where both θ and e_i are represented by shifts v and e_i in the group lattice. The group law has to be satisfied and this leads to certain restrictions as we can see from the consideration of the example discussed in the last section. On the SU(3) lattice we have $\theta e_1 = e_2$ and the group law requires therefore $a_1 = a_2$, i.e. we have only three independent Wilson lines. Moreover, we have to realize that for any element X of the space group we have X^3 =identity. This then implies that 3a is a lattice vector. Similar conditions are obtained from the requirement of modular invariance which we shall not discuss here. It is important to point out that these restrictions lead to a *quantization of the value of a*. The string model is only consistent for a discrete set of values of the background fields (at least in this case of the abelian embedding). This quantization is a necessary condition for the change of the chiral structure (e.g. the number of families) by a change of the background fields. Index theorems guarantee that such changes cannot be achieved through a smooth mechanism.

What are the properties of these models with nontrivial background fields? First of all, we can have a further breakdown of the gauge group. This happens because of an additional projection in the untwisted sector requiring, for example, $p \cdot a$ to be of integer value (p being an E_8 lattice vector).

The most important change, however, concerns the twisted sectors: they no longer have identical spectra and the 27-fold degeneracy is broken. To see this, let us again consider Fig.4, where we have three fixed points. The fixed point at the origin is invariant under the rotation. In the corresponding twisted sector we have therefore the same situation as before, p being shifted by v, the shift vector representing the rotation. The other fixed points, however, are not invariant under the rotation alone; instead this rotation has to be accompanied by a shift e_1 in one case and $e_1 + e_2$ in the second case. Since we have now gauge background fields attached to these directions, we have in the corresponding sectors to shift p not only by v but by $v + a_1$ or $v + a_1 + a_2$, respectively, and the spectrum in these twisted sectors will be different.

Let us illustrate the mechanism with a simple example[18]. Take the Z-orbifold with the standard embedding $v = \frac{1}{3}(1,1,2,0,\ldots,0)$ and consider one Wilson line $a_1 = a_2 = \frac{1}{3}(0,\ldots,0,1,1)(2,0,\ldots,0)$ which satisfies the requirements from group law and modular invariance. The projection $p \cdot a \in Z$ in the untwisted sector leads to the gauge group $SU(3) \times SU(6) \times U(1) \times O(14) \times U(1)$. From the $3(\mathbf{3},\mathbf{27})$ matter states in the untwisted sector only $3(\mathbf{3},\mathbf{15})$ survive (the $\mathbf{15}$ being the antisymmetric tensor in $SU(6)$). With one Wilson line the 27 twisted sectors split into three groups of nine. The first nine are independent of the presence of a_1 and give the same spectrum as before; with the difference that the $\mathbf{27}$ of E_6 is decomposed into $\mathbf{27} = \mathbf{15} + \bar{\mathbf{6}} + \bar{\mathbf{6}}$ of $SU(6)$. In the other twisted sectors we obtain a different result since we have to shift p to $p + v + a_1$ or $p + v + a_1 + a_2$, respectively. Here we have $2a_1 = -a_1$ and these 18 sectors give the same massless spectrum consisting of $(\bar{\mathbf{3}},\mathbf{1}) + (\mathbf{1},\bar{\mathbf{6}})$. In total we therefore have a model with 18 families of $SU(6)$ to be compared with the 36 E_6-families previously found. A closer inspection of models with Wilson lines reveals the fact that millions of 3-family models can be constructed quite easily[18,9]. We shall not discuss this search in detail and refer to the original literature. It is actually interesting to observe that it is here much easier to construct 3-family models than, for example, 4-family models. This has its origin in the fact that we compactify three complex dimensions.

TOWARDS $SU(3) \times SU(2) \times U(1)$

Amongst this large number of 3-family models we can now search for those that ressemble most closely to the standard model. The first question concerns the gauge group, where we want to see how close we could get to $SU(3) \times SU(2) \times U(1)$. As already discussed, our method of constructing models based on the abelian embedding of the space group into the gauge group does not allow us to obtain models with lower rank. Starting with the observable E_8 we should therefore try to break it to $SU(3) \times SU(2) \times U(1)^5$. A variety of such models can be found[22] and let us here discuss one example. We choose $v = \frac{1}{3}(1,1,1,1,2,0,1,1)(1,1,0,0,0,1,1,2)$, $a_1 = a_2 = \frac{1}{3}(1,1,1,2,1,0,0,0)(0,\ldots,0,2,0,0)$, $a_3 = a_4 = \frac{1}{3}(0,\ldots,0,2)(1,1,1,2,0,0,0,1)$ and obtain the desired gauge group. Apart from additional singlets we get $3[(\mathbf{3},\mathbf{2}) + (\mathbf{1},\mathbf{1})]$ in the untwisted sector. In the $(v + a_1)$-twisted sector as well as in the $(v + a_1 - a_3)$-twisted sector we obtain three $[(\bar{\mathbf{3}},\mathbf{1}) + (\mathbf{1},\mathbf{2})]$ each, whereas in the $(v - a_1 + a_3)$-sector there are nine additional SU(2) doublets. As a result we have three families of quarks and leptons plus 12 Higgs doublets plus singlets of $SU(3) \times SU(2) \times U(1)^5$. Observe first that this models does not contain additional so-called Higgs-triplets which could mediate fast proton decay; we have just six $(\bar{\mathbf{3}},\mathbf{1})$ states correponding to the right-handed quark superfields. Thus the potential problem of fast proton decay is solved in the must natural way, since, because of the quantization of the a_i, there is no fine tuning involved. There are other models[22,9] which contain additional color triplets and we shall discuss those in the next section.

In this example it is also interesting to note the special pattern in which quark and leptons are distributed over twisted and untwisted sectors. While the lefthanded quarks $(\mathbf{3},\mathbf{2})$ come from the untwisted sector we find the right-handed ones $(\bar{\mathbf{3}},\mathbf{1})$ in different twisted sectors. Since also the potential Higgs-bosons are in twisted sectors we arrive at the situation, that quark masses cannot be generated through renormalizable (i.e. trilinear) Yukawa couplings. This question of masses, however, should be postponed till we have discussed the breakdown of the additional U(1)'s in these models. This breakdown will require vev's of some of the singlet fields, and with the inclusion of these states we shall also be able to generate quark and lepton masses through nonrenormalizable Yukawa couplings.

Let us therefore first discuss the question concerning the rank of the gauge group. In the early times of a phenomenological discussion of superstring models there appeared a certain tendency to proclaim that additional $U(1)$ gauge symmetries should exist. Our example seems to confirm this prejudice. We should, however, not forget that our method of construction does not allow a reduction of the rank of the original gauge group. This comes from the fact that we embedded the space group into the gauge group in an abelian way. The shift on the gauge lattice has no effect on the oscillator states $N_L = 1$ in compactified dimensions and therefore preserves the rank of the gauge group.

A method to lower the rank can be obtained[23] by realizing that the space group is always nonabelian, even if the point group is abelian. The rotations of the point group simply do not commute with the shifts that define the torus. To exploit this fact we have to embed the space goup into the gauge group in a nonabelian way: we represent rotations by rotations and shifts by shifts. This involves a quite technical discussion of the way to perform rotations of the $E_8 \times E_8$ lattice. We do not have the time here to present this method in detail and have to content ourselves to explain the main idea.

Let us therefore consider (θ, e_i) in the space group represented by (R, a_i) in $E_8 \times E_8$, where R is a rotation of the lattice. This rotation has to correspond to an automorhism of the lattice and in our example it will correspond to a 120° rotation. A classification of the possible embeddings involves a knowledge of the Z_3 automorphisms of the lattice. Here it turns out to be useful to decompose E_8 into $SU(3)^4$ and for a detailed explanation we refer to the original literature[23]. We again observe in this example, that the group law requires $a_2 = Ra_1$ and that therefore we only have three independent Wilson lines. We also know that $(\theta, e_1)^3 = 1$ and in the case of the abelian embedding this required the quantization of the Wilson lines. But here $(R, a_1)^3 = 1$ is satisfied for any value of a_1. This implies that the Wilson line is no longer quantized and there is also no further restriction from modular invariance. Please keep in mind that with such a mechanism we are not able to change the chiral structure of the model, but that it breaks the gauge group in a smooth way. We can continuously change the value of the a_i and this corresponds to a continuous change of the mass of the gauge bosons corresponding to the broken generators of the gauge group. At certain critical values of the Wilson lines (typically $p \cdot a_i \in Z$ with p a vector in the E_8 lattice) we shall find enhanced gauge groups. Changing a_i will break the gauge group: accompanied by a smooth Higgs mechanism. It actually turns out that these simple models exhibit a Three-Higgs-Rule[9] (T$_H$R) in such a way that with one gauge boson three Higgs bosons acquire a mass: one combination playing the rôle of the longitudinal component of the gauge boson. Apart from this fact it is important to remember that this mechanism proved the existence of a *smooth* mechanism to break E_8 in a model with N=1 supersymmetry. We believe that such a mechanism should eventually describe the breakdown of $SU(2) \times U(1)$ and be responsible for the masses of the weak gauge bosons. It would therefore be interesting to see, how far T$_H$R remains valid under more general circumstances.

Why can we actually lower the rank of the gauge group with such a mechanism? Consider the root lattice of $SU(3)$ as in Fig.6 and perform 120° rotations. From the original 8 states (6 lattice vectors and 2 oscillator states) only two are invariant under this rotation, and they correspond to a certain combination of the lattice states;

$SU(3)$ is broken to $U(1)^2$. But now the states corresponding to the gauge bosons of the two U(1)'s are no longer oscillator states and they feel the presence of Wilson lines, represented through the shifts a_i on the lattice. As a result we observe that $U(1)^2$ is broken if $p \cdot a_i$ is not an integer and this will reduce the rank of the gauge group. We also see that for all values of a_i with integer $p \cdot a_i$ the rank of the gauge group will be preserved.

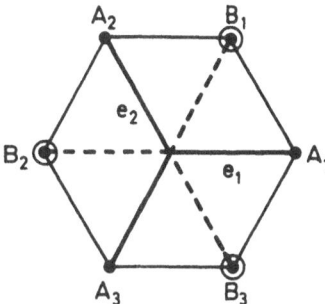

Fig.6. The two invariant combination of the lattice states.

This mechanism allows us to lower the rank of the gauge group arbitrarily and even break $E_8 \times E_8$ completely. We therefore conclude that strings do not necessarily predict gauge groups of higher rank. We still have to investigate, however, whether models that contain $SU(3) \times SU(2) \times U(1)$ have the tendency to contain additional U(1)'s.

The explicit construction of string models using this mechanism is quite cumbersome and a detailed study of these models has not yet been performed. We have no time here to explain these technicalities and refer the reader to the original paper[23]. Recently, a simpler way to discuss such models has been proposed and we shall discuss this in the following section.

DEGENERATE ORBIFOLDS[24]

Our final goal would be a classification of all consistent string theories in four extended dimensions. This is supposed to be equivalent to a classification of superconformal field theories on the two-dimensional world sheet. There are various ways to construct such 4-dimensional string models and we have here presented the method based on orbifold compactification. The question remains, whether such methods cover the space of all consistent string vacua. Maybe there are additional consistent models which at the moment, because of technical difficulties, cannot be constructed explicitly? New methods are required to answer such questions.

For the well-known reasons we here concentrate on models with N=1 space-time supersymmetry, i.e. either (2,2) or (2,0) world-sheet supersymmetry. We also know that the vacuum structure of supersymmetric theories is quite special. In the case of global supersymmetry all symmetric vacua are degenerate at vanishing vacuum energy, implying vanishing vev's for the so-called D- and F-terms. With D one denotes the auxiliary fields of the gauge supermultiplets

$$D_\alpha = g_\alpha \Phi_i^* (T_\alpha)_j^i \Phi^j \tag{21}$$

where T_α are the group generators and Φ_i the chiral matter superfields. The F_i denote the highest components of those chiral superfields and can be represented by the derivatives of the superpotential W

$$F_i = \frac{\partial W}{\partial \Phi_i} \tag{22}$$

and the contribution to the vacuum energy is given by

$$\frac{1}{2} \sum_\alpha D_\alpha D^\alpha + \sum_i F_i^* F^i \tag{23}$$

In the case of local supersymmetry (i.e. supergravity) the vacuum structure is a bit more complicated, but since we here discuss string theories with classical scale invariance, we are essentially in the same situation; we only have the additional constraint that the vev of the superpotential should vanish as well: all the symmetric vacua are degenerate and $F = D = W = 0$.

This special property of theories with $N = 1$-supersymmetry provides us with a method to search efficiently for a new class of consistent string theories known as *degenerate orbifolds*. Just use the following recipe. Start with a string model as constructed in the previous section and identify the spectrum of massless fields. Compute all the vertex operators and all the interactions of these fields and determine the corresponding scalar potential. Then consider nonvanishing vev's for these massless fields and find all the degenerate vacua. All the different groundstates will describe consistent string models. In the language of the underlying 2-dimensional conformal field theory these nontrivial vev's of the massless fields will correspond to so-called marginal operators.

This new method of searching for previously unknown consistent string models relies mainly on a field theoretic argumentation for the massless states and one might argue that string effects are not properly taken into account. This is not true since the information about string theory is contained in the interaction between the massless fields. It should, however, be stressed here that we have to compute *all* the relevant couplings. In the superpotential, for example, we have to compute all couplings and cannot just consider the trilinear (so-called renormalizable) ones. The described method, on the other hand, is reliable only pertubatively in the vev's of the massless fields. In the case of "large" vev's this method might not give us some models in so-called multicritical points where additional stringy modes (like e.g. winding states) become massless. This is quite similar to the enhancement of gauge groups in the framework of torus compactification. Given the remarks, it is clear that this method allows us to find only those new models that are continuously connected with previously known ones by a smooth variation of parameters. This then provides us with a new way to break gauge symmetries smoothly, accompanied by a Higgs mechanism.

To follow the recipe let us start with the D-terms, where the discussion is quite simple since this is essentially an exercise in supersymmetric field theory[25]. One of the methods uses the construction of gauge invariant functions $I(\Phi)$. Then

$$\delta I_\alpha = \frac{\partial I}{\partial \Phi_i} (T_\alpha)_i^j \Phi_j = 0 \tag{24}$$

and if we choose the vev's of the Φ-fields to satisfy

$$\frac{\partial I}{\partial \Phi^i} = c\Phi_i^* \qquad (25)$$

we obtain a vanishing D-term for arbitrary values of c. Thus the search for vanishing D-terms is just an exercise in group theory. In the case of the presence of an anomalous U(1) symmetry we have to face the slight complication to consider also the presence of a Fayet-Iliopoulos term and obtain

$$D \cong \frac{g}{192\pi^2} TrQ + q_i \Phi_i^* \Phi^i \qquad (26)$$

indicating a nonvanishing vev for some combination of the Φ_i. For the discussion of this case we should remember that the gauge coupling g is determined dynamically through the vev of the dilaton field. The various vev's adjust themselves to give a vanishing D-term and supersymmetry remains unbroken.

Next we have to consider the F-terms and here all the information about string theory enters. To answer the questions concerning the F-terms we have to compute the full superpotential including terms with an arbitrary large number of external fields. The scattering amplitudes for the physical states propagating on orbifolds can be computed using the powerful tools of conformal field theory[20,21]. We have no time here to explain these calculations in detail. The geometric structure of orbifolds, however, allows us to formulate some simple selection rules which we shall describe briefly.

We first have the so-called point- and space-group selection rules. To each field entering an amplitude there is an element of the space group associated to. The amplitude vanishes unless the product of these group element contains the identity. These rules are at the origin of certain discrete symmetries in the field theory of the massless states.

Next we have a rule that might be called lattice selection rule. The conservation of the lattice momentum in the $E_8 \times E_8$ lattice is equivalent to the requirement of gauge invariance. The conservation of lattice moment in the O(10) lattice of the bosonized NSR-fermions, however, gives new restrictions corresponding to discrete R-symmetries.

A third rule takes care of additional invariances under phase rotations of the internal coordinates and gives rise to new discrete symmetries.

Rule 4 is more difficult to understand. It is a special rule telling us which combination of fixed points appears in the allowed couplings. This rule seems to be really "stringy" and it cannot be understood entirely in terms of the symmetries of the field theory of the massless states[26]. We shall discuss this rule later in detail.

Equipped with these selection rules we can now determine the superpotential for the groundstates of an orbifold model. Such an investigation reveals some surprising results. One finds an enormous number of new models and flat directions, many of them not yet constructed as a full string model. As a matter of fact we find back all the previously discussed models constructed through the nonabelian embedding of Wilson lines; the parameters that can be smoothly varied are in one to one correspondence with the vev's of the massless fields in the untwisted sectors. New models are found along flat directions that involve nonvanishing vev's of fields in the twisted sectors[24]. A small subset of those had been considered previously in the discussion of the blowing-up procedure to remove the orbifold singularities[27].

The most important result from this discussion, apart from the surprise about the huge number of new models, is the observation that the rank of the gauge group can be lowered efficiently. The presence of gauge groups of high rank in previously constructed models seems to be an artifact of the construction method used. This, of course, has consequences for the discussion whether superstrings predict additional $U(1)$ gauge groups or not.

We would like to stress here again that the models obtained by this mechanism correspond to a smooth symmetry breakdown and that it is valid pertubatively in the vev's of the massless fields. Knowledge about the behaviour for large vev's cannot always be obtained by this method.

Let us illustrate the method by a simple example. Consider the Z_3-orbifold with the nonstandard embedding

$$v = \frac{1}{3}(1,1,1,1,-2,0,0,0)(-2,0,\ldots,0) \qquad (27)$$

with gauge group $SU(9) \times O(14) \times U(1)$. The charged matter content consists of three copies of

$$(84,1)_0 + (1,14)_{-2} + (1,64)_1 \qquad (28)$$

in the untwisted sector and 27 copies of $(\overline{9},1)_{4/3}$ in the twisted sectors. In Ref.23 it was realized that representing the embedding by a Weyl rotation Θ in $E_8 \times E_8$, instead of a shift, one can add Wilson lines represented by arbitrary shifts in the E_8 lattice. In particular, choosing a representation of E_8 in terms of $SU(3)^4$, we could consider the Wilson line

$$a_1 = \lambda(d,0,0,0)(0,0,0,0) \qquad (29)$$

where d is an SU(3)-weight and λ an arbitrary parameter. When we vary λ continuously away from zero, the SU(9) gauge group is broken down to $SU(3)^3$. Points in λ space where λ reaches integer values are so-called multicritical points, where the gauge symmetry is enhanced to $SU(3)^3 \times U(1)^2$, which incidentally has the same rank as the original gauge group.

We can also obtain this model by consideration of the original SU(9)-model, giving nontrivial vev's to some components of one of the 84-dimensional representations in the untwisted sector[24]. Denoting this 84 by X^{ijk}, completely antisymmetric in i,j and k, we can see that a flat direction in the potential corresponds to

$$X^{123} = X^{456} = X^{789} \neq 0 \qquad (30)$$

and all other components vanishing. It is straightforward to show that (30) does not induce a D-term. Selection rule 2 implies the absence of F-terms, the only allowed couplings being $(XYZ)^n$, where Y and Z represent the other two 84's. In the vacuum, these terms vanish as well as their first derivatives with respect to the fields. The vev's in (30) break the gauge group to $SU(3)^3$ with a smooth Higgs mechanism. The number of broken generators in the coset $SU(9)/SU(3)^3$ is 56 and one 84 leaves us with $28 = 84 - 56$ fields in the representation $[(3,3,3)+(1,1,1)]$ of $SU(3)^3$, the same as in the model with the Wilson line given in (29). We have found the same model by two different methods. Observe, however, that in the second case we are not able to uncover the multicritical points at $\lambda \in Z$ with enhanced gauge group $SU(3) \times U(1)^2$.

It can be shown that all the models with the nonabelian embeddings of Wilson lines correspond to models with nontrivial vev's of massles fields in the *untwisted* sector. New models are obtained when one considers vev's of twisted sector fields. In the model discussed, for example, it is possible to break the gauge symmetries down to SO(12), while having vanishing D-terms, by giving vev's to nine of the $(\bar{9}, 1, 4/3)$ and to one $(1, 14, -2)_1$, where the subscript refers to the first (complex) internal coordinate. The only dangerous F-term could arise from the coupling

$$(84)_1 \times (\bar{9})^3 \times (14)_1^2 \tag{31}$$

but this vanishes if we consider the $\bar{9}$'s with vev's to be at the same fixed point in the first (complex) internal coordinate. A more detailed discussion of these degenerate orbifolds can be found in the original paper[24].

SU(3)×SU(2)×U(1) MODELS

Being equipped with these methods we can now ask the question whether one can obtain a realistic $SU(3) \times SU(2) \times U(1)$ model and such investigations have started recently. The previously described models[22] with gauge group $SU(3) \times SU(2) \times U(1)^n$ serve as a good starting point and have been considered by Casas and Muñoz[28] as well as our group[29]. In a first step one tries to find flat directions that allow a breakdown of $U(1)^n$ to $U(1)_Y$ where Y denotes the correct hypercharge. It should be stressed that this is already a nontrivial constraint and it is just a consequence of the huge number of flat directions that this constraint can be fulfilled. Although such investigations are straightforward they turn out to be very tedious, since one has to identify many U(1)-charges for many fields and we have no time here to discuss the details. The corresponding tables can be found in the literature[28,29].

Within these $SU(3) \times SU(2) \times U(1)^n$ models we can distiguish two classes: those models which contain additional triplets and antitriplets (and therefore could suffer from fast proton decay) and those where these Higgs triplets are absent. In our discussion earlier, of course, we preferred the latter for the obvious reasons. But now we should also reconsider the former ones since the breakdown of the extra U(1)'s might provide a mechanism to make these triplets massive. As a matter of fact, it turned out that model 3 of Ref.22 (with additional triplets) is a very promising starting point and most of the investigations up to now have been performed within this model. There are various flat directions in the potential that allow a breakdown of $U(1)^n$ and give mass to the additional triplets, avoiding the problem of fast proton decay.

The next question concerns the number of Higgs-doublets. Do they also become heavy when the triplets gain a mass? The answer is no and this is really a nontrivial result. It could be considered as a potential solution of the μ-problem of super-symmetric theories. This fact has certainly a connection with the constraints from selection rule 4, which seems to require a breakdown of the concept of naturalness in the framework of these theories[26], and we shall come back to this discussion in the next section.

We can now ask the question concerning the quark and lepton masses, i.e. the Yukawa couplings between quarks, leptons and the candidate Higgs-doublets. As in any supersymmetric model, we have first to find out in a given model which of the SU(2)-doublet superfields represent Higgses and which of them represent left-handed leptons. This distinction can be made with the help of an R-symmetry and such R-symmetries can appear as a consequence of selection rule 2.

Concerning the question of fermion masses we could, of course, now stop our discussion and just assume that all the terms allowed by the symmetries exist, and we could then proceed to adjust them at our will. Fortunately we are not allowed to do so, since we can really compute these couplings. We have to do this even at the risk that certain models could be ruled out by these considerations. Let us for example consider model 1 of Ref.22. Here the left-handed quarks $Q = (3, 2)$ are in the untwisted sector, while antiquarks $\bar{q} = (\bar{3}, 1)$ and the Higgs $H = (1, 2)$ are in twisted sectors. The model is built with a Z_3 twist and rule 1 requires the number of twisted fields to be a multiple of 3. As a result, quark masses through the trilinear couplings $Q\bar{q}H$ are not allowed because they are of type UTT. The model does not allow the generation of quark masses. Not all is lost, however. In the process of breaking $U(1)^n$ to $U(1)_Y$ we have some $SU(3) \times SU(2)$ singlet fields X that receive a nonvanishing vev, and they could very well come from the twisted sector. The coupling

$$\frac{1}{M_P} Q\bar{q}HX \tag{32}$$

could then be an allowed term in the superpotential since it is of type $UTTT$. With a nontrivial vev for X (32) would induce an effective trilinear Yukawa coupling and allow a generation of quark masses. Preliminary investigations actually show that these so-called nonrenormalizable Yukawa couplings play a crucial rôle in the generation of quark and lepton masses, at least in the currently considered models. A careful and tedious examination of the vacuum structure of the models is necessary in order to reveal all these mechanisms. This situation is in fact quite desireable, since these nonrenormalizable Yukawa couplings all contain exponential factors[27]. Together with the nonvanishing vev's of the fields that break the additional U(1)'s, these factors might be just what we need to explain large fermion mass ratios as we observe in nature, although a model with the correct pattern still needs to be constructed.

Such a pattern has been worked out for the model under consideration, and the results are very encouraging for a further study of of these questions. A detailed explanation of the structure of the Yukawa couplings requires many technicalities and we have to refer the reader to the original literature[29].

Let us close this section with some comments concerning the general structure of these models. The model under consideration has three families and no antifamilies as it comes from an orbifold with a prime number twist. One could, of course, also consider models with $n + 3$ families and n antifamilies. Such models in general allow a greater flexibility to adjust Yukawa couplings, while the former have more predictive power. Actually the model proposed in the framework of Calabi-Yau compactification has 9 families as well as 6 antifamilies and allows a parametrization of realistic quark and lepton masses[30]. It remains to be seen, however, whether this is still true after one has *computed* the Yukawa couplings. The same applies to the superstring inspired model[31] based on $SU(5) \times U(1)^4$. These approaches and ours differ in some important fundamental point of view. We try to see whether it is possible to construct a model based on $SU(3) \times SU(2) \times U(1)$ directly at the string level, while the others aim at some larger unified group structure that is broken at an intermediate scale, which incidentally turns out to be quite close to the string scale[32]. The breakdown of gauge groups at this "intermediate" scale needs more input and cannot yet be understood in terms of the string theory. It ussually assumes the existence of some semiflat directions, i.e. directions that are flat if one only takes into account the trilinear couplings in the superpotential. This has to be compared with the approach using really flat directions and the considerations of degenerate supersymmetric vacua. It is not clear at the moment which of the approaches will be more useful and further research is needed in both directions. One should try to minimize the number of input assumptions that cannot be derived from the string itself. Ultimately a better understanding of supersymmetry breakdown at the string level is required for the answer of these questions. But it should also be clear that progress can only be made if the Yukawa couplings are really computed. We cannot just rely on the symmetries of the field theory of the massless modes, as we shall see explicitly in the next section.

NATURALNESS AND STRING THEORIES[26]

A central issue in the discussion of string theory should be be an effort to isolate stringy effects which cannot be understood in the framework of field theories at low energies. The consequences of selection rule 4 seem to have such properties and we wish to discuss them now in detail. While we can understand the other rules in terms of the symmetries (discrete symmetries, discrete R-symmetries, gauge symmetries) of the field theory, such an understanding is not possible for rule 4.

To compute the Yukawa couplings, we usually consider correlation functions of vertex operators of the type

$$V_{-1/2}V_{-1/2}V_{-1}V_0 \ldots V_0 \tag{33}$$

where the subscript denotes superconformal ghost number which should equal to -2 for tree level correlation functions. You need not understand all the details to follow the discussion; it is enough to know that half integer ghost number corresponds to space time fermions and integer ghost number to bosons. V_0 includes factors proportional to (∂X). Such factors thus always appear in correlations of type (33) with more than three fields, and it becomes important to determine the classical solutions for (∂X). In the case of twisted sector fields a classical solution for (∂X) is typically given by the distance v between fixed points, and the correlator (33) is proportional to $v^n \exp(-v^2/\alpha')$ if $n + 3$ fields participate in the interaction. What happens now if all the fields sit at the same fixed point? The trivial solution is then $v = 0$ and we have to see whether there are different classical solutions for (∂X). Consider a field in a twisted sector defined through the rotation θ and fixed point x_f. Possible solutions are then[21]

$$v = (1 - \theta)[x_f + ne_i] \tag{34}$$

where the e_i are lattice vectors. As an example let us take again the Z_3 orbifold and display the solutions for $x_f = 0$ in Fig.7. We see that these solutions themselves form a lattice and that

$$\sum_i v_i^{2k+1} = 0 \tag{35}$$

at least for integer k. As a result of this, we observe that couplings tend to vanish if there are several fields from the same fixed point participating in the interactions. This is actually even true when the corresponding fixed points have the same entry in a 2-dimensional sublattice. Such a behaviour of the couplings is very peculiar and "stringy". As a matter of fact, it cannot be understood in the framework of the field theory of the massless states. Let us explain this in the Z_3 example. We denote the twisted sectors by (n_1, n_2, n_3) with $n_i = 0, \pm 1$. Consider fixed point $(0,0,0)$ and take a **27** of E_6, which we denote by Φ. Checking selection rules 1 through 3 we remain with Φ^{3+9k} as allowed couplings. According to rule 4 we have the addtional constraint that for $n \in Z$, Φ^{12+18n} is forbidden. Let us now try to understand this in terms of the symmetries of the field theory of the massless states. If this cooresponds to a symmetry, it must be an R-symmetry since Φ^3 is allowed. This would imply $R(\Theta) = 1$ and $R(\Phi) = 2/3$. Consider now Φ^m with large $m = 12 + 18n$ which is a forbidden coupling if all the fields sit at the same fixed point. Now add fields from other fixed points in such a way to obtain an allowed coupling $\Phi^m Y^k$. We immediately conclude that also $\Phi^m Y^{2k}$ is allowed. But this cannot be a result of a symmetry. Φ^m differs from $\Phi^m Y^k$ by ΔR, but this is the same as the difference between $\Phi^m Y^k$ and $\Phi^m Y^{2k}$. Thus we see that rule 4 cannot be understood in terms of symmetries. This might have far reaching consequences in the sense that it might enable us to isolate "stringy" effects. Such effects can be seen to play a major rôle in model building and they might be responsible for the presence of light Higgs doublets, solving the μ-problem in supergravity models[33]. This result of rule 4 shows us that it is not enough to consider the low energy field theory, but instead has to do calculations in string theory. It might happen in some models that some quark and lepton masses vanish although they are allowed by all the symmetries of the low energy theory. Arguments based on naturalness do no longer apply and explicit string calculations have to be performed. With a more optimistic attidude, of course, we might also hope that this strange behaviour dictated by rule 4 might give us a key for measurable predictions from string theory.

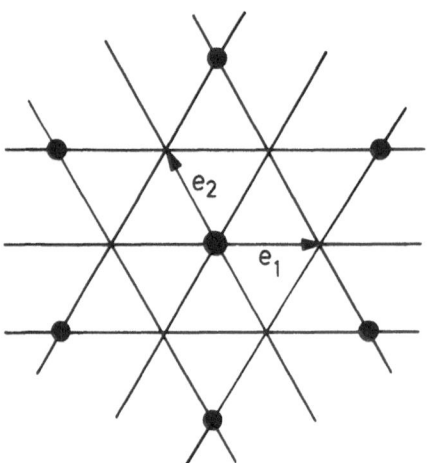

Fig.7. The solutions of equation (34)

FINAL REMARKS

It should be clear that there is still a far way to go until we obtain a satisfactory string model that explains the parameters of the $SU(3) \times SU(2) \times U(1)$ standard model. We first have to classify all the consistent string vacua in 4 extended dimensions. Although tremendous progress has been made during recent years, we have not yet reached this goal. A central question remains the understanding of supersymmetry breakdown in the framework of string theories. Unfortunately I had no time here to discuss this topic, but I shall come back to this point at another occasion. Such an understanding of supersymmetry breakdown would enable us to discuss explicitely how the classical vacuum degeneracy is lifted and how the true vacuum is obtained. But we then also have to face the problem to explain a vanishing cosmological constant. In the field theory at low energies we have a dynamical mechanism to break supersymmetry based on the conjectured formation of gaugino condensates in the hidden sector. Such a formation of condensates is of nonpertubative nature and not yet understood in the framework of string theories. Many definite statements about the phenomenological prospects of string theory have to wait till we understand these issues more deeply.

REFERENCES

1. For a review see: M.B. Green,J.H. Schwarz and E. Witten, Superstring Theory, Cambridge Monographs on Mathematical Physics, Cambridge University Press 1987
2. P. Candelas, G. Horowitz, A. Strominger and E. Witten, Nucl. Phys. B258 (1985) 46
3. For a review see: H.P. Nilles, Physics Reports 110 (1984) 1
4. H.P. Nilles, Phys. Lett. 115B (1982) 193; Nucl. Phys. B217 (1983) 366
5. S. Ferrara, L. Girardello and H.P. Nilles, Phys. Lett. 125B (1983) 457
6. R. Barbieri, S. Ferrara and C.A. Savoy, Phys. Lett. 119B (1982) 343
7. J.P. Derendinger, L. Ibáñez and H.P. Nilles, Phys. Lett. 155B (1985) 65
8. M. Dine, R. Rohm, N. Seiberg and E. Witten, Phys. Lett. 156B (1985) 55
9. L. Ibáñez, J. Mas, H.P. Nilles and F. Quevedo, Nucl. Phys. B301 (1988) 157
10. M. Green and J. Schwarz, Phys. Lett. 149B (1984) 117
11. D.J. Gross, J.A. Harvey, E. Martinec and R. Rohm, Phys. Rev. Lett. 54 (1985) 502; Nucl. Phys. B256 (1985) 253; Nucl. Phys. B267 (1986) 75
12. K.S. Narain, Phys. Lett. 169B (1986) 41
13. K.S. Narain, M.H. Sarmadi and E. Witten, Nucl. Phys. B279 (1987) 369
14. R. Rohm, Nucl. Phys. B237 (1984) 553
15. L. Dixon, J. Harvey, C. Vafa and E. Witten, Nucl. Phys. B261 (1985) 678 and B274 (1986) 75
16. W. Lerche, D. Lüst and A.N. Schellekens, Nucl. Phys. B287 (1987) 477
17. H. Kawai, D.C. Lewellen and S.H.H. Tye, Nucl. Phys. B288 (1987) 1; I. Antoniadis, C. Bachas and C. Kounnas, Nucl. Phys. B289 (1987) 87
18. L. Ibáñez, H.P. Nilles and F. Quevedo, Phys. Lett. 187B (1987) 25
19. K.S. Narain, M.H. Sarmadi and C. Vafa, Nucl. Phys. B288 (1987) 551
20. S. Hamidi and C. Vafa, Nucl. Phys. B279 (1987) 465
21. L. Dixon, D. Friedan, E. Martinec and S. Shenker, Nucl. Phys. B282 (1987) 13
22. L. Ibáñez, Jihn E. Kim, H.P. Nilles and F. Quevedo, Phys. Lett. 191B (1987) 282
23. L. Ibáñez, H.P. Nilles and F. Quevedo, Phys. Lett. 192B (1987) 332
24. A. Font, L. Ibáñez, H.P. Nilles and F. Quevedo, Nucl. Phys. B307 (1988) 109
25. F. Buccella, J.P. Derendinger, S. Ferrara and C.A. Savoy, Phys. Lett. 115B (1982) 375; R. Gatto and G. Sartori, Comm. Math. Phys. 109 (1987) 327
26. A. Font, L. Ibáñez, H.P. Nilles and F. Quevedo, On the concept of naturalness in string theories, CERN preprint TH-5034 (1988), to appear in Physics Letters.
27. M. Cvetič, Phys. Rev. Lett. 59 (1987) 2829
28. J.A. Casas and C. Muñoz, Phys. Lett. 209B (1988) 214

29. A. Font, L. Ibáñez, H.P. Nilles and F. Quevedo, Phys. Lett. 210B (1988) 101
30. B. Greene, K.H. Kirklin, P.J. Miron and G.G. Ross, Nucl. Phys. B278 (1986) 667
31. For a review and a list of references see: J. Ellis, Aspects of superstring model building, CERN preprint TH-5103 (1988)
32. F. del Aguila and G.D. Coughlan, Very large intermediate scales in three generation models, CERN preprint TH-5143 (1988)
33. Jihn E. Kim and H.P. Nilles, Phys. Lett. 138B (1984) 150

Chairman: H. P. Nilles

Scientific Secretaries: F. Freire and Z. Lalak

DISCUSSION I

- *Brandt:*

As I understand, the idea of compactification is to go from 10 to 4 dimensions. You gave us the example of torus–compactification to go from 10 to 9 dimensions. Would this method then be repeated to get to 4 dimensions or is it generalized to dispose of 6 dimensions at once?

- *Nilles:*

I just presented the simple example to go from $d = 10$ to $d = 9$ to illustrate the mechanism of string compactification. There are many possibilities to reach d=4 and such a mechanism could also appear in steps, combining different methods of compactification.

- *Lalak:*

For some values of the radius of compactified dimension the resulting gauge groups are bigger than for other ones. How do you choose the radius? Is there any dynamical principle behind this?

- *Nilles:*

In the framework of the supersymmetric models discussed in the lecture all these values of the radii are equally probable as minima of a potential with vanishing vacuum energy. This degeneracy, however, will be lifted in the presence of supersymmetry breakdown.

- *Volkas:*

My question is closely related to the previous one. Is such a choice for R or the background field set once and for all, or is it subject to dynamical corrections?

- *Nilles:*

As said before, such a question can only be answered once we have an understanding of the mechanism of supersymmetry breakdown. This would then hopefully fix R and changes could be understood through dynamical corrections to the vacuum solution.

- *Horne:*

You mentioned a couple of methods for the breakdown of supersymmetry. What are these methods, and do they depend on the details of the compactification?

Methods to break supersymmetry fall into two distinct categories, the smooth ones obtained through a continuous variation of a parameter and those which do not correspond to such a smooth variation. The latter are used in string theory in connection with the breakdown of $N = 4$ to $N \leq 1$ needed for a chiral theory. Geometrically such a breakdown can be understood through the twists of boundary conditions and the holonomy group of the compact space. Maximal holonomy group ($SO(6)$ for six compactified dimensions) would lead to a nonsupersymmetric model in $d = 4$ while $N = 1$ is obtained through $SU(3)$ holonomy.

The smooth method to break supersymmetry could play a role in the $d = 4, N = 1$ theory. While such mechanisms exist in field theory it is not clear, whether they have a string theory analogue. We also still need to understand into which category the mechanism based on gaugino condensation will eventually fall.

– *Giannakis:*

Do you think that a realistic theory of interacting strings propagating on a background torus should be independent of the radius of the torus which is not the case with the Frankel–Kac (FRS) mechanism where R is fixed?

– *Nilles:*

The Frankel–Kac mechanism works for arbitrary values of the radii. At the critical values one obtains massless gauge bosons, while at other values of R a Higgs mechanism gives rise to masses for these gauge bosons.

– *Rahal:*

In what way does the CP–violation come out of a given string model?

– *Nilles:*

There are various ways in which this can happen for example through the Yukawa couplings in a model with three generations.

– *Skarke:*

Suppose people had invented superstrings 50 years ago, they would have looked for a low energy limit predicting electrons, protons, neutrons, neutrinos as fundamental particles. Does that not shed a bad light on the chances of deriving the Standard Model from strings?

– *Nilles:*

Looking at string theory it is difficult to imagine that it could give protons and neutrons as fundamental particles. A consideration of these questions now could shed some light on the question whether quarks and leptons are fundamental. In any case, we cannot stop working on a theory because it has not existed 50 years ago.

– Lalak:

One can compactify from 10 to 4 dimensions in two ways: at the level of string or at the level of 10–dimensional field theory. Are the two approaches equivalent?

– Nilles:

They are two complementary methods to tackle the same problem and we should try to use them both. We have a lot of intuition in field theory argumentations and this is quite useful for the discussion of e.g. the relevance of the Green–Scwarz counterterms in $d = 4$. On the other hand we certainly need explicit string results. An example is the appearance of winding states in the massless sector. If string calculations would be very easy we could of course just be happy with this way of discussing the problem, but we have not yet gotten so far and still need hints from the field theory argumentation.

– Sarid:

Can the effective cosmological constant actually be calculated in string theory?

– Nilles:

In principle yes, but we would really need a full understanding of a given theory. The question is closely related to that of supersymmetry breakdown, and this mechanism is poorly understood in the framework of string theory. The candidate mechanism of gaugino condensation is based on nonperturbative effects and can at the moment only be discussed in the framework of a field theoretic approximation. Amazingly such a mechanism indicates also a possibility for a vanishing cosmological constant: the dilaton can adjust its vacuum expectation value in such a way as to render $E_{vac} = 0$. This would at the same time fix the value of the gauge coupling constant.

– Warr:

Is not the "microscopic condition" for the vanishing of the cosmological constant in string models now irrelevant, in the light of Coleman's arguments, namely that topology changing processes on the space–time cause $\Lambda = 0$?

– Nilles:

Besides the "macroscopic condition" also a "microscopic condition" for the vanishing of the cosmological constant seems to be very relevant for the finiteness and therefore the consistency of the string theory. A nonvanishing cosmological constant at a given loop level would indicate a divergency at the next loop as discussed in the lecture by M. Green.

– *Colas:*

Would it be possible that some of the superstring particles of the mass of the order of Planck scale be stable and thus contribute to the present density of the universe?

– *Nilles:*

This could very well be, but I am not aware of an exhaustive discussion of the subject. It is certainly a question that depends quite strongly on the model under consideration.

– *Giannakis:*

Why do we want to compactify the string itself and not the field theory limit of the theory on Calabi–Yan manifolds?

– *Nilles:*

One of our goals is to isolate predictions of string theory that cannot be understood within the framework of the low energy field theoretic approximation. It is clear that such an investigation needs explicit string calculations.

DISCUSSION II

– *Lalak:*

At the end of your talk you mentioned "stringy zeros" and you have told us that there are no symmetries associated with them. Are these effects not destroyed by radiative corrections?

– *Nilles:*

This is a very interesting question and closely related to a discussion of "decoupling" in string theory. The "zeros" came from certain properties of the classical solutions of the two–dimensional system and could therefore be regarded as an "on–shell result" that could be perturbed by radiative corrections. Nobody has worked out this, but I think it is true. In the given framework of supersymmetric theories we can, however, rely on a stability of the "stringy zeros" due to nonrenormalization theorems. Eventually these "zeros" will then become a measure of supersymmetry breakdown. This would then also imply a solution of the μ–problem.

– *Volkas:*

In ordinary field theory GUTS supersymmetry is introduced in order to stabilize the gauge hierarchy. Some of the string models you described had no GUT group after compactification. Is space–time supersymmetry then really necessary?

– *Nilles:*

At the moment we have every reason to believe that a consistent string model needs supersymmetry. In the $d = 4, N = 1$ models supersymmetry is supposed to play the same role as before. It has to explain why the weak scale is so small compared to the Planck scale, independent of the presence of grand unification.

– *Horne:*

You mentioned that within families it was possible for different particles to have masses differ by an exponential factor. Could you explain how this factor arises?

– *Nilles:*

The centre of mass of a string in a twisted sector is located at one of the fixed points. For the ground state this implies that the string really collapses to the fixed point. In terms of field theory you would then consider two fields at different points and they would not interact with each other. You have to excite the string so that it can virtually extend to reach out and meet the other one and this gives an exponential suppression. In terms of a world sheet σ–model calculation such contributions arise from world sheet instantons.

– Freire:

Does any value of a Wilson line correspond to a minimum of the effective potentials?

– Nilles:

The models I have discussed so far are all supersymmetric, so all the vacua of consistent models are degenerate. In some cases, however, we found that the Wilson lines are quantized and only discrete gauge background values lead to a consistent model. In terms of an effective potential these discrete values would correspond to the minima while the interpolating values correspond to nonminimal potential.

– Lalak:

Are there any global symmetries in models of the type you described? For example, in Calabi-Yau compactification one has residual symmetries of the compactified manifold. Is there something like that here?

– Nilles:

Yes, there is. Discrete global symmetries appear in the selection rules for the Yukawa couplings through space and point group selection rules and they are a consequence of the geometrical symmetries of the orbifold.

– Warr:

Do the Higgs bosons in your string models satisfy bounds on mass analogous to the Weinberg–Linde bound (lower) and the unitarity bound (upper)?

– Nilles:

There is no special difficulty in string models to fulfil these bounds.

SUPERSYMMETRIC PARTICLES

R. Barbieri

Dipartimento di Fisica, Università di Pisa
INFN, Sezione di Pisa

INTRODUCTION

Supersymmetry[1] is among the best options for physics beyond the standard model. It has both "fundamental" and "pragmatic" motivations. The fundamental motivations are related to the general role of symmetries in field theory and to the connection of local supersymmetry[2] to gravity. The pragmatic motivations depend upon the relevance of fundamental scalar fields in the present day description of particle physics.

On the basis of these motivations, an effort has been made to formulate a consistent supersymmetric extension of the standard model. Such an extension exists, based on N=1 supergravity[3,4]. This general framework actually defines a class of models which can be usefully compared with experiments. This comparison has not given, up to now, any positive result. Nevertheless, on one side, the present negative searches are far from being able to rule out the general idea. On the other side, some forthcoming experiments will provide a crucial test for a significant category of supersymmetric models.

On these matters, various review articles already exist[5]. In these lectures I will address a few issues, singled out because of their special importance and because they have not yet received adequate attention.

THE "STANDARD SUPERGRAVITY MODEL" DEFINED

As said, N = 1 supergravity offers a framework to describe supersymmetric extensions of the standard model, which are at the same time internally consistent and fully compatible with the present experimental informations. On a general basis, the success of N = 1 supergravity rests on the possibility of incorporating in it a softly broken globally supersymmetric theory. In turn this gives a neat way to understand the separation between the standard particle masses and the spectrum of their superpartners. This separation appears geometrical in nature, being related to the invariance of the superpartner

The Superworld III
Edited by A. Zichichi
Plenum Press, New York, 1990

masses under the full SU(3) x SU(2) x U(1)group, unlike the case of all standard particle masses.

The building blocks of a general N = 1 supergravity model can be summirized as follows:

a) Take a gauge group G with matter supermultiplets transforming as a representation $\underline{r}$ of G, and couple this system to the graviton-gravitino supermultiplets[3,6].

b) Introduce a "hidden" superHiggs sector[3] where supersymmetry is spontaneously broken at a scale M_S. This sector is called "hidden" since it couples to the standard matter supermultiplets only via gravitational couplings.

c) After supersymmetry breaking in the classical vacuum, turn off gravity by taking the $M_{Pl} \to \infty$ limit at fixed $M^2{}_S/M_{Pl}$ [3]. In this way one obtains a softly broken globally supersymmetric theory.

d) Scale the Lagrangian so obtained at a grand scale, of order M_{Pl}, down to low energies, $E < G_F{}^{-1/2}$, according to the Renormalization Group Equations, (RGE), and the particle content of the theory[7].

Admittedly,this general scheme gives rise to a large class of "low energy" Lagrangians. In turn, the related experimental signatures can vary in an important way, depending on specific assumption. A significant class of models is singled out by this set of assumptions:

i) Following a criterium of simplicity, the gauge group G is taken to be SU(3) x SU(2) x U(1), acting on the "minimal" set of supermultiplets, quarks,leptons and a pair of Higgs doublets H_1, H_2.

ii) The R-parity[8], positive on all standard particles and negative on all superpartners, is taken to be unbroken. The consequent stability of the lightest supersymmetric particle gives a very interesting and natural connection between supersymmetry and the dark matter problem, to be illustrated below.

iii) The theory does not contain any significant breaking of flavour and/or CP other than in supersymmetric Yukawa couplings. Even though radical and even plausible alternatives are possible in this case, it is unlikely that they modify the specific experimental signatures to be discussed in the following.

Along the lines outlined, from these hypotheses, a unique, softly broken globally supersymmetry Lagrangian follows at a grand scale, before rescaling by RGE's:

$$L = L_{supersymm} + L_{breaking}$$

$$L_{supersymm} = L_{supersymm} [SU(3) \times SU(2) \times U(1); f]$$
$$f = f_y + \mu H_1 H_2$$
$$f_y = \lambda^u{}_{ab} Q_a u^c{}_b H_2 + \lambda^d{}_{ab} Q_a d^c{}_b H_1 + \lambda^e{}_{ab} L_a e^c{}_b H_1$$

$$L_{breaking} = - m^2 \Sigma_i |\phi_i|^2 - M \Sigma_\alpha \lambda_\alpha \lambda_\alpha$$
$$+ (A m f_y + B \mu H_1 H_2 + h.c.) \tag{1}$$

where:

- the supersymmetric piece of L is specified, other than by the standard gauge group, by a superpotential f containing the Yukawa couplings, f_y, and a mass coupling, $\mu H_1 H_2$, between the two Higgs douplets;

- the breaking piece of L contains a universal mass term for all scalars, ϕ_i, of the theory, an equally universal mass term for all gauginos, λ_α, and an "analytic" breaking contribution to the scalar potential, dependent on the superpotential terms as shown;

- the 5 parameters μ, m, M, A, B, upon which L depends other than the standard gauge and Yukawa couplings, can all be taken real with a suitable (re-)definition of the various fields.

This Lagrangian can be meaningfully confronted with experiments. Before doing that, however, one needs a handle on the space of the 5 parameters μ, m, M, A, B, which replace in some sense the parameters appearing in the Higgs potential of the standard model. They fix, other than the enlarged Higgs spectrum of the theory, the masses and the couplings of all the super-partners.

In principle it would be possible to choose these parameters in such a way as to let all superpartners become arbitrarily heavy, relative to the W and Z masses which are kept fixed at their physical value. Quite clearly however, this could be done at the price of introducing increasingly precise tunings among the various parameters themselves. But the absence of unnatural fine tunings is the very "raison d'être" of low energy supersymmetry. In turn an upper bound is obtained on the dimensionful parameters μ, m, M and on the physical masses of the various superpartners, by tolerating cancellations among parameters <u>at most</u> of a given relative amount. By giving a precise mathematical meaning to this criterium, it turns out that these bounds depend on the value at the top quark mass m_t. For $m_t \approx 100$ GeV and cancellations allowed of at the most one order of magnitude, one obtains[9]

m < 250 GeV, M < 200 GeV, μ < 120 GeV

and, for the physical masses,

m (gluino) < 700 GeV

m (squark) < 700 GeV

m (slepton) < 300 GeV

m ($\chi^\pm$) < 180 GeV

m (χ) < 100 GeV.

$\chi^\pm$ and χ refer respectively to the lightest "chargino" and "neutralino", superpositions of higgsinos and weak-gauginos. I shall refer in the following to these limits as "naturalness bounds"on the parameter space and on the superparticle masses.

THE MOST RELEVANT SEARCHES IN e^+e^- COLLISIONS

Among the variety of reactions which can be usefully studied to get positive signatures or constraints on the Lagrangian (1),a few appear of special significance.In fact,the following statements seem to be well justified:

(i) The combined effort to search for gauginos and higgsinos (charginos, neutralinos and gluinos) at SLC/LEP I, LEP II and TEVATRON will allow to explore almost all of the "natural region" of parameter space, as defined above.

ii) At e^+e^-, up to LEP II energies, an appropriate Higgs search should allow to uncover part of the supersymmetric Higgs structure.

Let me motivate and,at the same time,qualify these statements in turn.

As said, the charginos (2) and the neutralinos (4) are appropriate superpositions of weak gauginos ($\tilde{\omega}^{\pm},\tilde{z},\tilde{\gamma}$) and higgsinos ($\tilde{h}^{\pm},\tilde{h}^0_1,\tilde{h}^0_2$)

$$\chi = \alpha\tilde{z} + \beta\tilde{\gamma} + \gamma\tilde{h}^0_1 + \delta\tilde{h}^0_2$$

$$\chi^{\pm} = \sigma\tilde{\omega}^{\pm} + \tau\tilde{h}^{\pm}$$

As their masses,the chargino and neutralino compositions are determined by the parameters of the Lagrangian. Their interactions are correspondingly fixed.Due to their weak interactions,charginos and neutralinos can best be studied in e^+e^- collisions,especially via

$$e^+e^- \rightarrow \chi^+\chi^-$$
$$ \llcorner\!\lrcorner\!\rightarrow \chi f\bar{f}$$

$$Z \rightarrow \chi\chi'$$
$$ \llcorner\!\rightarrow \chi f\bar{f}$$

where χ is the lightest,stable,undetected neutralino and χ' is an "excited" neutralino. The rates for these reactions depend crucially upon the parameters μ and M and only weakly on the remaining ones. Infact chargino production mainly depends on the chargino mass alone.On the other hand, the branching ratio for $Z\rightarrow\chi\chi'$ is also influenced by the actual composition of the neutralinos, as it is clear from the fact that the Z boson couples to higgsinos but not to the neutral gauginos.

The significance[10] of these reactions is illustrated in Fig.1) which shows the region in the (M_2-μ_R) plane that will be probed in future e^+e^- experiments.M_2 and μ_R are the M,μ parameters after rescaling down to low energy by

$$M_2 = Z_2 M \quad , \quad \mu_R = Z_\mu\mu$$

$$Z_2 \sim 0.8 \quad , \quad Z_\mu = 1.4 (1-2.4 \times 10^{-5} M^2_t/\text{Gev}^2)^{1/4}$$

$$M_t = \sqrt{2} \ (h_t/g) \ M_W$$

where g is the SU(2) coupling constant, M_W is the W-mass and h_t is top-Yukawa coupling. In the shaded region B $(Z \to \chi\chi') \geq 10^{-5}$. This figure can be considered as a realistic sensitivity limit (after more than 10^6 Z's are produced) in view of the very distinctive signature associated with these decays, with no conventional background. χ' decays predominantly into $\chi q\bar{q}$ or $\chi l\bar{l}$. ($\chi' \to \chi\gamma$ is possible in a region of the parameter space)[10]. The resulting events are then characterized by a pair of jets or leptons on one side, with a large amount of missing energy carried away by the undetected stable neutralinos. Fig.1 also shows the contour plot corresponding to the lightest chargino masses $m(\chi^{\pm})=22.5,40,80$ GeV. From $\chi^{\pm} \to \chi qq'$, $\chi l\nu$, The sensitivity limit for chargino pair production at LEP II is $m(\chi^{\pm}) \sim 80$ GeV, where the background from W-pair production takes over. Notice that the region of the $(M_2-\mu_R)$ plane shown in Fig.1 is also the "natural region" for the parameters M,μ, as defined in the previous Section. The remaining parameters only affect the ratio v_2/v_1 among the vacuum expectation values of the two Higgs fields; in turn the weak dependence on this ratio of the rates for chargino and neutralino production is also illustrated in Fig.1.

The Higgs structure of the models under examination is reacher than the minimal one. Instead of the single Higgs scalar of the one doublet model, one has here two neutral scalars (H^0_1,H^0_2), one neutral pseudoscalar (H^0_3) and a charged Higgs $(H^{\pm})$. The masses and the couplings of this system are determined by two effective parameters, which can be chosen as the mass m_{H2} of the lightest scalar (H_2) and, as before, the ratio v_2/v_1.

For the purposes of the present discussion, the most interesting features of the Higgs system, as described by the Lagrangian (1), are:

i) There is always at least one neutral scalar (H_2) which is lighter than the Z boson[11]. For $v_2/v_1<3$, (H^0_2) couples to the Z boson much in the same way as the standard Higgs (H^0) does, so that the processes

$$Z^0 \to H^0_2 \ \mu^+\mu^-$$

$$e^+e^- \to H^0_2 \ Z^0 \tag{2}$$

keep their significance in the Higgs search, up to $m(H^0_2)<40$ GeV (LEP I) and $m(H^0_2)<80$ GeV (LEP II).

ii) The strength of the ZZH^0_2 coupling drops for large v_2/v_1. In this case, however, the pseudoscalar H^0_3 becomes degenerate with H^0_2, with full-strength $H^0_2 H^0_3 Z$ coupling, giving rise to a large branching ratio for $Z \to H^0_2 H^0_3$ up to the border of the phase space[12]. One has

$$B \ (Z \to H^0_2 \ H^0_3) = 3 \times (10^{-3}+ 10^{-2})$$

for $m^0_{H2} \sim m^0_{H3}$ ranging from 40 to 10 GeV. Since both H^0_2 and

H^0_3 decay predominantly into $b\bar{b}$ pairs (for $m^0_{H2} > 10$ GeV),one would have a relatively large amount of Z decays into $b\bar{b}$ $b\bar{b}$,as well as $b\bar{b}$ $\tau\bar{\tau}$, down by a factor of 20.

Although the importance of the background for these re-actions is still under study,their consideration[12] together with the conventional processes,(2),gives a great chance of uncovering the Higgs structure of the theory.

DARK MATTER NEUTRALINOS

The Lagrangian (1) implies, in almost all of the natural region of parameter space, the existence of dark matter in the form of stable neutralinos.This observation, together with the intrinsic interest of the dark matter problem, has stimulated a great amount of work on the subject, with the focus centered on the direct or indirect methods of dark matter detec-tion.Quite clearly the detection of dark matter particles in the laboratory is,in absolute terms, an exciting prospect.This possibility is in fact being discussed[13] and actually already attempted[14] for the various kinds of stable neutral weakly-interacting massive particles (WIMPs), which could form mas-sive halos around galaxies.

A neutralino is defined to be "cosmologically (and astro-physically) interesting" if,for its relic density, $\Omega_\chi = \rho_\chi/\rho_c$, relative to the critical density ρ_c,

$$0.025 \leq h^2\Omega_\chi \leq 1 \tag{3}$$

where h denotes the Hubble parameter in units of 100 km/sec/ Mpc.The lower bound in (3) corresponds to a relic neutralino density exceeding the baryon cosmic density, as inferred from the nucleosynthesis limit $\eta_B = n_B/n_\gamma < 7 \times 10^{-10}$ [15].On the other hand the upper bound saturates the limit of the total rela-tive density Ω, as obtained from the dating of the oldest stars[16] and $h > 0.4$. Correspondingly a "cosmologically inter-esting" region of the parameter space is singled out by re-quiring that (3) be satisfied.

The present relic abundance of cosmic neutralinos, is given[17] by the standard formula

$$\rho^0_\chi = 5 \times 10^{-40} \left(\frac{T^0_\chi}{2.7^0 \text{ K}}\right)^3 \sqrt{n_{eff}\left(T_f\right)} \left(\frac{\text{GeV}^2}{ax_f + \frac{1}{2} bx_f^2}\right) \text{ g/cm}^3 \tag{4}$$

where T^0_χ is the effective temperature of the χ's today[18]

$$\left(\frac{T^0_\chi}{2.7^0 \text{ K}}\right)^3 = \frac{n_{eff}\left(1 \text{ MeV}\right)}{n_{eff}\left(T_f\right)} \times \frac{4}{11} \tag{5}$$

$n_{eff}(T)$ is the effective number of degrees of freedom at temper-ature T.T_f is the neutralino decoupling temperature, at which the

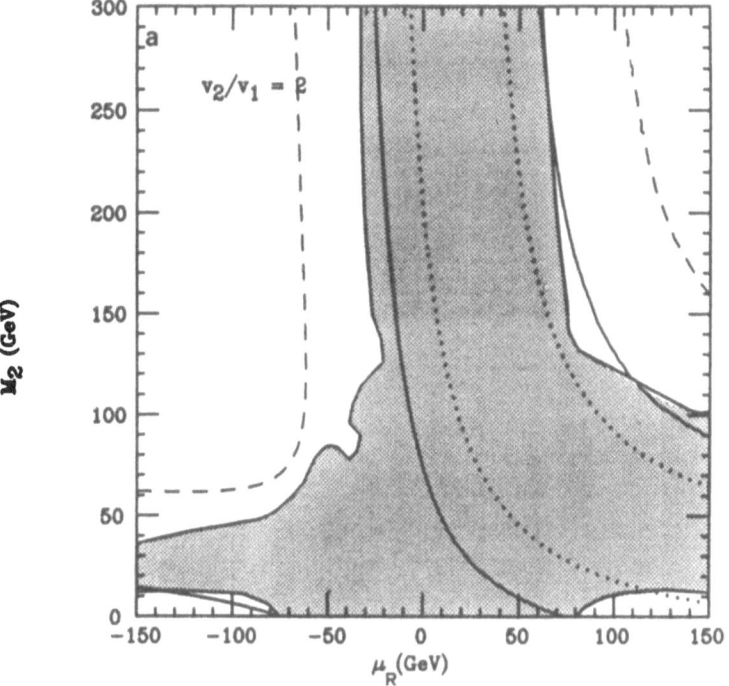

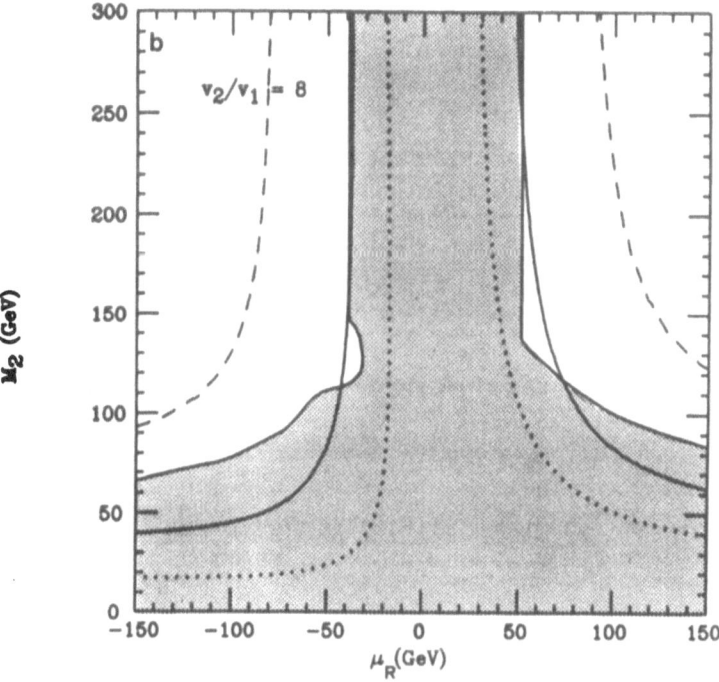

Fig.1.Regions in the $(M_2-\mu_R)$ plane that will be probed in future e^+e^- experiments. In the shaded region $BR(Z^0 \to \chi\chi') \geq 10^{-5}$. Also shown are the regions where charginos are lighter than 22.5 GeV (dotted line), 40 GeV (solid line) and 80 GeV (dashed line). In Fig.1a $v_2/v_1 = 2$, in Fig. 2b $v_2/v_1 = 8$.

expansion rate equals the annihilation rate

$$3H\ (T_f)\ =\ \langle\sigma_{ann}\ v_{rel}\rangle\ n^{eq}_\chi \tag{6}$$

and the thermally averaged annihilation cross section of a χ pair, for small χ velocity v, is given by[17]

$$\sum_f\ \langle\sigma(\chi\chi \to f\bar{f})\ v_{rel}\rangle\ \equiv\ \langle\sigma_{ann}\ v_{rel}\rangle\ \sim\ a\ +\ bx \tag{7}$$

$$x\ =\ \frac{\langle v^2\rangle}{6}\ =\ \frac{kT}{m_\chi}$$

The annihilation amplitude is mediated by a t-channel sfermion exchange or a s-channel Z-boson exchange.

Given the sfermion masses, contour plots of the relic neutralino density, Eq.(4), can therefore be drawn in the plane of the parameters (M_2,μ), for representative values of the ratio v_2/v_1, as it was originally done in Ref.17. In this same plane we can get the "cosmologically interesting region" corresponding to (3).

To this end, we draw the contour line corresponding to the upper bound $h^2\Omega_\chi \leq 1$, by minimizing ρ_χ with respect to the sfermion masses. In turn, this requires the maximization of the annihilation cross section, which is obtained by setting the sfermion masses to their minimum values allowed by the present experimental informations[19]

m(sleptons) > 25 GeV
m(squarks) > 55 GeV

Conversely, the contour line corresponding to the minimal χ density, $h^2\Omega_\chi = 0.025$, is drawn for infinite sfermion masses or minimal annihilation cross section[*].

The resulting cosmologically relevant area in the $(M_2-\mu_R)$ plane is shown in Fig.1 for two representative values of v_2/v_1 = 2,8. All the shaded area in this plane may give rise to a significant relic neutralino density, depending on the value of the sfermion masses. The white "excluded" regions are understood as follows. Along both the $M_2 = 0$ and the $\mu_R = 0$ axes, as well as on a line crossing the $\mu_R > 0$ half-plane, the neutralino mass is too small to give rise to a significant annihilation cross section. Correspondingly, the relic neutralino density is too high, $h^2\Omega_\chi > 1$, to be acceptable. On the contrary, the "excluded" regions in the $\mu_R < 0$ half-plane and in the upper $\mu_R > 0$ half-plane correspond to a too high annihilation cross section, or a too low density $h^2\Omega_\chi < 0.025$, due to the closeness of the Z pole to the c.m.s. energy $\sqrt{s} \sim 2m_\chi$.

──────────

[*]We bar the possibility of a distructive interference between the Z-exchange and the sfermion exchange amplitudes, since this requires a tuning of the sfermion mass parameters and it could only suppress the neutralino annihilation cross section into one specific $f\bar{f}$ channel.

In Fig.2 we also show the contour lines corresponding to $h^2\Omega_\chi = 0.16$ for infinite sfermion masses. These lines, together with those corresponding to $h^2\Omega_\chi = 1$, delimit the region in the parameter space consistent with $\Omega_\chi = 1$ for $0.4 < h < 1$. In this case the relic neutralinos could be the dominant component of the present density in an inflationary cosmology.

As mentioned, if neutralinos form a massive halo around our galaxy, direct or indirect detection methods can be attempted. The dark matter, in other words, may turn out not to be so dark. One possibility consists in the direct identification of the recoil energy deposited when a dark matter particle scatters off a nucleus in a low-background, low-threshold detector[13]. More indirectly, one can try to detect the annihilation products of halo dark matter particles gravitationally trapped in the sun, or in fact also in the earth[20]. Experimental work[21,22] has already been made and is being made, with negative results up to now, along both directions.

Quite clearly, in both cases, a relevant physical parameter is the elastic neutralino-nucleus non relativistic cross section $\sigma(\chi N \to \chi N)$. In turn for a heavy nucleus, this cross section is likely to be dominated by a t-channel Higgs exchange[23], more precisely the light scalar H_2, discussed in the previous Section. This is because the scalar exchange, unlike the Z exchange, gives rise, in the non relativistic limit, to a spin independent interaction of the neutralino with the nucleons and allows in this way a sizeable coherent cross section of the neutralino with a heavy nucleous. This pattern is related to the Majorana nature of the neutralino, or the vanishing of $\chi\gamma_\mu\chi$, as opposed, for example, to the case of a massive Dirac neutrino ν_D, which acquires a coherent interaction from the Z-exchange amplitude. For the elastic neutralino cross section, $\sigma_H(\chi)$, out of a nucleous of mass M, mediated by the Higgs exchange, one has infact[23]

$$\sigma_H(\chi) = \left(\frac{16\ M\ M_W}{27\ m_{H_2}^2}\right)^2 \frac{m_\chi^2\ M^2}{8\pi\ (m_\chi + M)^2}\ G_F^2\ F^2$$

(8)

where m_χ is the neutralino mass and the factor F, typically of order 1, contains all the information about the neutralino and H_2 couplings. As such, F depends upon the usual parameters $M, \mu, v_2/v_1$, as the neutralino mass does. In turn, on the nuclei of a suitable detector, the expected counting rate for the elastic scattering of neutralinos making the halo of our galaxy, with a mass density ρ and a mean velocity v is ($A = M/m_N$)

$$R = \frac{950\ events}{kg \times day}\ \frac{4\ m_\chi\ M}{(m_\chi + M)^2}\ \left(\frac{10\ GeV}{m_{H_2}}\right)^4 \left(\frac{A}{100}\right)^2 F^2 \times$$
$$\times \frac{\rho}{10^{-24}g/cm^3}\ \frac{v}{300\ km/sec}\ .$$

(9)

In both Eqns.(8,9), a presumably lower estimate[24] is used for the Higgs-nucleon coupling, which neglects the light quark contribution (u,d,s).

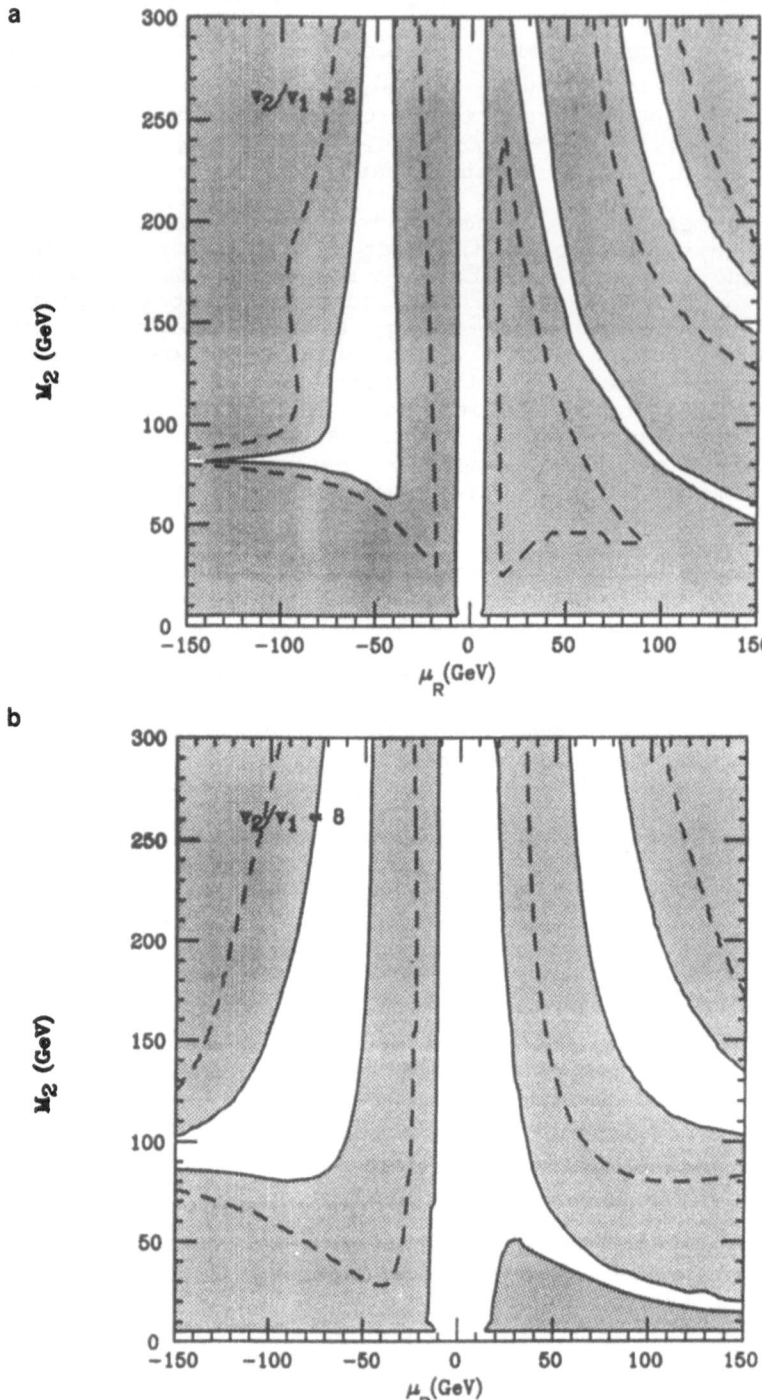

Fig.2. The "cosmologically interesting" region (shaded area) in a general supergravity theory with minimal particle content, corresponding to $0.025 \le h^2\Omega_\chi \le 1$, for $v_2/v_1 = 2$ (Fig.2a) and 8 (Fig.2b). Also shown are the contour lines for $h^2\Omega_\chi = 0.16$ (dashed lines), which delimit the region consistent with $\Omega_\chi = 1$.

164

The relatively large amount of events expected, makes it worth discussing whether the existing experiments[21] using ^{76}Ge detectors are already able to constrain the supersymmetric parameter space from the negative dark matter searches. We recall that these experiments are able to rule out, as dark matter candidates, Dirac neutrinos with masses ranging from 10 GeV to about 1 TeV. The negative experimental results[21] can infact be interpreted as giving a mass dependent upper bound on the elastic cross section ranging typically from 10^{-33} up to 10^{-31} cm^2, in the range 10 GeV ÷ 1 TeV for the dark matter particle mass. These bounds allow us to exclude the parameter ranges shown in Fig.3 for m_{H2} = 10, 20 and 30 GeV. Notice that the excluded regions reflect in a rather direct way the trivial dependence of the cross section on the neutralino mass. The shape of the boundaries resembles the contour lines, in the (M_2-μ_R)-plane, of the neutralino mass itself. Even more than actually showing a range of excluded parameters, Fig.3 is meant to visualize the sensitivity of the direct methods for dark matter neutralino searches. It is also useful to compare Fig.3 with Fig.1, which shows how much of the (M_2-μ_R) plane can be probed experimentally by studying chargino and neutralino pair production at e^+e^- colliders.

A final comment concerns the trapping of neutralinos in the sun and on earth, with their subsequent annihilation producing high energy neutrinos, in principle detectable in proton-decay-type experiments. Quite clearly the trapping rate is again determined by the neutralino nucleus elastic cross section. Eq.(8) is therefore also relevant to this problem, with some effect on the trapping in the sun, where the heavy nuclei are relatively few, and a much more important effect on the trapping in the earth[25]. As a consequence, for a Higgs exchanged of 10 GeV mass, depending on the neutralino mass, m_χ = 10 ÷ 100 GeV, one expects[25] a number of neutrino events per kiloton, per year between 1 and 100 from the annihilation in the earth and between 2 and 5 from the annihilation in the sun. A somewhat higher number of muon events per 100 m^2, per year is expected from neutrino interactions in the rock surrounding the detector.

CONCLUSIONS

The class of supergravity models that have been described in these lectures offers a sensible, concrete guide to experiments in the search for supersymmetry. The implementation of the usual "naturalness" criterium constrains the space of the parameters of the basic Lagrangian, Eq.(1). As a consequence, a few especially significant experimental tests are singled out, which appear to be able to probe the theory in basically all this constrained parameter region. Even the possible connection between the dark matter problem with the existence of a stable massive neutralino can be subjected to experimental verification in a quite definite way.

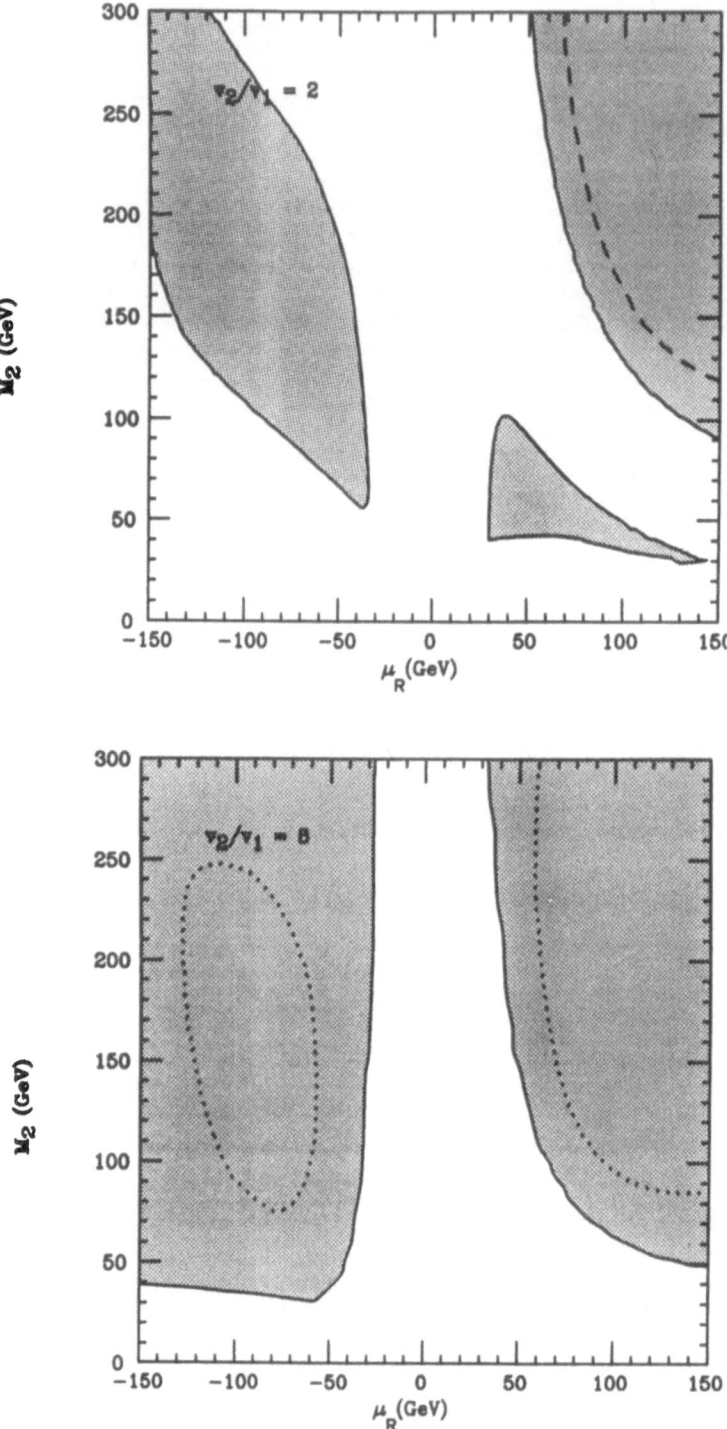

Fig.3. Regions in the (M_2-μ_R) plane that are excluded by negative dark matter searches with ^{76}Ge detectors for Higgs mass $m_{H2} = 10$ GeV (solid line), 20 GeV (dashed line), 30 GeV (dotted line). v_2/v_1 is chosen as in Fig.1. The experimental results are taken from Ref.21. A halo density of 0.4 GeV/cm^3 is assumed.

REFERENCES

1. Y.Gol'fand and E.Likhtam,JETP Lett.13(1971)323;
    D.Volkov and V.Akulov, Phys.Lett.46B(1973)109;
    J.Wess and B.Zumino, Nucl.Phys.B70(1974)39.
2. D.Freedman, P.van Nieuwenhuizen and S.Ferrara,Phys.Rev.D13
    (1976)3214;
    S.Deser and B.Zumino, Phys.Lett.62B(1976)335.
3. R.Barbieri, S.Ferrara and C.Savoy, Phys.Lett.119B(1982)343;
    A.Chamseddine,R.Arnowitt and P.Nath, Phys.Rev.Lett.49
    (1982)970;
4. H.Nilles,Phys.Lett.115B(1982)193;
    L.Ibanez,Phys.Lett.118B (1982)73.
5. R.Barbieri and S.Ferrara, Survey in high energy physics 4
    (1983)33;
    H.Nilles, Phys.Rep.110(1984)1;H.Haber and G.Kane,
    Phys.Rep.117(1985)75;
    M.Chen,C.Dionisi,M.Martinez and X.Tata, Phys.Rep.C159
    (1988) 202
    R.Barbieri, La Rivista del Nuovo Cimento 11 (1988).
6. E.Cremmer et al.,Phys.Lett.79B(1978)231;Nucl.Phys.B147(1982)
    231.
7. K.Inoue,A.Kakuto,H.Komatsu and S.Takeshita, Prog.Theor.Phys
    68(1982)927;ibid71(1984)413;
    J.Ellis,D.Nanopoulos and K.Tamvakis, Phys.Lett.121B(1983)
    123;
    L.Alvarez-Gaumè, J.Polchinsky and M.Wise,Nucl.Phys.B221
    (1983)495;
    C.Kounnas, A.Lahanas,D.Nanopoulos and M.Quiros,Phys.Lett.
    B132(1983)95;Nucl.Phys.B236(1984)438;
    L.Ibanez, Nucl.Phys.B218(1983)514;
    L.Ibanez and C.Lopez,Phys.Lett.126B(1983)54;Nucl.Phys.B233
    (1984)511;
    L.Ibanez C.Lopez and C.Munoz, Nucl.Phys.B256(1985)218;
    A.Bouquet,J.Kaplan and C.Savoy,Nucl.Phys.B262(1985)299.
8. P.Fayet, in Unification of the Fundamental Particle Interac-
    tions,eds.S.Ferrara et al.(Plenum Press,New York,1980)
    p.587.
9. R.Barbieri and G.Giudice, Nucl.Phys.B306(1988)63.
10. J.M.Frère and G.Kane,Nucl.Phys.B233(1983)331;
    J.Ellis,J.Hagelin,D.Nanopoulos and M.Srednicki,
    Phys.Lett.177B(1986)201; R.Barbieri,G.Gamberini,G.Giudice
    and G.Ridolfi,Phys.Lett. 195B(1987)500;1B296(1988)75.
11. H.Nilles and M.Nusbaumer,Phys.Lett.145B(1984)73;
    P.Majumdar and P.Roy, Phys.Rev.D30(1984)2432.
12. G.F.Giudice,Phys.Lett.208B(1988)315.
13. M.Goodman and E.Witten,Phys.Rev.D31(1985)3059.
14. G.Gelmini, Telemark IV Conference Proceedings,eds.V.Barger
    et al.,World Scientific (1987); D.Caldwell,ibidem.
15. See,e.g.,J.Young et al.,Astrophys.J.281(1984)483.
16. See,e.g.,I.Iben and A.Renzini,Phys.Rep.105(1984)329.
17. J.Ellis et al.,Nucl.Phys.B238(1984)453; see also H.Goldberg,
    Phys.Lett.50(1983)1419.
18. K.Olive,D.Schramm and G.Steigman,Nucl.Phys.B180(1981)497.
19. W.Bartel et al., Phys.Lett.146B(1984)126;
    B.Adeva et al.,Phys.Lett.152B(1985)439;
    I.J.Kroll (UA1 coll.), CERN preprint EP/87.103(1987).
20. G.Steigman,C.Sarasin,H.Quintana and J.Faulkner,Ap.J.83(1978)
    1050;
    W.Press and D.Spergel,Ap.J.296(1985)679;
    J.Silk,K.Olive and M.Srednicki, Phys.Rev.Lett.55(1985)257.

21. S.Ahlen et al.,Phys.Lett.195B(1987)603;
    D.Caldwell et al.,Phys.Rev.Lett.61(1988)510.
22. B.Kurinik, Orsay preprint LAL-87-21(1987);
    J.Losecco et al.,Phys.Lett.188B(1987)388;
    Y.Totsuta, Tokyo preprint UT-ICEPP-87-02(1987).
23. R.Barbieri,M.Frigeni and G.Giudice,to appear in Nucl.Phys.B.
24. M.Shifman,A.Vainshtein and V.Zacharov,Phys.Lett.78B(1978)443.
25. G.Giudice and E.Roulet,ISAS-Trieste preprint 103 EP(1988).

Chairman: R. Barbieri

Scientific secretaries: C. Gonzalez-Garcia and U. Sarid

DISCUSSION I

– Gonzalez-Garcia

Could you explain the concept of naturalness in more detail and how it leads to the upper bounds on the lightest chargino and neutralino masses?

– Barbieri:

In short, the concept of naturalness is the statement that hierarchies of physical parameters in a basic Lagrangian should not be the result of any fine tuning. This should also be the case for the electroweak scale (the Higgs mass) relative to any higher physical mass scale (M_P, ...). In turn, this can only be achieved by embedding the standard model into a supersymmetric theory, with the effective SUSY breaking scale (the typical mass of the superpartners) close to the Fermi scale itself. In the supersymmetric theories under examination, the W mass and the superpartner masses are related, being determined by the same set of parameters. To let the superpartner masses become arbitrarily high while fixing at the same time the W mass at its physical value is possible, but only at the price of reintroducing the fine tunings among the parameters that one wanted to avoid. For the precise Lagrangian defined in this text, for any given amount of fine tuning that one is willing to tolerate, an upper bound correspondingly arises for any of the superparticle masses.

– Gonzalez-Garcia:

In order to obtain, for instance, the neutralino mass, you must diagonalize the 4 X 4 mass matrix. In this matrix the parameters M, m and μ etc. appear, but the VEV of the Higgs also appears. In your figures, where are all these parameters hidden?

– Barbieri:

These parameters do not appear because the bounds are precisely obtained by maximizing the superpartner masses with respect to these same parameters. It is important in fact to realize that every individual bound is obtained by maximizing the corresponding mass irrespective of what the other masses do. In other words, the bounds could not all be saturated at the same time since that would require choosing different values for the same parameters.

– Warr:

In order to solve the naturalness problem, there are theories with only fundamental fermions and gauge bosons. Can you say why we are forced to consider such theories with a gauge group U(45), that is, what happened to technicolour theories?

– Barbieri:

I did not go so far as to say that one needs to go to a U(45) gauge group to make a purely fermionic theory work. The mention of the U(45) group – 45 being the number of Weyl spinors occurring in the standard model Lagrangian – is a way of pointing out that purely fermionic renormalizable gauge theories have generally unwanted global symmetries associated with the reducibility of the fermionic representation of the basic gauge group. In U(45) one would have all known fermions occurring in only one irreducible representation of the gauge group, with only one global charge conserved, the fermion number itself. A priori this is a much better situation than having, for example, all the global symmetries of the standard model, with the Yukawa couplings turned off. Of course, one has traded in this way the global symmetries with local symmetries that will have to be broken spontaneously. The known fermions and the known forces cannot however give rise to this breaking, at least as far as we can tell. Other forces (technicolour) and other fermions (technifermions) are needed. The group must be enlarged further. Does this have an end?

– Calafiura:

What are the consequences of the possible observation of the reaction $K_L \to \pi_0 \ell^+ \ell^-$ that has been studied at Fermilab?

– Barbieri:

Very interesting, if it were true.

– Calafiura:

Why does the evidence of deviations from the KM matrix in flavour mixing decays favour the purely fermionic model?

– Barbieri:

A more precise statement is that no evidence of deviation from the Cabibbo-Kobayashi-Maskawa description of flavour changing phenomena disfavours the purely fermionic models relative to the Higgs models. This is my opinion, for the reasons explained in the text. On the contrary, evidence of deviations would put the two alternatives on the same footing.

– Lalak:

You have restricted yourself to discuss a model with the neutralino as the lightest supersymmetric particle (LSP). Could we arrange the sneutrino to be the LSP? If so, what is wrong with such a model?

– *Barbieri:*

The possibility cannot be excluded, but it is unlikely to occur in the full space of the parameters of the Lagrangian under examination.

– *Rius:*

Why cannot one of the Higgs doublets be the supersymmetric partner of one of the left-handed fermions?

– *Barbieri:*

From the pure point of view of the gauge quantum numbers, there is obviously no obstacle to this possibility. Again, however, this does not appear as a likely solution in the full parameter space.

Scientific Secretaries: C. Gonzalez-Garcia and U. Sarid

DISCUSSION II

– *Gonzalez-Garcia:*

Is there any reason why neutralinos are better candidates than heavy majorana neutrinos for dark matter?

– *Barbieri:*

Technically the answer is no, as far as I know. The difference appears in their raison d'être. As I have explained, dark matter neutralinos are a natural consequence of supergravity models.

– *Pollak:*

If there is no evidence for a Higgs lighter than $W^{+,-}, Z^0$ at LEP, which assumptions would you give up first?

– *Barbieri:*

The assumption of minimal particle content. Enlarging the particle supermultiplets (adding, e.g., a group singlet) allows us to evade the bound. It will, however, generally remain true that a relatively light Higgs ($m_H \lesssim 150$ GeV) exists, due to the requirement of not having any Yukawa-like coupling getting too large before the grand scale. The experimental study of such a Higgs will not be an easy task, however.

– *Pollak*

How big would an e^+e^- collider need to be to rule out SUSY?

– *Barbieri:*

I have explained quite in detail the relevance of LEP II ($\sqrt{s}$ up to 200 GeV) to study the class of models under consideration. Since my favourite particles to be looked for are neutralinos and charginos, I attribute in general the greatest significance to e^+e^- collisions in the search for SUSY. More than that I am not willing to say.

– *Volkas:*

Going back to Pollak's question, I think that the following might be the easiest way to evade the light Higgs phenomenon: Since the fundamental reason one has a light Higgs is that the quartic terms in the potential only come from D terms, what one really wants is an F term contribution. Having only Higgs doublets is not enough since you need Higgs trilinear terms in the superpotential. Now $2 \otimes 2 = 3 \oplus 1$ so you need either a Higgs singlet or triplet.

– *Barbieri:*

This is well known. See my answer to Pollak.

– *Mannel:*

Will SUSY be ruled out by the new accelerators if no superpartners are found?

– *Barbieri:*

As a general idea, the answer is no. But the problem becomes philosophical in nature. These lectures are the best I can say about the problem of testing SUSY as a low energy phenomenon. If nothing is found, everyone will have to make his or her own judgment.

– *Fordham*

You mentioned only the $b\bar{b}$ decay modes for your neutral Higgs boson. Do you exclude masses for the neutral Higgs which are lighter than the $b\bar{b}$ production threshold?

– *Barbieri:*

In general no. The point is that I mentioned the $b\bar{b}$ mode in connection with the study of $Z \rightarrow H_2 H_3$ which becomes relevant for $v_2/v_1 \gg 1$. In this regime the coupling of H_2 to $b\bar{b}$ gets large, since it scales as v_2/v_1 relative to the coupling of the standard Higgs to $b\bar{b}$. As such, if H_2 is also lighter than $2m_b$, it should have been discovered in $\Upsilon \rightarrow H_2 \gamma$ at Cornell.

– *Horne:*

If one assumes that WIMPs are the lightest supersymmetric particles, can one use this to set any limits on these particles?

– *Barbieri:*

If you want to know if neutralinos can play the role of cosmions, solving at the same time the dark matter and solar neutrino problems, the answer is no. The annihilation and scattering cross–sections of neutralinos are not right for that purpose. In general I prefer to keep the two issues separated. My personal opinion is that the solar neutrino problem is not manifestly a problem. This without reducing in any way the importance of looking as best as possible at solar neutrinos.

– Lalak:

You discuss neutralinos having only spin-dependent interactions. But there might easily be coherent interaction, say through left-right squark mixing. Could this change your quoted limits?

– Barbieri:

I said that neutralinos have spin-dependent interactions in the non- relativistic limit only if these interactions are mediated by vector (Z^0) exchange, due to the majorana nature of neutralinos. Scalar exchanges give rise to spin-dependent interactions. I discuss Higgs exchange and not squark exchange because, in the models under consideration, the lightest Higgs is lighter than Z^0, whereas this is unlikely for squarks. The effect of squark exchange in these matters has been recently discussed by Griest.

– Lalak

If there is no R-parity, how fast would the LSP decay?

– Barbieri:

As fast or as slow as you want. If the LSPs have to decay, I prefer them to do it with a lifetime longer than the age of the universe, so as to be able to make the dark matter. Even this is conceivable.

– Sarid:

How would an axion fit into your SUSY standard model and could you predict anything about it?

– Barbieri:

Neither better nor worse than in the non-SUSY models. Normally a very light axino, the fermionic partner of the axion, is implied, which may become the LSP.

– Servoli:

What is the energy spectrum (or the momentum distribution) for the $q\bar{q}$ pairs in $\chi' \rightarrow \chi q\bar{q}$ decay?

– Barbieri:

What counts is the χ, χ' masses, especially the mass difference δ. Again these are determined by the usual parameters. For detection in Z-decays, it's important to know that in general $\delta \gtrsim (10 - 20)\%$ of the average χ, χ' masses in the range of 10-50 GeV. We have been told by Martin Perl that no major obstacles exist to go down to low δ in searching for $Z \rightarrow \chi\chi' \rightarrow \chi\chi q\bar{q}$.

FRACTALS IN PHYSICS

L. Pietronero

Dipartimento di Fisica, Università di Roma "La Sapienza"
P.le A.Moro 2, 00185 Roma, Italy

ABSTRACT

We give a brief overview of the impact of fractal geometry on physical sciences. In particular we will describe the prototype of fractal growth models and the recent developments in the direction of the formulation of an analytical theory that allows to understand why natural phenomena often give rise to fractal structures. Finally we will show that the large scale properties of the matter distribution in the universe need a radical revision on the of light of these new concepts.

1. INTRODUCTION AND GENERAL CONCEPTS

Mathematical methods in physics have been usually based on analytical functions. This has produced a search for regualrity and simplicity in natural phenomena. At the same time systems that do not satisfy these requirements of regularity have in general been disregarded. This situation has changed substantially in the last few years with the new concepts of fractal geometry[1] and deterministic chaos[2].

Fractal geometry describes systems in which increasing detail is revealed by increasing magnification, and the newly revealed structure looks similar to what one can see at lower magnification. Concepts of this type were defined as purely mathematical entities long ago by Poincarè, Hausdorff and others. However it was Mandelbrot who realized that many familiar structures in nature could be described with these methods[1]. This concept was clearly lacking for the description of complex structures in nature, that are actually the majority, and this is one of the reasons of its present success. In this respect it is interesting to compare this field with that of particle physics. The situation in just the opposite: in particle physics often theoretical results are available and enormous experimental efforts have to be made in order to confirm or reject them. In the case of fractals the "experimental data" have always been around but, given the absence of an appropriate concept, they have not been considered part of the scientific problematic. Now that such a concept has been introduced one is able to give a precise characterization to these phenomena.However this

The Superworld III
Edited by A. Zichichi
Plenum Press, New York, 1990

does not mean that one has a theory that explains how these structures actually arise from physical laws. A deepening of the interrelation between fractal geometry and physical phenomena is what may be called the theory of fractals and forms the objective of the present activity in the field[3,4].

One should note that the concept of self-similarity has already appeared in the critical properties of thase transitions and it has been instrumental to the formulation of the renormalization group approach that has essentially solved this problem[5]. In that case however self-similarity was considered a peculiarity of the competition between order and disorder at a particular temperature for equilibrium phase transitions[3,4]. We can now see that this property is much more common and it appears in many equilibrium and non equilibrium phenomena apparently unrelated to the problem of phase transitions. In fact the theory developed for critical phenomena cannot be applied to these new problems.

Self-similarity is not compatible with analiticity because, since at smaller scales the same complexity appears, curves never become regular or differentiable. The possibility of describing these systems with a non integer, fractal dimension arises from a generalization of the mass or generalized volume (N)-lenght (R) relation[1]

$$N(R) \simeq R^D. \tag{1.1}$$

One can see that a self-similar structure can be described by such a power law relation with a non integer exponent D. This generalization of the concept of dimension is only valid in the metric sense but not in the topological one. Therefore it does not have the same general validity as in euclidean geometry.

It is easy to see why self-similarity gives rise to power law correlations. Let us consider the correlation function for the occupation number $n(\vec{r})$ of a given structure

$$\Gamma(r) = \langle n(\vec{r}_0)n(\vec{r}_0+\vec{r})\rangle_0 . \tag{1.2}$$

This function defines how many points of our structure can be found at a distance r (scalar for simplicity) from an occupied point at $\vec{r}_0$. The average refers to all possible choices of $\vec{r}_0$. For random fractals one may also average over different realizations of the whole pattern. By integrating Eq (1.2) we obtain a relation between average volume and lenght as Eq. (1.1)

$$N(R) = \int_0^R d\vec{r} \langle n(\vec{r}_0)n(\vec{r}_0+\vec{r})\rangle_0 \simeq R^D. \tag{1.3}$$

We can now consider the mathematicxal restrictions imposed on the function $\Gamma(r)$ (or N(R)) by the property of self - similarity. This implies that the system is invariant under scale trasformation

$$R \rightarrow R' = bR, \tag{1.4}$$

in the sense that

$$N(R') = N(b \cdot R) = A(b) \cdot N(R) \qquad (1.5)$$

where the function A depends only on b and not on R. It is easy to see that power laws are the natural solutions of this funtional equation.

From the mass length relation (Eq. (1.1)) one can see that the average conditional density is

$$\rho(R) = \frac{N(R)}{V(R)} \simeq \frac{R^D}{R^d} \simeq R^{-(d-D)} \qquad (1.6)$$

where d is the enclidean dimension of the space in which the fractal structure is defined. It is clear from Eq. (1.6) that the density is a function of length - scale considered (R). Therefore in a fractal the average density cannot be properly defined. The system is non analytic in every point and the average conditional density decays, from each of these points, with a power law whose exponent is the codimension (d-D).

2. FRACTAL STRUCTURES FROM PHYSICAL MODELS

In the past few years a great deal of activity has been devoted to the study of fractal structures in relation to physical phenomena[3,4]. The prototype fractal growth model is based on a combination of the Laplace equation and a stochastic field. The first model of this class to be formulated was Diffusion Limited Aggregation (DLA)[6]. A few years later the more general Dielectric Breakdown Model (DBM)[7] was introduced. This model used the relation between random walk and potential theory and made clear that growth could also occurr "from inside". In addition to their intrinsic theoretical interest, these models are now believed to capture the essential features necessary to describe pattern formation in seemingly different phenomena like electrochemical deposition, dendritic growth, dielectric breakdown, viscous fingering in fluids, fracture propagation and others.

From the point of view of dielectric breakdown this growth model assumes that the already grow pattern at a given time is equipotential. One then computers the local field around this structure by solving Laplace equation

$$\nabla^2 \phi = 0 \qquad (2.1)$$

with the above boundary condition. The growth probability p_j for each bond(j) connected to the structure is then related to the local field

$$p_j = \frac{|\nabla\phi|_i^\eta}{\sum_j |\nabla\phi|_j^\eta} \qquad (2.2)$$

where η is a parameter that modulates the amount of randomness in the process.

This probability distribution is used to select one bond that becomes part of the pattern. The process then continues by iterating this procedure.

This model, apparently very simple, produces random fractals whose dimension depends continuously on the parameter η[7]. The mathematical process described in however very complex because it is long ranged in both space and time and it cannot be analyzed within the known mathematical methods.

Most of the activity in this field has been based on computer simulation and the analytical attempts have not been very succesful. Recently we have proposed a new approach in this direction that appears reasonably promising[8,9]. This method clarifies the origin of fractal structures in these models and provides a systematic method for the calculation of the fractal dimension. Here we are just going to sketch the basic ideas. For a detailed description the reader should consult the original papers[8,9].

The key points of this new approach are the following: first, one should identify the set of basic configurations that correspond to the procedure of fine (or coarse) graining of the structure. For the DLA or DBM models (in two dimensions) these diagrams are just two whose occurrence is defined by the probability distribution (C_1, C_2). The fractal dimension D can be directly related to this distribution

$$D = 1 + \frac{\ln(C_1 + 2C_2)}{\ln 2}.$$

(2.3)

The next step is to find an iteration process whose fixed point provides the asymptotic distribution (C_1, C_2). In view of the scale invariance of Laplace equation, the structure is self-similar under scale transformation as well as under growth. This allows to introduce a Fixed Scale Transformation (FST) with respect to growth. for this problem such a transformation results to be more convenient than a Renormalization Transformation under coarse graining. The matrix elements Mij (i(j) = 1,2) of the FST correspond to the conditional probability that a frozen (not modified by further growth) configuration of type i is followed, in the growth direction, by a frozen configuration of type j. This defines the interaction process

$$\begin{pmatrix} C_1^{(k+1)} \\ C_2^{(k+1)} \end{pmatrix} = \begin{pmatrix} M_{11} & M_{21} \\ M_{12} & M_{22} \end{pmatrix} \begin{pmatrix} C_1^{(k)} \\ C_2^{(k)} \end{pmatrix}$$

(2.4)

where k is the index of the iteration.

In order to compute the matrix elements Mij one has to consider, in principle, an infinite number of growth steps conditional to the existence of a frozen configuration of preassigned type. In practice a rapid convergency of the series is ensured by the screeming properties of the Laplace equation.

The boundary conditions appear to play an important role in the conditional growth processes that define M_{ij}. It is important therefore to consider the effect of fluctuations in the boundary conditions. In this case the matrix elements become non linear functions of the distribution (C_1, C_2)itself and the iteration process is much more complex.

This approach allows a reasonably systematic analytical calculation of the fractal dimension for these models. The method can easily be extended to other types of fractal growth models and we are now trying to apply similar ideas to the problem of fully developed turbulance[10].

3. GALAXY DISTRIBUTION; IS THE UNIVERSE FRACTAL?

In this section we describe a field in which fractal geometry is used phenomenologically in a problem that may have important relations with particle physics. The question concerns in fact the large scale properties of matter distribution in the universe.

The usual point of view is that this distribution is, at large scale, homogeneons[11]. All cosmological theories are based in fact the on Friedmann metric that corresponds to a constant density in the equations of general relativity[12]. The case for homogeneity is based in part on questionable statistical analysis, but even more on the fact that it is considered as a sort of principle, more or less equivalent to the cosmological principle in its varions formulations. Now by assuming a cosmological principle one would like to avoid the possibility that we sit in the center of the universe or someting like this. Such a reasonable assumption is often confused with homogeneity that is a much stronger one. In fact what one really should impose is a principle of local isotropy, in the sense that form every galaxy the universe looks statistically the same. Only if the additional (and arbitrary) assumption of analiticity is made then local isotropy also implies homogeneity[12]. As we have seen in Sect 1 for a fractal the principle of local isotropy is well satisfied (Eq. 1,6) however the structure is not analytic and therefore not homogeneous. This means that the question of whether the large scale distribution of matter in the universe is homogeneous or not is a progmatic question to be answered by experiments[13] instead of a sort of unquestionable principle as it been until now considered. Even the statistical correlation analysis of the galaxy distributions in fact assumes homogeneity a priori and adopts mathematical concepts that are only valid if the distribution is indeed homogeneous[11,14].

One can see from this discussion how the introduction of a new concept may open important questions even with respect to the information derived from experiments. From the traditional correlation analysis that implicitaly assumes homogeneity one obtains the result that the galaxy distribution becomes homogeneous at a langth-scale $r_0 \simeq 5$ Mpc[11,14]. Comparing this length the Hubble radius $R_H > 5000$ M_{pc} the usual conclusion is that the universe is homogeneous at large scale and dishomogenities only appear at rather small scales.

The previous discussion however motivated us to reconsider the evidence for homogeneity in a more critical way. We have therefore reanalized the usual galaxy catalogs without assunning homogeneity but instead in order to test it[15]. This implies the use of a different method of analysis that is however standard in modern statistics. The results are quite surprising. First no tendency to homogeneity is present at all. This means that the usual result $r_0 \simeq 5$ Mpc was an artifact of the implicit (and incorret) a priori assumption of homogeneity. Second the galaxy distributions show fractal behavior up to the present limits of observations.

This result reconciles the statistical analysis of galaxy distributions with the observations of large voids and superclusters[13,15]. This means however that no evidence for homogeneity is present in the available data at any length. If confirmed by further analysis this result may have a profound impact on our ideas about the matter distribution in the universe[13,15].

REFERENCES

1. B.B Mandelbrot, "The Fractal Geometry of Nature", (Freeman, New York, 1982).

2. See e. g.: P. Cvitanovic Ed., "Universality in Chaos", Adam Hilger Ltd, Bristol, (1984).

3. L.Pietronero and E.Tosatti eds., "Fractals in Physics", North-Holland, Amsterdam (1986).

4. H.E.Stanley and N.Ostrowsky, "On Growth and Form", Nijhof, Dororecht (1986).

5. D.J.Amit: "Field Theory, the Renormalization Group and Critical Phenomena", Mc Graw Hill (1978).

6. T.A.Witten and L.M.Sander, Phys. Rev.Lett. $\underline{47}$, 1400 (1981).

7. L.Niemeyer, L.Pietronero and H.J.Wiesmann, Phys. Rev. Lett. $\underline{52}$, 1033 (1984).

8. L.Pietronero, A.Erzan and C.Evertsz, Phys, Rev. Lett., $\underline{61}$, 861 (1988).

9. L.Pietronero, A.Erzan and C.Evertsz, Physica A $\underline{151}$, 207 (1988).

10. G.Paladin and A.Vulpiani, Phys. Rep. $\underline{156}$, 147 (1987).

11. P.J.E.Peebles, "The Large-Scale STructure of the Universe", Princeton Univ. Princeton (1980).

12. S.Weinberg, "Gravitation and Cosmology", Wiley, New-York (1972).

13. L.Pietronero, Physica A $\underline{144}$, 257 (1987).

14. M.Davis and P.J.E.Peebles, Ap. J. $\underline{267}$, 465 (1983)

15. P.H.Coleman, L.Pietronero and R.H,Sanders, Astron. Astrophys. $\underline{200}$, L32 (1988).

DISCUSSION

– *Warr:*

What has the lack of an upper critical dimension for the "laplacian model", as distinct for the "polymer model", got to do with the inapplicability of Wilson's RG (renormalization group).

– *Pietronero:*

The absence of an upper critical dimension prevents the applicability of the ϵ– expression method. From the point of view of Real Space Renormalization instead this is not a limitation. However, also this approach is not very appropriate to the present problem for other reasons (see Ref. (9)).

– *Horne:*

How does the observation that the galaxy distribution appears to be fractal and not homogeneous coexist with the measured isotropy of the microwave background radiation?

– *Pietronero:*

Experimentally, the data show that the microwave background radiation is isotropic whereas the distribution of galaxies is not homogeneous. My remarks refer only to the analysis of the data. I have no theory to explain them. One may note however that, if you take away the assumption of analyticity, one may have more possibilities to reconcile these observations. A fractal is in some sense locally isotropic, but what lacks is analyticity (homogeneity). Current theories are all in the framework of analyticity, which might be a bit narrow.

– *Rahal:*

Which kind of statement can you make which would lead to an observable effect for the universe being non–homogeneous?

– *Pietronero:*

People thought, there is homogenization at 5 Megaparsecs. I have shown by a better data analysis, that this is not the case. The same holds (although we have less data) at the level of clusters and at the level of superclusters. The new analysis just shows a fractal distribution to all lengths. If there is eventually homogeneity, its scale is much larger than previously thought.

– *Rahal:*

What are the new phenomena about irreversibility and boundary conditions mentioned at the beginning of the talk?

– *Pietronero:*

In the most well known fractal problems like Laplacian growth and turbulence, the phenomenon of the generation of self–similarity is very dependent on boundary conditions and it is irreversible. This appears to be a general feature of fractals. The phenomena are very different from equilibrium phase transition.

– *Polychronakos:*

For the fractal dimension of a set to be well–defined, either the set should have "infinite detail" (short–distance limit) or infinite size (large–distance limit). Neither is true for the universe (intergalactic distance and Hubble radius set cutoffs). Thus, how well–defined is the fractal dimension of the universe?

– *Pietronero:*

Indeed, these cutoffs alway exist for self–similar structures. This means that in the range of scales between the two cutoffs the distribution of galaxies consists of a fractal distribution with approximate dimension $D=1.2$.

– *Warr:*

a) Can you calculate, in analogy to Wilson's method for critical exponents, the value of η in the Laplacian model appropriate for a given system (e.g. lightning)?

b) More generally, is there evidence of universality in classes of fractal phenomena in nature?

– *Pietronero:*

a) The parameter η is given a priori in the model and the theory (analogous to the Renormalization Group) provides the value of the fractal dimension $D(\eta)$ that corresponds to a certain value of η. The microscopic origin of η can be found by analyzing the detailed process occurring in a particular type of dielectric breakdown process.

b) In this respect fractals seem to be different than critical phenomena. In most of the models of fractal growth one can observe a lower degree of universality as compared to the critical properties of phase transitions.

– *Giavarini:*

When people try to explain phenomena using fractal techniques it is not clear to me where the purely geometric description stops and where the dynamics come in.

– Pietronero:

Of course the geometrical characterization of a particular phenomenon does not explain the phenomenon. Nevertheless a good understanding of the geometrical structures and properties of a phenomenon is essential for constructing any model that can explain it.

– Mincer:

You have shown computer simulations for lightnings. Has any similar calculation been done for the universe which would show that a fractal structure can arise from a set of expanding mutually attracting points.

– Pietronero:

Calculations have been done, but they tried to reproduce the presumed homogeneous structure of the universe. In this respect one should note that the input of these calculations is not at all unique and one cannot hope to show by computer simulation what the real evolution has been.

– Sarid:

You showed that a decaying correlation function is implied by a fractal distribution, but does the converse hold, i.e. does the measured galaxy correlation function necessarily imply self–similarity?

– Pietronero:

The new analysis shows a conditional density that decays as a power law to all scales. This is a necessary and sufficient condition to conclude it is a fractal.

STATUS OF PC VIOLATION

Michel Gourdin

Université Pierre et Marie Curie
Paris — France

INTRODUCTION

The aim of these notes is to provide a general view of the status of the violation of PC in the middle of 1988 from both the experimental and the theoretical sides. We concentrate our comparison between theory and experiment on the minimal standard model and we develop the necessary material for such a comparison emphasizing the defects of both theory and experiment.

The full set of available experimental data is compatible with calculations based on the minimal standard model with three generations of quarks and leptons. Therefore it is rather natural to start a comparison between theory and experiment for the violation of PC by using the framework provided by the standard model and the Cabibbo-Kobayashi-Maskawa 3×3 matrix. In this case the violation of PC is described by one phase, δ, which has to be the same in the $K^0 \overline{K}^0$, $D^0 \overline{D}^0$, $B^0 \overline{B}^0$, $T^0 \overline{T}^0$, etc... systems. Indeed this is a very strong constraint which has to be satisfied by experiment. If it happens that a unique δ is not able to describe all the types of observed violations of PC then it will be justified to look carefully at other possible mechanisms producing a violation of PC.

Concerning the violation of PC in the $K^0 \overline{K}^0$ system we shall give only an account of the phenomenological analysis of experimental data already discussed in details in one of our recent paper and we shall concentrate on the predictions of the minimal standard model.

For the two $B^0 \overline{B}^0$ systems the experimental material is unfortunately very poor and very preliminary. But it happens that positive signals of PC violation have been detected and they look very promizing. Interesting pieces of information are expected in a near future.

No effect of PC violation in the $D^0 \overline{D}^0$ has been experimentally seen and for such a reason we shall not discuss the $D^0 \overline{D}^0$ system.

Determinant progresses in our knowledge of the standard model in the PC violating sector go through a determination of the top quark mass m_t. When the top quark will be experimentally observed it will be possible to obtain reliable informations on the phase δ of the minimal standard model. For the time being the data can only be analyzed in terms of correlations between m_t and δ. Before to proceed in the discussion of PC violation let us keep in mind the following fact. Calling as M^0 a neutral, pseudoscalar, flavoured meson, it happens that the $M^0 \overline{M}^0$ systems are very convenient places to discover new phenomena in physics. This has been the case in the past with the violation of PC in weak decays. This might be the case in the future with the discovery of new physics.

RECENT REVIEW PAPERS

A. Ali   Proceedings of the International Symposium on Production and Decay of Heavy Hadrons, Hamburg (1986), Preprint DESY 87/083 (July 1987) to appear in the Proceedings of the UCLA Workshop on BB factory.

G. Altarelli and P. Franzini
Z Phys. C **37** 271 (1988), Preprint CERN TH 4914/87

G. Altarelli   Proceedings of the International Europhysics Conference on High Energy Physics Uppsala Sweden (July 1987)

S. Stone
Cornell preprint CLNB 87/103 (October 1987) to appear in CP Violation, ed. C.Jarlskog (World Scientific).

PART A - THE $K^0 \overline{K}^0$ SYSTEM
EXPERIMENT - PHENOMENOLOGY AND THEORETICAL FRAMEWORK

I - EXPERIMENTAL OBSERVATION OF THE VIOLATION OF PC IN NEUTRAL K MESON DECAY

1) There exist two states having a definite lifetime
K_S^0 is the short lived component with $\tau_S \simeq .9\,10^{-10}$s.
K_L^0 is the long lived component with $\tau_L \simeq 5\,10^{-8}$s.
It has been observed in 1964 that both states K_S^0 and K_L^0 decay into $\pi^+\pi^-$ and $\pi^0\pi^0$ which are eigenstates of PC with PC=+1. On the other hand $\pi^0\pi^0\pi^0$ is one of the important decay modes of K_L^0 and it has PC=-1. The direct conclusion is that PC is violated in the decay of neutral K mesons and in order to describe such a violation we introduce two complex parameters associated to the 2π modes.

$$
\begin{aligned}
\eta_{+-} &= \frac{\langle \pi^+\pi^- \mid H_W \mid K_L^0 \rangle}{\langle \pi^+\pi^- \mid H_W \mid K_S^0 \rangle} = |\eta_{+-}| e^{i\phi_{+-}} \\
\eta_{00} &= \frac{\langle \pi^0\pi^0 \mid H_W \mid K_L^0 \rangle}{\langle \pi^0\pi^0 \mid H_W \mid K_S^0 \rangle} = |\eta_{00}| e^{i\phi_{00}}
\end{aligned}
\tag{1}
$$

The particle data table results on the four real quantities are

$$
\begin{aligned}
|\eta_{+-}| = (2,275 \pm 0,021)10^{-3} \qquad & \phi_{+-} = 44°,6 \pm 1°,2 \\
|\eta_{00}| = (2,299 \pm 0,036)10^{-3} \qquad & \phi_{00} = 54° \pm 5°
\end{aligned}
\tag{2}
$$

Therefore the order of magnitude of the violation of PC in both modes is $\mathcal{O}(10^{-3})$. A recent experiment at CERN has improved our knowledge of the moduli $|\eta_{+-}|$ and $|\eta_{00}|$

$$
\left| \frac{\eta_{00}}{\eta_{+-}} \right|^2 = 0,980 \pm 0,004 \pm 0,003
\tag{3}
$$

Adding the errors in quadrature and combining the two sources of information we obtain

$$
1 - \left| \frac{\eta_{00}}{\eta_{+-}} \right| = (1,01 \pm 0,32)10^{-2}
$$
$$
|\eta_{+-}| - |\eta_{00}| = (2,27 \pm 0,74)10^{-5}
\tag{4}
$$

Both quantities differ from zero by 3 standard deviations.

2) The violation of PC has also been measured in the semi-leptonic decay mode of the K_L^0 component. Defining the charge asymmetry δ by

$$\delta = \frac{\Gamma(K_L^0 \to \pi^- e^+ \nu_e) - \Gamma(K_L^0 \to \pi^+ e^- \bar{\nu}_e)}{\Gamma(K_L^0 \to \pi^- e^+ \nu_e) + \Gamma(K_L^0 \to \pi^+ e^- \bar{\nu}_e)} \tag{5}$$

we have the experimental value quoted in the particle data tables

$$\delta_{\text{exp}} = (0,333 \pm 0,014)10^{-2} \tag{6}$$

Assuming the validity of the $\Delta S = \Delta Q$ rule for neutral K meson decay a non zero value of δ means a violation of PC invariance.

3) A violation of PC is also expected in the 3π modes. Unfortunately the decay of the K_S^0 into 3π has not been experimentally observed.

II - ISOTOPIC SPIN ANALYSIS OF THE 2π MESON MODES

1) Because of the Bose-Einstein statistics a 2π S state can only have a total isotopic spin $I = 0$ or $I = 2$. For theoretical purposes it is usual to define two new PC violating parameters

$$\epsilon_0 = \frac{\langle 0 \mid H_W \mid K_L^0 \rangle}{\langle 0 \mid H_W \mid K_S^0 \rangle}$$

$$\epsilon_2 = \frac{1}{\sqrt{2}} \frac{\langle 2 \mid H_W \mid K_L^0 \rangle}{\langle 0 \mid H_W \mid K_S^0 \rangle} \tag{7}$$

Of course ϵ_0 and ϵ_2 are related to η_{+-} and η_{00} via Clebsch-Gordan coefficients.

2) Experimentally the final state with $I = 0$ strongly dominates over the final state with $I = 2$. This result is known as the empirical $|\Delta \vec{I}| = \frac{1}{2}$ rule. Let us define, for $K_S^0 \Rightarrow 2\pi$ decay, the ratio R

$$R = \frac{\Gamma(K_S^0 \Rightarrow \pi^0 \pi^0)}{\Gamma(K_S^0 \Rightarrow \pi^+ \pi^-)} \tag{8}$$

With an exact $|\Delta \vec{I}| = \frac{1}{2}$ rule and taking into account phase space corrections due to the $\pi^\pm - \pi^0$ mass difference we predict

$$R_{TH} = 0,507 \tag{9}$$

Experimentally the particle data tables give

$$R_{\text{exp}} = 0,457 \pm 0,005 \tag{10}$$

As a conclusion the $|\Delta \vec{I}| = \frac{1}{2}$ rule is slightly violated with a 10 standard deviations effect. Such a violation will be described by the complex parameter ω defined by

$$\omega = \frac{1}{\sqrt{2}} \frac{\langle 2 \mid H_W \mid K_S^0 \rangle}{\langle 0 \mid H_W \mid K_S^0 \rangle} = |\omega| e^{i\phi_\omega} \tag{11}$$

3) The physical states $|K_S^0\rangle$ and $|K_L^0\rangle$ can be written as linear superpositions of the basic states $|K^0\rangle$ and $|\overline{K}^0\rangle$. Assuming TCP invariance we introduce only one complex parameter ϵ as follows

$$|K_S^0\rangle = \frac{1}{\sqrt{2(1+|\epsilon|^2)}}\left[(1+\epsilon)|K^0\rangle + (1-\epsilon)|\overline{K}^0\rangle\right]$$
$$|K_L^0\rangle = \frac{1}{\sqrt{2(1+|\epsilon|^2)}}\left[(1+\epsilon)|K^0\rangle - (1-\epsilon)|\overline{K}^0\rangle\right] \tag{12}$$

and the hermitian product of K_S^0 and K_L^0 is a real number

$$\langle K_S^0 \mid K_L^0 \rangle = \frac{2\,\mathrm{Re}\,\epsilon}{1+|\epsilon|^2} \tag{13}$$

On the other hand the analysis of the experimental data is conveniently made after introducing reduced amplitudes A_I free of the strong interaction phases δ_I associated to the $\pi\pi$ final state with isotopic spin I and energy m_K.

$$\langle I \mid H_W \mid K^0 \rangle = A_I e^{i\delta_I} \qquad I = 0, 2 \tag{14}$$

Assuming TCP invariance we have

$$\langle I \mid H_W \mid \overline{K}^0 \rangle = A_I^* e^{i\delta_I} \qquad I = 0, 2 \tag{15}$$

With our choice of phases the complex nature of the reduced amplitudes A_I is due to the violation of PC in $\Delta S = 1$ transitions. In what follows we shall use the ratios

$$\xi_I = \frac{\mathrm{Im}A_I}{\mathrm{Re}\,A_I} \qquad I = 0, 2 \tag{16}$$

4) From equations (11), (14) and (15) it is clear that the phase of ω is simply

$$\phi_\omega = \delta_2 - \delta_0 \tag{17}$$

The information for the phase difference $\delta_2 - \delta_0$ is coming from $\pi\pi$ phase shift analysis and K_{e_4} decay. A weighted average quoted by Devlin and Dickey is

$$\delta_2 - \delta_0 = -41°, 4 \pm 8°, 1 \tag{18}$$

5) As pointed out previously the violation of PC in neutral K meson decay is very small, $O(10^{-3})$ and a first order calculation seems to be justified. The relevant relations between the previous parameters have the simple form

$$\epsilon_0 \simeq \epsilon + i\xi_0$$
$$\frac{\epsilon_2}{\omega} \simeq \epsilon + i\xi_2 \tag{19}$$

and we get the constraints due to TCP invariance

$$\mathrm{Re}\,\epsilon_0 = \mathrm{Re}\,\frac{\epsilon_2}{\omega} = \mathrm{Re}\,\epsilon \tag{20}$$

In practice instead of the parameter ϵ_2 we use the parameter ϵ' defined by $\epsilon' = \epsilon_2 - \omega\epsilon_0$. From equations (19) we get

$$\epsilon' = i\omega(\xi_2 - \xi_0) \tag{21}$$

In this approximation the parameter ω is independent of the violation of PC and it has the form

$$\omega = \frac{1}{\sqrt{2}}\frac{\mathrm{Re}\,A_2}{\mathrm{Re}\,A_0}e^{i(\delta_2 - \delta_0)} \tag{22}$$

Let us emphasize that the parameters ϵ_0 and ϵ' are phase convention independent. This is not the case for ϵ whose real part only has a physical meaning.

III - PARAMETER ω AND THE $|\Delta \vec{I}| = \frac{1}{2}$ RULE

1) Consider first the $K_S^0 \Rightarrow 2\pi$ data and more specifically the ratio R whose experimental value is given in equation (10). Taking into account the phase space corrections we arrive to the following relation for the complex parameter ω

$$\left| \frac{1 - 2\omega}{1 + \omega} \right|^2 = 0,902 \pm 0,010 \tag{23}$$

which is nothing but a relation between $|\omega|$ and ϕ_ω represented on Figure 1 where the two curves correspond to one standard deviation as given in equation (23). Let us remember that ϕ_ω is simply the difference of two strong interactions phases $\phi_\omega = \delta_2 - \delta_0$.

2) We use for ϕ_ω the weighted averaged value (18) and we get the dashed region of Figure 1 and for $|\omega|$ the result

$$|\omega| = (2,3^{+0,7}_{-0,5})10^{-2} \tag{24}$$

The decay of an isotopic spin $I = \frac{1}{2}$ K-meson into a 2π final state of total isotopic spin $I = 0$ or $I = 2$ can proceed via $|\Delta \vec{I}| = \frac{1}{2}$, $|\Delta \vec{I}| = \frac{3}{2}$, or $|\Delta \vec{I}| = \frac{5}{2}$ transitions. Using experimental data for $K_S^0 \Rightarrow \pi^+\pi^-$, $K_S^0 \Rightarrow \pi^0\pi^0$ and $K^+ \Rightarrow \pi^+\pi^0$ we can compute the three reduced amplitudes describing these three types of transitions. The expected result is the dominance of the $|\Delta \vec{I}| = \frac{1}{2}$ transition over the two other ones. However with the previous value (18) of ϕ_ω we obtain a small but non zero value of the $|\Delta \vec{I}| = \frac{5}{2}$ transition.

3) In most of the previous analysis the absence of $|\Delta \vec{I}| = \frac{5}{2}$ transitions was assumed at the beginning. With such an assumption it becomes possible to derive the reduced amplitude A_2 from $K^+ \Rightarrow \pi^+\pi^0$ decay and we get, for $|\omega|$ a more accurate value

$$|\tilde{\omega}| = (3,17 \pm 0,02)10^{-2} \tag{25}$$

Such a value has been represented on Figure 1 by an horizontal bar and the phase ϕ_ω takes now the value

$$\tilde{\phi}_\omega = -56°,5 \pm 4° \tag{26}$$

IV - THE INCONSISTENCY PROBLEM

1) We can consider the TCP constraint as given in equation (20)

$$\text{Re}\,\epsilon_0 = \text{Re}\,\frac{\epsilon_2}{\omega}$$

Using the experimentally measurable quantities η_{+-}, η_{00} and ω we obtain

$$\text{Re}\left[\frac{(1 + \omega)(1 - 2\omega)}{\omega}(\eta_{+-} - \eta_{00}) \right] = 0 \tag{27}$$

Defining Φ as

$$\Phi = \text{Phase}\left[\frac{\omega}{(1 + \omega)(1 - 2\omega)} \right] + \frac{\pi}{2} \tag{28}$$

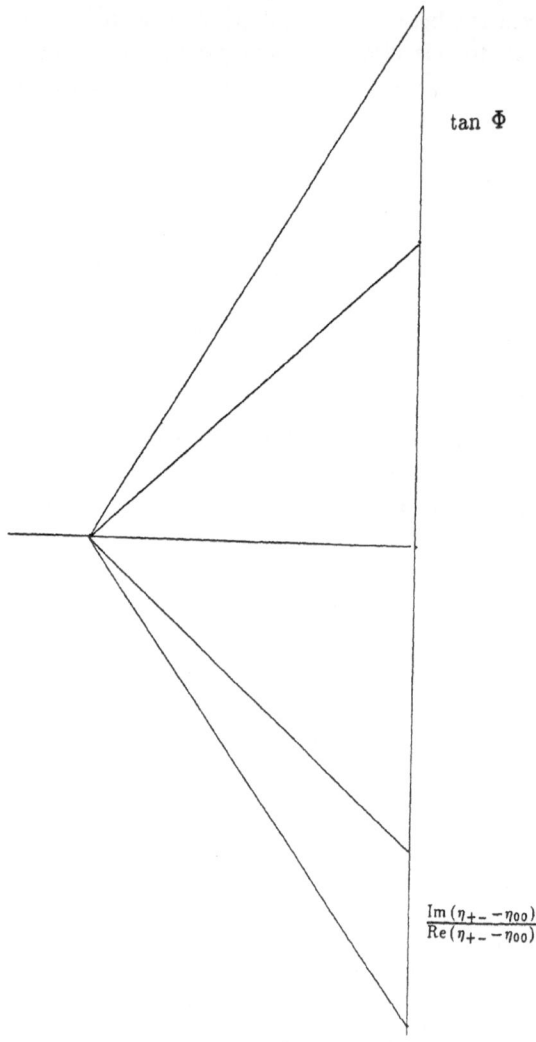

Figure 1 The two domains for $\frac{\text{Im}\,(\eta_{+-} - \eta_{00})}{\text{Re}\,(\eta_{+-} - \eta_{00})}$ and tan Φ allowed within one standard deviation limits.

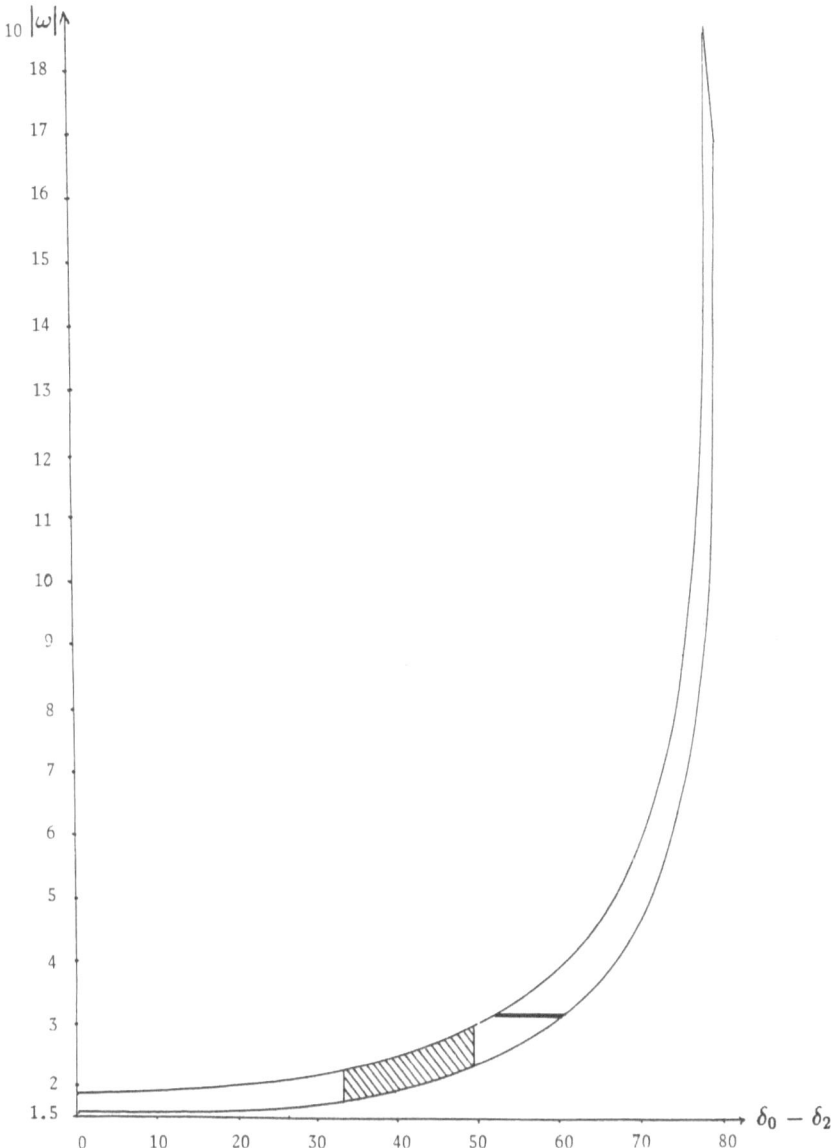

Figure 2 The modulus of ω as a function of the phase difference $\delta_0 - \delta_2$; the two curves correspond to one standard deviation limit for the ratio R.

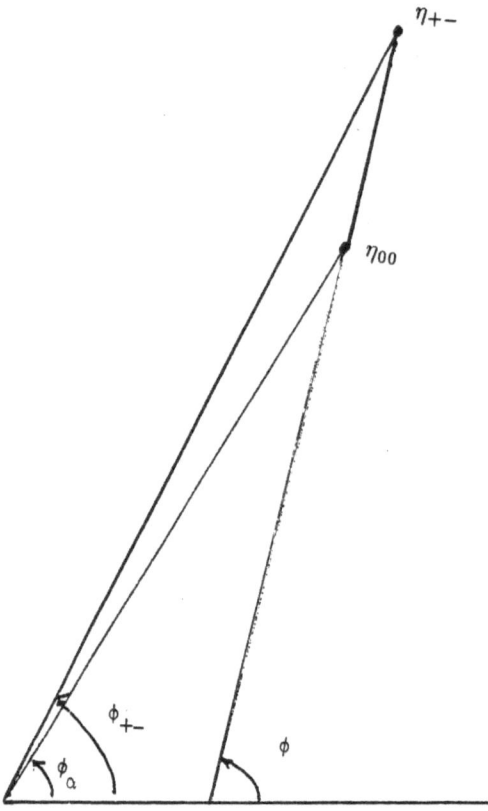

Figure 3 The vectors η_{+-}, η_{00} and $\eta_{+-} - \eta_{00}$ in their complex plane.

the TCP constraint takes the simple form

$$\text{Phase}[\eta_{+-} - \eta_{00}] = \Phi \qquad \text{modulo } \pi \tag{29}$$

Using the particle data tables (2) we obtain

$$\frac{\text{Im}(\eta_{+-} - \eta_{00})}{\text{Re}(\eta_{+-} - \eta_{00})} = -0,98 {\,}^{+0,13}_{-0,36} \tag{30}$$

From the previous determination of ω we have

$$\text{tg } \Phi = 1,10 {\,}^{+0,37}_{-0,27} \tag{31}$$

In equations (30) and (31) the errors correspond to one standard deviation for η_{+-}, η_{00} and $\delta_2 - \delta_0$. As shown on Figure 2 the two numbers (30) and (31) are definitively different. It is the inconsistency problem.

Two conclusions are possible
   i) the TCP constraint is not satisfied by experiment and we have a violation of TCP invariance.
   ii) the experimental data are not accurate enough to test the constraint (27) and, in particular, the difference $\eta_{+-} - \eta_{00}$ is not well measured.

In what follows we investigate the second explanation of the inconsistency problem and we try to derive constraints on the phase difference $\phi_{+-} - \phi_{00}$ such that the inconsistency problem disappears.

2) Consider the difference $\eta_{+-} - \eta_{00}$ written in the form coming from TCP invariance

$$\eta_{+-} - \eta_{00} = \Delta e^{i\Phi} \tag{32}$$

We shall choose for Φ the numerical value

$$\Phi = 47°,6 \pm 8°,1 \tag{33}$$

as coming from the phase difference (18) and from the corrections to the $|\Delta \vec{I}| = \frac{1}{2}$ rule obtained with the results of the previous section. The two complex parameters η_{+-} and η_{00} are represented on Figure 3 as two vectors and they form a triangle whose third side is precisely $\eta_{+-} - \eta_{00}$. We then use the elementary relations in a triangle between the length of one side and the sine of the opposite angle. The angles ϕ_{+-}, ϕ_{00} and Φ having similar values the triangle is extraflat and we get

$$\Delta = |\eta_{+-}| - |\eta_{00}| \tag{34}$$

$$\phi_{+-} - \phi_{00} = \left[\frac{|\eta_{+-}|}{|\eta_{00}|} - 1 \right] (\Phi - \phi_{+-}) \tag{35}$$

Using the experimental values of ϕ_{+-} and $|\eta_{+-}|/|\eta_{00}|$ and the estimated value (33) of Φ we obtain the following range of allowed values for the phase difference $\phi_{+-} - \phi_{00}$ due to the TCP invariance.

$$-0°,08 < \phi_{+-} - \phi_{00} < +0°,16 \tag{36}$$

Comparing with the particle data tables result

$$\phi_{+-} - \phi_{00} = -9°,6 \pm 6°,2 \tag{37}$$

we see that our estimate (36) differs by less than two standard deviations from the experimental value. It is now clear that an accurate measurement of $\phi_{+-} - \phi_{00}$ is urgently needed and it could solve in a very satisfactory way the inconsistency problem.

V - PARAMETERS ϵ_0 AND ϵ' FROM EXPERIMENT

1) The parameter ϵ_0 is related to measured quantities by the formula

$$\epsilon_0 = \eta_{+-} - \frac{1 - 2\omega}{3}(\eta_{+-} - \eta_{00}) \tag{38}$$

Inserting the experimental data for the 2π decay mode we get

$$0 < \phi_{\epsilon_0} - \phi_{+-} \leq 0°,02 \tag{39}$$

$$\text{Re}\,\epsilon = (1,615 \pm 0,034)10^{-3} \tag{40}$$

Using now the experimental value (6) for the charge asymmetry we obtain for $\text{Re}\,\epsilon$ a value in very nice agreement with the result (40)

$$\text{Re}\,\epsilon = (1,665 \pm 0,070)10^{-3} \tag{41}$$

and which, as a by product, produces a nice test of the $\Delta S = \Delta Q$ rule.

2) The parameter ϵ' is given, in terms of physical quantities, by the expression

$$\epsilon' = \tfrac{1}{3}(1 + \omega)(1 - 2\omega)(\eta_{+-} - \eta_{00}) \tag{42}$$

The phase of ϵ' is predicted by the TCP constraint (27)

$$\phi_{\epsilon'} = \delta_2 - \delta_0 + \frac{\pi}{2} \tag{43}$$

The determination of the modulus of ϵ' comes from the data on the 2π-mode and it is given by

$$|\epsilon'| = (7,44 \pm 2,42)\,10^{-6} \tag{44}$$

VI - GENERAL FORMALISM FOR $K^0\overline{K}^0$ MIXING

1) Basic equations

We use the Wigner-Weisskopf formalism with an Hamiltonian H represented as a 2×2 matrix and decomposed, as usual, into a mass matrix M and a decay matrix Γ

$$H = M - i\frac{\Gamma}{2} \tag{45}$$

Both M and Γ are hermitian matrices and assuming TCP invariance we obtain, in the $K^0, \overline{K}^0$ basis

$$M = \begin{vmatrix} m & m_{12} \\ m_{12}^* & m \end{vmatrix} \qquad \Gamma = \begin{vmatrix} \gamma & \gamma_{12} \\ \gamma_{12}^* & \gamma \end{vmatrix} \tag{46}$$

The observed states $|K_S^0\rangle$ and $|K_L^0\rangle$ defined in equation (12) are the eigenstates of the hamiltonian H and the eigenvalues $m_S, m_L, \gamma_S, \gamma_L$ are measured quantities. We define the mass difference and the width difference

$$\Delta m = m_L - m_S \qquad \Delta\gamma = \gamma_L - \gamma_S \tag{47}$$

From experiment $\Delta m > 0$ and $\Delta\gamma < 0$. The constraints arising from the eigenvalue equations are given in the form of two complex equations

$$\left(\frac{1 + \epsilon}{1 - \epsilon}\right)^2 = \frac{m_{12} - \frac{i}{2}\gamma_{12}}{m_{12}^* - \frac{i}{2}\gamma_{12}^*} \tag{48}$$

$$\left(\Delta m - \frac{i}{2}\Delta\gamma\right)^2 = 4\left(m_{12} - \frac{i}{2}\gamma_{12}\right)\left(m_{12}^* - \frac{i}{2}\gamma_{12}^*\right) \tag{49}$$

194

2) Solution for Δm and $\Delta \gamma$

Consider first the equation (49). We can write

$$m_{12} - \frac{i}{2}\gamma_{12} = \left(\operatorname{Re} m_{12} - \frac{i}{2}\operatorname{Re} \gamma_{12}\right)(1 + Y)$$

$$m_{12}^* - \frac{i}{2}\gamma_{12}^* = \left(\operatorname{Re} m_{12} - \frac{i}{2}\operatorname{Re} \gamma_{12}\right)(1 - Y)$$

where the PC violating complex parameter Y is defined by

$$Y = \frac{\frac{1}{2}\operatorname{Im} \gamma_{12} + i\operatorname{Im} m_{12}}{\operatorname{Re} m_{12} - \frac{i}{2}\operatorname{Re} \gamma_{12}} \tag{50}$$

The equation (49) becomes

$$\left(\Delta m - \frac{i}{2}\Delta\gamma\right)^2 = 4\left(\operatorname{Re} m_{12} - \frac{i}{2}\operatorname{Re} \gamma_{12}\right)^2 (1 - Y^2) \tag{51}$$

For the differences Δm and $\Delta \gamma$ the violation of PC occurs only at the second order. When it is neglected we simply have

$$\Delta m = 2\operatorname{Re} m_{12} \tag{52}$$

$$\Delta\gamma = 2\operatorname{Re} \gamma_{12} \tag{53}$$

3) Solution for ϵ

We now consider equation (48). Introducing the parameter Y it takes the form

$$\left(\frac{1+\epsilon}{1-\epsilon}\right)^2 = \frac{1+Y}{1-Y} \tag{54}$$

The solution of this second order equation in ϵ is

$$\epsilon = \frac{Y}{1 + \sqrt{1 - Y^2}} \tag{55}$$

To first order in the violation of PC we simply have

$$\epsilon = \frac{1}{2}Y \tag{56}$$

We now introduce the angle α_K defined by

$$\alpha_K = -\operatorname{Arc\,tg}\frac{2\Delta m}{\Delta\gamma} \tag{57}$$

Using the experimental data on Δm and $\Delta \gamma$ the numerical value of α_K is close to $\pi/4$

$$\alpha_K = 43°,72 \pm 0°,12 \tag{58}$$

After some elementary algebraic manipulations we obtain from equations (50),(56) and (57) an interesting form for ϵ

$$\epsilon = \frac{1}{2}e^{i\alpha_K}\left[\sin \alpha_K \frac{\operatorname{Im} m_{12}}{\operatorname{Re} m_{12}} + i\cos \alpha_K \frac{\operatorname{Im} \gamma_{12}}{\operatorname{Re} \gamma_{12}}\right] \tag{59}$$

195

4) We now compute the non diagonal element of the decay matrix γ_{12} for the particular case of K mesons. We have to make a summation over the physical states occurring in both the decay of K^0 and $\overline{K}^0$ mesons. Because of the $\Delta S = \Delta Q$ rule these state are hadronic and strongly dominated by the 2π meson states. Retaining only these contributions we get

$$\gamma_{12} = \sum \left[\langle \pi^+\pi^- \mid H_W \mid K^0 \rangle^* \langle \pi^+\pi^- \mid H_W \mid \overline{K}^0 \rangle \right.$$
$$\left. + \langle \pi^0\pi^0 \mid H_W \mid K^0 \rangle^* \langle \pi^0\pi^0 \mid H_W \mid \overline{K}^0 \rangle \right]$$

or equivalently by using states of definite isotopic spin

$$\gamma_{12} = \sum \left[\langle 0 \mid H_W \mid K^0 \rangle^* \langle 0 \mid H_W \mid \overline{K}^0 \rangle + \langle 2 \mid H_W \mid K^0 \rangle^* \langle 2 \mid H_W \mid \overline{K}^0 \rangle \right]$$

Assuming TCP invariance we obtain γ_{12} in term of reduced amplitudes

$$\gamma_{12} = \sum \left[(A_0^*)^2 + (A_2^*)^2 \right] \tag{60}$$

Retaining only the first order contribution in the violation of PC we obtain the ratio of the imaginary part to the real part of γ_{12} in the form

$$\frac{\operatorname{Im}\gamma_{12}}{\operatorname{Re}\gamma_{12}} = -2\,\frac{\xi_0 + 2|\omega|^2\xi_2}{1 + 2|\omega|^2} \tag{61}$$

Neglecting now second order terms in the violation of the $|\Delta \vec{I}| = 1/2$ rule $|\omega|^2 = 0(10^{-3})$ we obtain for ϵ the following expression

$$\epsilon = e^{i\alpha_K}\left\{ \sin\alpha_K \frac{\operatorname{Im}m_{12}}{2\operatorname{Re}m_{12}} - i\cos\alpha_K\xi_0 \right\} \tag{62}$$

However as pointed out previously the parameter ϵ depends on the phase convention chosen for $|K^0\rangle$, $|\overline{K}^0\rangle$ and only parameters like ϵ_0 ϵ_2 or ϵ' are observable quantities. In particular using the relation (19), $\epsilon_0 = \epsilon + i\xi_0$, we obtain the following expression for ϵ_0

$$\epsilon_0 = e^{i\alpha_K}\sin\alpha_K\left[\frac{\operatorname{Im}m_{12}}{2\operatorname{Re}m_{12}} + \xi_0 \right] \tag{63}$$

A second equivalent expression of ϵ_0 is obtained by replacing $2\operatorname{Re}m_{12}$ by Δm as given by the equation (52)

$$\epsilon_0 = e^{i\alpha_K}\sin\alpha_K\left[\frac{\operatorname{Im}m_{12}}{\Delta m} + \xi_0 \right] \tag{64}$$

5) As a consequence of the equation (54) the phase of ϵ_0 is predicted to be α_K up to correction of the order $O(|\omega|^2)$

$$\phi_{\epsilon_0} = \alpha_K + O(|\omega|^2)$$

As obtained in the phenomenological analysis of the experimental data the two phases ϕ_{ϵ_0} and ϕ_{+-} differ by less than $0°,02$. Comparing now the experimental values of ϕ_{+-} and α_K we get

$$\alpha_K - \phi_{+-} = 2°,32 \pm 1°,4 \tag{65}$$

which is consistent with zero in the 1,5 standard deviation limit.

VII - CONCLUSIONS

1) We have a good experimental knowledge of ϵ_0 at the 2% level.

2) It is necessary to improve the measurement of $\left|\frac{\eta_{+-}}{\eta_{00}}\right| - 1$ presently known at the 30% level.

3) The inconsistency problem could be solved by measuring the phase difference $\phi_{+-} - \phi_{00}$ at the 1° level and if possible better.

4) The situation concerning the experimental value of the phase difference $\delta_2 - \delta_0$ and the (non) existence of $|\Delta \vec{I}| = 5/2$ transitions in $K \Rightarrow 2\pi$ decay has to be clarified.

REFERENCES FOR EXPERIMENTAL DATA

J.H. Christenson, J.W. Cronin, V.L. Fitch and R. Turlay
    Phys. Rev. Lett. **13**, 138 (1964).

R.H. Bernstein et al.   Phys. Rev. Lett. **54**, 1631 (1985).

J.K. Black et al.   Phys. Rev. Lett. **54**, 1628 (1985).

M. Woods et al.   Phys. Rev. Lett. **60**, 1695 (1988).

H. Burkhart et al.   Phys. Lett. **B206**, 169 (1988).

Particle Data Tables   Phys. Lett. **B170** (1986).

REFERENCES FOR PHENOMENOLOGICAL ANALYSIS

T.J. Devlin and J.O. Dickey   Rev. Mod. Phys. **51**, 237 (1979).

V.V. Barmin et al.   Nucl. Phys. **B247**, 293 (1984).

N.N. Biswas et al.   Phys. Rev. Lett. **47**, 1378 (1981).

M. Gourdin   University of Paris preprint LPTHE 88-27 (July 1988).

PART B – THE $K^0 \overline{K}^0$ SYSTEM
COMPARISON OF EXPERIMENT WITH THE MINIMAL STANDARD MODEL

I - STANDARD MODEL AND THE CABIBBO-KOBAYASHI-MASKAWA MATRIX

1) The standard model is based on the gauge group

$$SU(3)_c \otimes SU(2)_L \otimes U(1)$$

The first component $SU(3)_c$ is the colour group for quantum chromodynamics (QCD) where the strong interaction forces are mediated by 8 coloured massless gluons g. The second part $SU(2)_L \otimes U(1)$ is the electroweak group of the Glashow-Salam-Weinberg model where the weak interaction forces are mediated by 3 massive bosons $W^\pm$ and Z° and the electromagnetic interactions by the massless photon γ. In the 100 GeV region the original gauge symmetry is broken and only the exact symmetry

$$SU(3)_c \otimes U(1)_{em}$$

survives. One popular candidate for such a breaking is the so called Higgs mechanism but, from the experimental side, the situation is open.

2) By minimal standard model we mean only 3 generations of quarks and leptons and only 1 complex doublet of scalar Higgs fields. However, in the quark sector, the t quark of electric charge $Q = +2/3$ forming a doublet with the b quark has not yet been experimentally observed. A lower bound on its mass comes from $e^+ e^-$ experiments, $m_t > 27 \text{GeV}$ and an upper bound can be derived by using radiative corrections arguments for $\sin^2 \theta_w$. For the coming numerical computations we shall take

$$30 \text{ GeV} < m_t < 200 \text{ GeV} \tag{1}$$

3) It is known from a very long time that, in the quark sector, the states defined in strong interactions and those occuring in weak interactions are not the same. More technically we shall say that the flavours are not conserved in weak interactions involving charged currents associated to the W^+ and W^- bosons. The situation is different in weak interactions involving neutral currents associated to the Z° boson. These experimental facts are at the origin of mixing in the quark sector described, in the case of three generations of quarks, by an unitary 3×3 matrix called the Cabibbo-Kobayashi-Maskawa matrix (CKM).

$$V = \begin{vmatrix} V_{ud} & V_{us} & V_{ub} \\ V_{cd} & V_{cs} & V_{cb} \\ V_{td} & V_{ts} & V_{tb} \end{vmatrix} \tag{2}$$

The unitarity constraints are simply

$$\sum_i V_{ij} V_{ik}^* = \delta_{jk}$$

$$\sum_j V_{ij} V_{kj}^* = \delta_{ik} \tag{3}$$

Among the various parametrizations proposed for the CKM matrix we shall choose that due to Maiani

$$V = \begin{vmatrix} C_\theta C_\beta & S_\theta C_\beta & S_\beta e^{-i\delta} \\ -S_\theta C_\gamma - C_\theta S_\beta S_\gamma e^{i\delta} & C_\theta C_\gamma - S_\theta S_\beta S_\gamma e^{i\delta} & C_\beta S_\gamma \\ S_\theta S_\gamma - C_\theta S_\beta C_\gamma e^{i\delta} & -C_\theta S_\gamma - S_\theta S_\beta C_\gamma e^{i\delta} & C_\beta C_\gamma \end{vmatrix} \tag{4}$$

Here C means Cos and S means Sin for the three angles θ, β and γ. Of course θ is nothing but the Cabibbo angle θ_c.

4) The observation of relatively long lifetime for the B meson and of a small value for the ratio $\Gamma(b \to u)/\Gamma(b \to c)$ suggest a hierarchy in the 3 angles θ, β and γ. This is at the origin of a simple parametrization proposed by Wolfenstein who finds that $|V_{cb}|$ is of the order $\sin^2 \theta_c$ and $|V_{ub}|$ of the order $\sin^3 \theta_c$. Defining $\sin \theta_c = \lambda$ and

$$S_\gamma = A\lambda^2 \qquad S_\beta = Aa\lambda^3 \tag{5}$$

we obtain the Wolfenstein representation by making an expansion of the form (4) in power of λ retaining only terms of the order $O(\lambda^3)$. The result is

$$V = \begin{vmatrix} 1 - \frac{1}{2}\lambda^2 & \lambda & A\lambda^3 a e^{-i\delta} \\ -\lambda & 1 - \frac{1}{2}\lambda^2 & A\lambda^2 \\ A\lambda^3(1 - ae^{i\delta}) & -A\lambda^2 & 1 \end{vmatrix} \tag{6}$$

Estimates on the parameters λ, A and a coming from experiment are

$$\begin{vmatrix} \lambda = 0,221 \pm 0,002 \\ A = 1,05 \pm 0,17 \\ 0,3 < a < 0,6 \end{vmatrix} \tag{7}$$

The phase δ will describe the violation of PC and it is the object of experiments on PC violation to bring informations on δ.

II - BOX DIAGRAMS FOR m_{12}

1) The box diagrams for the $\Delta S = 2$, $K^0\overline{K}^0$ transition are represented on the Figure 1 where the virtual quarks u_m and u_l have the electric charge $Q = +2/3$: u, c or t.

2) The amplitude associated to the box diagrams involves the product of four elements of the CKM matrix. We shall adopt the following convention for the W $u_i d_j$ vertices

$$\begin{aligned} V_{u_i d_j} &\quad \text{if the outgoing quark is } u_i, \\ V_{u_i d_j}^* &\quad \text{if the outgoing quark is } d_j. \end{aligned}$$

The product of the four CKM matrix elements will be writen as $\lambda_k \lambda_l$ where

$$\lambda_k = V_{u_k d} \cdot V_{u_k s} \tag{8}$$

and the unitarity constraint (3) is simply

$$\lambda_u + \lambda_c + \lambda_t = 0 \tag{9}$$

Using (4), (6) and (8) we obtain

$$\begin{aligned} \operatorname{Re} \lambda_u &= \lambda(1 - \tfrac{1}{2}\lambda^2) + O(\lambda^5) & \operatorname{Im} \lambda_u &\equiv 0 \\ \operatorname{Re} \lambda_c &= -\lambda(1 - \tfrac{1}{2}\lambda^2) + O(\lambda^5) & \operatorname{Im} \lambda_c &= iA^2 a\lambda^5 \sin \delta \\ \operatorname{Re} \lambda_t &= -A^2\lambda^5(1 - a\cos\delta) & \operatorname{Im} \lambda_t &= -iA^2 a\lambda^5 \sin \delta \end{aligned} \tag{10}$$

3) The box diagrams of Figure 1 are computed by using an effective four-quark Fermi hamiltonian and the result is

$$m_{12}^{\text{box}} = -\frac{G_F^2}{16\pi^2} m_W^2 \Big[\sum_{u,c,t} \eta_{kl} F(z_k, z_l) \lambda_k \lambda_l \Big] \mathcal{O}_{12}^{\text{box}} \tag{11}$$

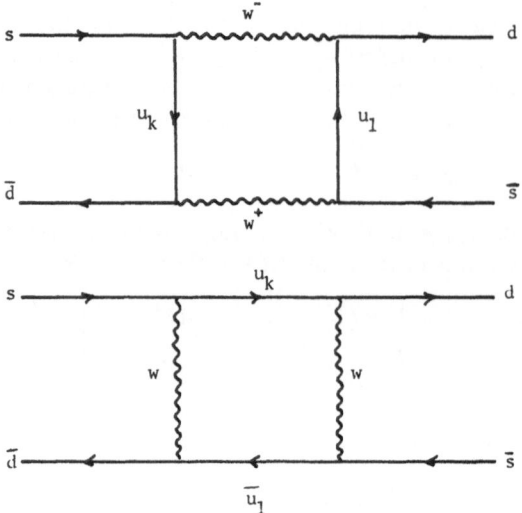

Figure 1 The box diagrams for the transition $\overline{K}^0 \Rightarrow K^0$ (m_{12})

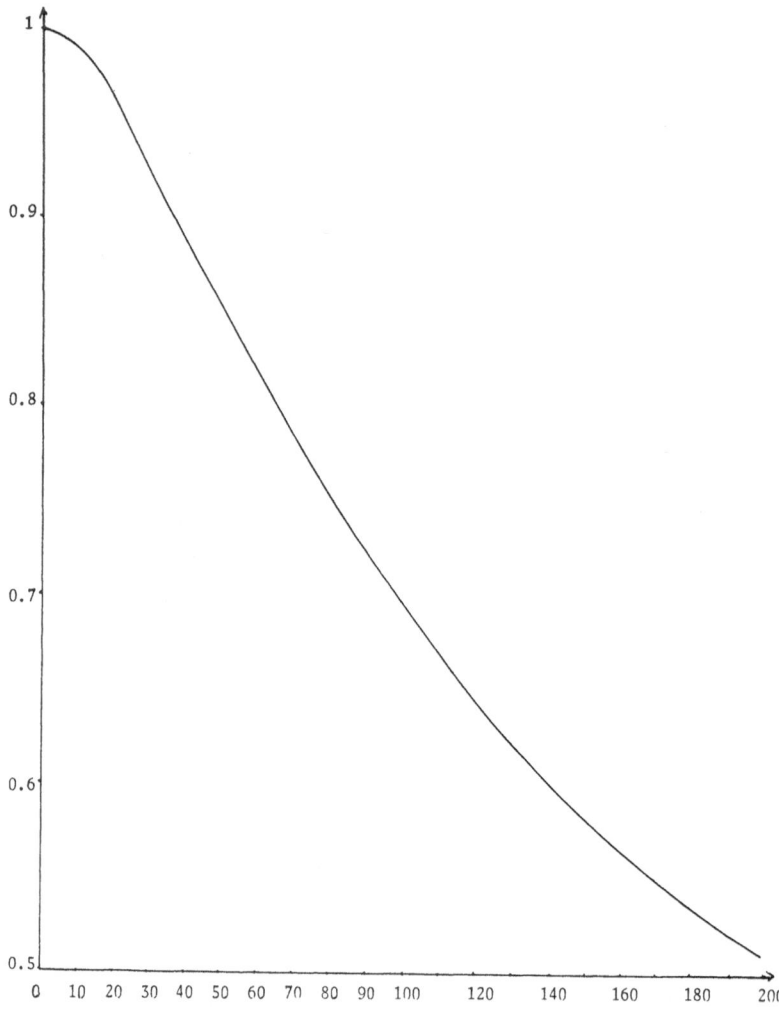

Figure 2 The function $F(z_t)$ for $0 \leq m_t \leq 200\,\mathrm{GeV}$

TABLE 1

METHOD	B_K	REFERENCE
Flavour Sum Rule	$0,33 \pm 0,33$	(1)
QCD Sum Rule, 2-point function	$0,33 \pm 0,09$	(2)
QCD Sum Rule, 3-point function	$1,2 \pm 0,1$	(3)
	$0,55 \pm 0,15$	(4)
	$0,84 \pm 0,08$	(5)
	$0,5 \pm 0,1 \pm 0,2$	(6)
$\frac{1}{N}$ expansion	$0,7 \pm 0,07$	(7)
QCD lattice	$0,9 \pm 0,2$	(8)

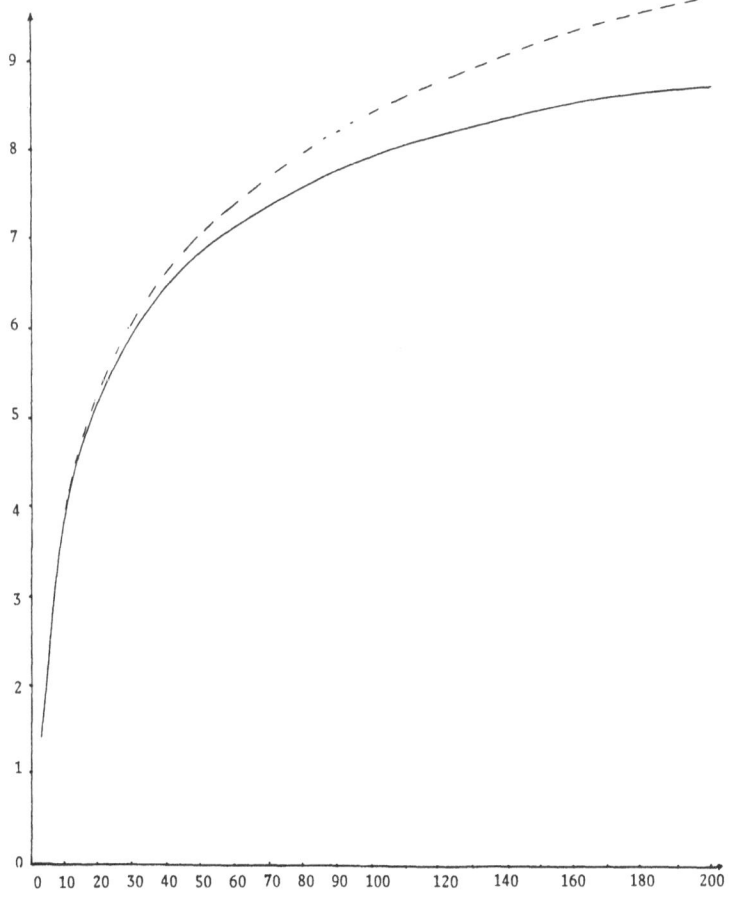

Figure 3 The function $G(z_c, z_t)$ for $0 \leq m_t \leq 200\,\text{GeV}$ and $m_c = 1,5\,\text{GeV}$. The dashed curve is simply the leading term $\log \frac{z_t}{z_c}$

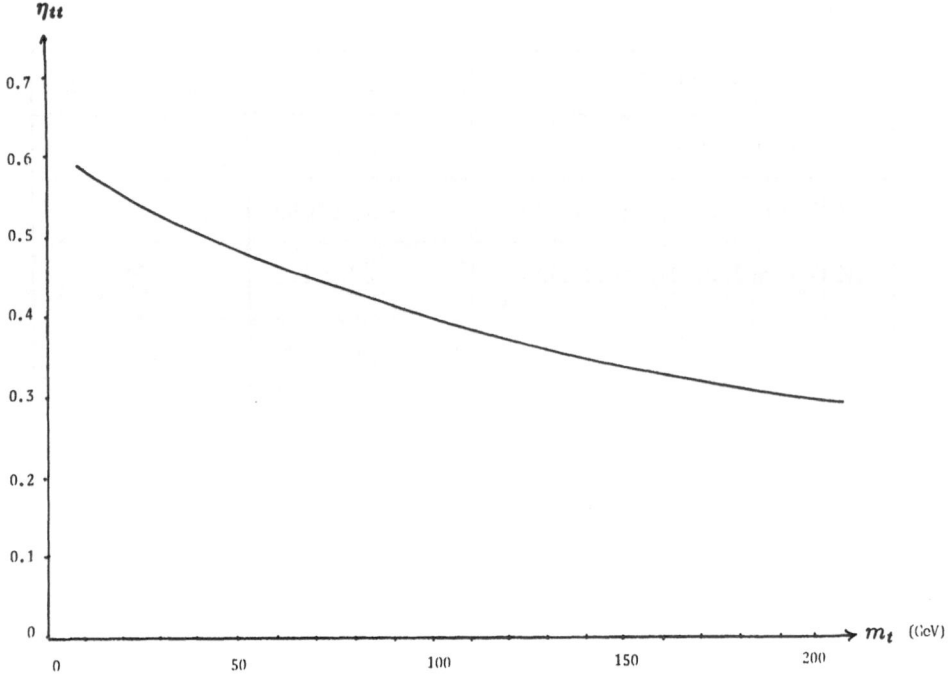

Figure 4 QCD coefficient η_{tt} as a function of m_t

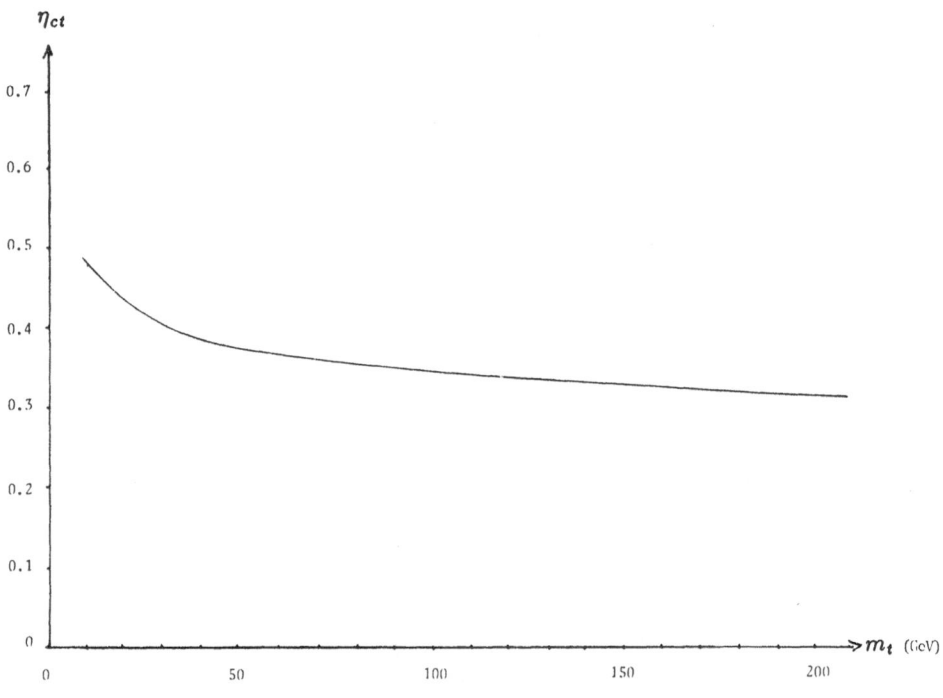

Figure 5 QCD coefficient η_{ct} as a function of m_t

where the coefficients η_{kl} are QCD corrections to the loop diagram.

The mass of the vertical quarks u_k and u_l appear through dimensionless parameters $z_k = (m_{u_k}/m_W)^2$ and the function $F(z_1, z_2)$ is given by

$$F(z_1, z_2) = z_1 z_2 \left\{ \left[1 - \frac{3z_1^2}{4(1-z_1)^2} \right] \frac{\log z_1}{z_1 - z_2} + \left[1 - \frac{3z_2^2}{4(1-z_2)^2} \right] \frac{\log z_2}{z_2 - z_1} \right. \tag{12}$$
$$\left. - \frac{3}{4} \frac{1}{(1-z_1)(1-z_2)} \right\}$$

The function $F(z_1, z_2)$ has a limit when $z_1 = z_2 = z$ and we write $F(z, z) \equiv zF(z)$ with

$$F(z) = 1 - \frac{3}{4} \frac{z(1+z)}{(1-z)^2} - \frac{3}{2} \frac{z^2}{(1-z)^3} \log z \tag{13}$$

The function $F(z)$ is represented on Figure 2 for quark masses between 0 and 200 GeV. When $z \ll 1$, $F(z)$ can be safely replaced by unity and such a situation occurs obviously for u and c quarks. For $z_1 \ll 1$ and $z_2 \gg z_1$ the expression of $F(z_1, z_2)$ can be replaced by $F(z_1, z_2) \simeq z_1 G(z_1, z_2)$ with

$$G(z_1, z_2) = \log \frac{z_2}{z_1} - \frac{3}{4} \frac{z_2}{1-z_2} \left[1 + \frac{z_2}{1-z_2} \log z_2 \right] \tag{14}$$

The function $G(z_c, z_t)$ is represented on Figure 3 with $m_c = 1, 5 \, \text{GeV}$ and $0 < m_t < 200 \, \text{GeV}$. The dashed line is simply the first logarithmic term in the right hand side of equation (14).

The matrix element $\mathcal{O}_{12}^{\text{box}}$ defined by

$$\mathcal{O}_{12}^{\text{box}} = < K^0 | \left[\bar{d}\gamma^\mu (1+\gamma_5)s \right] \left[\bar{d}\gamma_\mu (1+\gamma_5)s \right] | \overline{K^0} > \tag{15}$$

describes the quark content of the neutral K mesons and it is therefore associated to a non-perturbative part of QCD. A first estimate of $\mathcal{O}_{12}^{\text{box}}$ is obtained by using the vacuum insertion approximation (VI)and the result is

$$(\mathcal{O}_{12}^{\text{box}})_{\text{VI}} = -\frac{4}{3} f_K^2 m_K \tag{16}$$

where f_K is the decay coupling constant for the purely leptonic channel $K^+ \rightarrow l^+ + \nu_l$. In equation (16) the normalization is chosen so that the experimental value of f_K is $f_K = 169 \, \text{MeV}$.

The quantity $(\mathcal{O}_{12}^{\text{box}})_{\text{VI}}$ is now used as a reference scale for $\mathcal{O}_{12}^{\text{box}}$ and the parameter B_K defined by

$$\mathcal{O}_{12}^{\text{box}} = B_K (\mathcal{O}_{12}^{\text{box}})_{\text{VI}} \tag{17}$$

represents soft QCD corrections coming from long range contributions.

Inserting the expressions (16) and (17) of $\mathcal{O}_{12}^{\text{box}}$ we obtain the final form of m_{12}^{box}

$$m_{12}^{\text{box}} = \frac{G_F^2}{12\pi^2} m_K^2 f_K^2 B_K m_K \left[\sum_{u,c,t} \eta_{kl} F(z_k, z_l) \lambda_k \lambda_l \right]. \tag{18}$$

4) Various approaches have been used for estimating the parameter B_K and the results are summarized in the Table 1. The situation is somewhat confusing and we shall use a conservative range on the possible values for B_K.

$$\frac{1}{3} < B_K < 1 \tag{19}$$

5) The QCD correction factors have been computed by many authors and they exhibit a sensibility to large values of the unknown top mass m_t. We have represented on Figures 4 and 5 the coefficients η_{tt} and η_{ct} as estimated by Kaufman Steger and Yao for values of m_t up to $m_t = 200 \, \text{GeV}$. The coefficient η_{cc} turns out to be esssentially independent of m_t and we shall use in the numerical estimates $\eta_{cc} = 0, 7$.

6) Computation of $\operatorname{Re} m_{12}^{\text{box}}$

Neglecting second order effects in the violation of PC the mass difference $\Delta m = m_L - m_c$ is related to the real part of the non diagonal mass matrix element m_{12}

$$\Delta m = 2\operatorname{Re} m_{12} \tag{20}$$

Let us start first by evaluating $\operatorname{Re} m_{12}$ using the box diagram. From equations (10) the dominant contribution is due to cc exchange and we get

$$2\operatorname{Re} m_{12}^{\text{box}} \simeq \frac{G_F^2}{6\pi^2} f_K^2 \, B_K \, m_K \, m_c^2 \, \eta_{cc} \, \operatorname{Re} \lambda_c^2 \tag{21}$$

We compose this quantity with the experimental value of Δm and using

$$f_K = 169\,\text{MeV} \quad m_c = 1,5\,\text{GeV} \quad \lambda = 0,221 \quad \eta_{cc} = 0,7$$

we get

$$\frac{2\operatorname{Re} m_{12}^{\text{box}}}{\Delta m_{\exp}} = 0,71 \, B_K \left(\frac{m_c}{1,5\,\text{GeV}}\right)^2 \tag{22}$$

With a value of B_K less than unity it seems very difficult to reproduce the experimental value of Δm using only the box diagram contribution. The origin of such a difficulty might be the importance of long range effects associated to $\Delta S = 1$ transitions like $K \Rightarrow 2\pi$ or $K \Rightarrow \eta\pi$ which have not yet been accurately calculated.

7) Computation of $\operatorname{Im} m_{12}^{\text{box}}$

The quantity $\operatorname{Im} m_{12}$ enters in the determination of the PC violating parameter $\operatorname{Re} \epsilon$. The phases arising from the CKM matrix elements, with the usual parametrization where $\operatorname{Im} \lambda_u = 0$, only λ_c and λ_t have non-vanishing and opposite imaginary parts. As a consequence the virtual moments of the intermediate quarks are located in the $m_c - m_t$ range and for that reason the box diagram computation of $\operatorname{Im} m_{12}$ makes sense. We start with the expression

$$\operatorname{Im} m_{12}^{\text{box}} = \frac{G_F^2}{6\pi^2} m_W^2 \, f_K^2 \, B_K \, m_K \left[\sum_{c,t} \eta_{kl} \, F(z_k, z_l) \operatorname{Im} \lambda_k \lambda_l \right] \tag{23}$$

We eliminate λ_c by using the unitarity relation (9) and with the expressions (10) of λ_u and λ_t we get

$$\operatorname{Im} m_{12}^{\text{box}} = \frac{G_F^2}{6\pi^2} f_K^2 \, B_K \, m_K \, m_c^2 \, A^2 \, \lambda^6 a \sin\delta \, \{\} \tag{24}$$

where

$$\{\} = -\eta_{cc} + \eta_{ct} G(z_c, z_t) + A^2 \lambda^4 (1 - a\cos\delta)\eta_{tt} \left(\frac{m_t}{m_c}\right)^2 F(z_t) \tag{25}$$

The quantity $\operatorname{Im} m_{12}^{\text{box}}$ enters in $\operatorname{Re} \epsilon$ via the ratio $\operatorname{Im} m_{12}^{\text{box}}/2\Delta m$. Using as previously

$$f_K = 169\,\text{MeV} \quad m_c = 1,5\,\text{GeV} \quad A = 1,05 \pm 0,17 \quad \lambda = 0,221 \pm 0,002$$

and the experimental value of the mass difference Δm we get

$$\frac{\operatorname{Im} m_{12}^{\text{box}}}{2\Delta m} = 1,34 \, 10^{-3} \, \mathcal{E}_{\text{CKM}} \, B_K \, a \sin\delta \, \{\} \tag{26}$$

where the error factor $\mathcal{E}_{\text{CKM}}$ due to the uncertainties on A and λ varies in the range

$$0,63 < \mathcal{E}_{\text{CKM}} < 1,43 \tag{27}$$

The expression of $\operatorname{Re} \epsilon$ is made of two terms. It is remarkable that the magnitude of $\operatorname{Re} \epsilon$ as obtained from experiment is correctly reproduced by the box diagram calculation. The second term ξ_0 will be evaluated in the next section by using the so-called penguin diagram. Anticipating on the result we shall obtain the contribution to $\operatorname{Re} \epsilon$ due to ξ_0 turns out to be small and the box diagram term already dominates. Therefore $\sin_\delta$ has to be positive.

III - THE PENGUIN DIAGRAM

1) The penguin diagram describes $\Delta S = 1$ transitions and it is represented on Figure 6. The virtual quark u_k has the electric charge $Q = +\frac{2}{3}$: u,c,t.

The amplitude associated to the penguin diagram involves the product of two matrix elements of the CKM matrix in the form already defined for the box diagram amplitude

$$\lambda_k = V_{u_k d}^* V_{u_k s}$$

2) The penguin diagram of Figure 9 is computed by using the effective four quark Fermi hamiltonian and the result is

$$\mathcal{H}_{\text{eff}}^{\text{PENG}} = \sqrt{2}\, G_F \frac{\alpha_s}{12\pi} \{ \sum_{u,c,t} \lambda_k [\log z_k - \pi(z_k)] \} O^{\text{PENG}} \tag{28}$$

As previously $z_q = (m_q/m_W)^2$ and the function $\pi(z)$ is given by

$$\pi(z) = z \left[\frac{18 - 11z - z^2}{8(1-z)^3} + z \log z \frac{15 - 16z + 4z^2}{4(1-z)^4} \right] \tag{29}$$

For $z \ll 1$ we simply have $\pi(z) \simeq \frac{9}{4}z \ll 1$ and as a consequence we can neglect both $\pi(z_u)$ and $\pi(z_c)$ retaining only $\pi(z_t)$.

Using now the unitarity constraint of the CKM matrix $\lambda_u + \lambda_c + \lambda_t = 0$ we obtain an equivalent expression of $\mathcal{H}_{\text{eff}}^{\text{PENG}}$

$$\mathcal{H}_{\text{eff}}^{\text{PENG}} = \sqrt{2}\, G_F \frac{\alpha_s}{12\pi} \left\{ \lambda_t \left[\log\left(\frac{m_t}{m_u}\right)^2 - \pi(z_t) \right] + \lambda_c \log\left(\frac{m_c}{m_u}\right)^2 \right\} O^{\text{PENG}} \tag{30}$$

The operator O^{PENG} describes the quark structure of the involved mesons

$$O^{\text{PENG}} = 2 \left[\overline{d}\gamma^\mu (1 + \gamma_5) t^A s \right] \left[\sum_q \overline{q}\gamma_\mu t^A q \right] \tag{31}$$

where the 3×3 color matrices t^A are normalized so that

$$\text{Tr } t^A t^B = \frac{1}{2}\delta^{AB} \qquad A, B = 1, 2, \ldots, 8$$

3) The penguin diagram can be used for two main purposes

  i) a tentative understanding of the enhancement observed in the non leptonic $|\Delta \vec{I}| = \frac{1}{2}$ amplitude for K meson decay ; this refers to Re $\mathcal{H}_{\text{eff}}^{\text{PENG}}$.
  ii) an estimate of the PC violating amplitude Im A_0 and this aspect is obtained with Im $\mathcal{H}_{\text{eff}}^{\text{PENG}}$.

Of course both questions are related in magnitude and sign. However the short distance approach for $\mathcal{H}_{\text{eff}}^{\text{PENG}}$ is expected to work better for the imaginary part than for the real part.

We now restrict to the imaginary part of $\mathcal{H}_{\text{eff}}^{\text{PENG}}$. In the Maiani parametrization Im $\lambda_u = 0$ and therefore Im $\lambda_c = -\text{Im }\lambda_t$ and from equation (30) we get

$$\text{Im } \mathcal{H}_{\text{eff}}^{\text{PENG}} = \sqrt{2}G_F \frac{\alpha_s}{12\pi} \left\{ \text{Im } \lambda_t \left[\log\left(\frac{m_t}{m_c}\right)^2 - \pi(z_t) \right] \right\} O^{\text{PENG}} \tag{32}$$

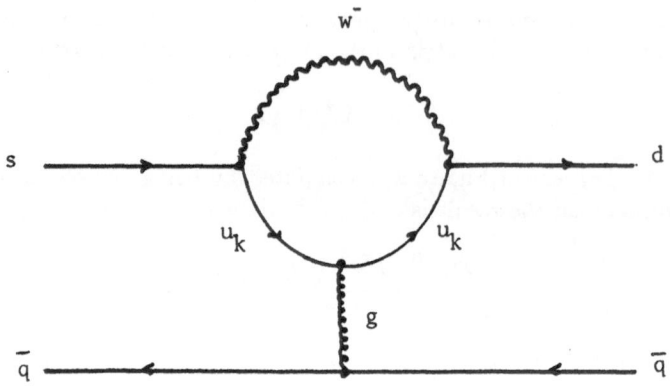

Figure 6 The penguin diagram

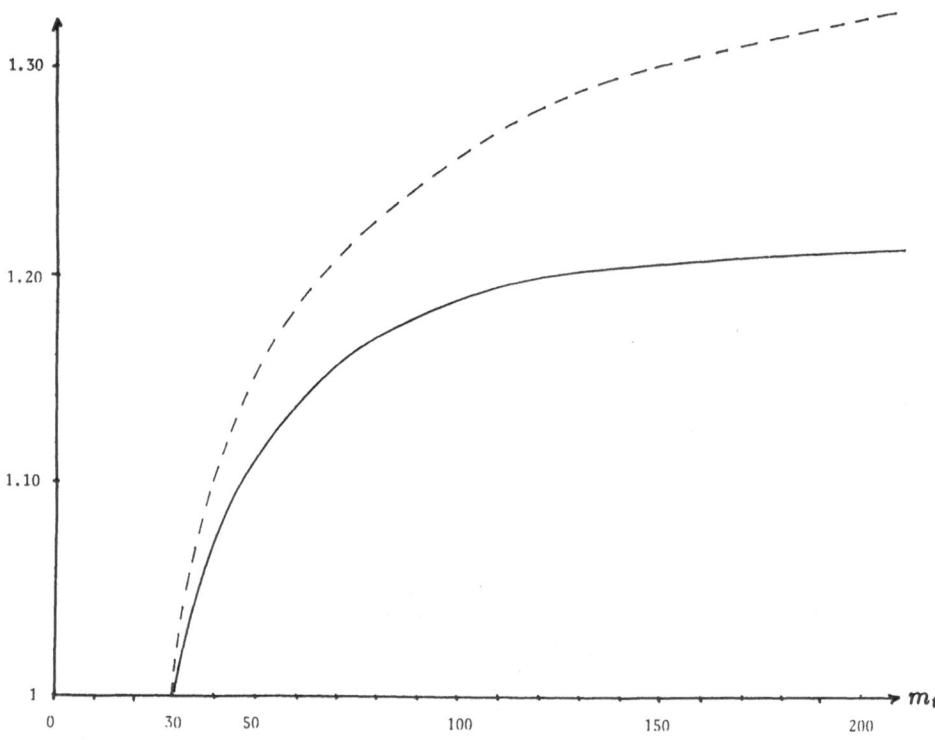

Figure 7 The function $C_I(m_t)$ for $30\,\text{GeV} < m_t < 200\,\text{GeV}$ normalized to unity at $m_t = 30\,\text{GeV}$. The dashed line is the function $\log(\frac{m_t}{m_c})^2 - \pi(z_t)$ with the same normalization

In the Wolfenstein parametrization we simply have

$$\text{Im}\,\lambda_t = -A^2\lambda^5\,a\,\sin\delta \tag{33}$$

4) Let us now discuss the QCD corrections to the expression (32). The mass of the top quark is expected to be large and the factor $\alpha_s/\pi \log\left(m_t/m_c\right)^2$ might be important. For such a reason it becomes necessary to make a resummation of the complete set of leading logarithms occuring in a more accurate QCD calculation. The result is taken from the work of Guberina and Peccei who introduce the coefficient a_I.

$$\begin{aligned}
a_I = &\left\{ K_t^{0,57}[0,0370\,K_c^{0,85} - 0,0308\,K_c^{0,42} - 0,0024\,K_c^{-0,13} - 0,0038\,K_c^{-0,35}] \right. \\
&\left. + K_t^{-0,29}[0,0124\,K_c^{0,85} + 0,0028\,K_c^{0,42} + 0,0140\,K_c^{-0,13} - 0,0292\,K_c^{-0,35}] \right\} \\
&\times \left\{ 0,8509\,K^{0,80} + 0,0091\,K^{0,42} + 0,1222\,K^{-0,12} + 0,00178\,K^{-0,30} \right\}
\end{aligned} \tag{34}$$

where

$$K = \frac{\alpha_s(\mu)^2}{\alpha_s(m_c^2)} \quad K_c = \frac{\alpha_s(m_c^2)}{\alpha_s(m_t^2)} \quad K_t = \frac{\alpha_s(m_t^2)}{\alpha_s(m_W^2)}$$

and

$$\alpha_s(q^2) = \frac{1}{1 + \dfrac{33 - 2n_f}{12\pi}\log\dfrac{q^2}{\Lambda_{QCD}^2}} \tag{35}$$

The various parameters are defined as usual

$\quad n_f$ number of flavours occuring at the energy $\sqrt{q}^2$

$\quad \mu$ substraction point parameter

$\quad \Lambda_{QCD}$ scale parameter

The expression of $\text{Im}\,\mathcal{H}_{\text{eff}}^{\text{PENG}}$ taking into account QCD corrections takes now the form

$$\text{Im}\,\mathcal{H}_{\text{eff}}^{\text{PENG}} = \sqrt{2}\,G_F\,O^{\text{PENG}}C_I(m_t)\text{Im}\,\lambda_t \tag{36}$$

where

$$C_I(m_t) = a_I\left[1 - \frac{\pi(z_t)}{\log(\frac{m_t}{m_c})^2}\right] \tag{37}$$

Of course C_I is also a function of m_c, μ and Λ_{QCD}.

The main effect of the QCD corrections is to decrease the importance of the top mass dependence as shown on Figure 7 where the function $C_I(m_t)$ and the original expression $\log(m_t/m_c)^2 - \pi(z_t)$ have been represented with an arbitrary normalization of unity at $m_t = 30\,\text{GeV}$.

Defining a new function $g(m_t)$ by

$$C_I(m_t) = C_I(30\,\text{GeV})g(m_t) \tag{38}$$

we have to compute $C_I(30\,\text{GeV})$ which depends on m_c, μ and Λ_{QCD}.

With $m_c = 1,5\,\text{GeV}$, $\mu = (0,5-1)\,\text{GeV}$ and $\Lambda_{QCD} = (100-200)\,\text{MeV}$, Altarelli and Franzini obtain

$$C_I(30\,\text{GeV}) = 0,075 \pm 0,031 \tag{39}$$

5) The penguin diagram of Figure 6 contributes only to the amplitude $|\Delta \vec{I}| = \frac{1}{2}$ hence to A_0. With the real part $\mathrm{Re}\, A_0$ taken from experiment and the imaginary part $\mathrm{Im}\, A_0$ from the penguin diagram we get

$$\xi_0 = \frac{\mathrm{Im}\, A_0}{\mathrm{Re}\, A_0} = \frac{\langle I = 0 \mid \mathrm{Im}\, \mathcal{H}_{\mathrm{eff}}^{\mathrm{PENG}} \mid K^0 \rangle}{\mathrm{Re}\, A_0} \qquad (40)$$

Using equation (36) the ratio ξ_0 is given in terms of the matrix element of the penguin operator O^{PENG}

$$\xi_0 = \sqrt{2}\, G_F \frac{\langle I = 0 \mid O^{\mathrm{PENG}} \mid K^0 \rangle}{\mathrm{Re}\, A_0} C_I(m_t) \mathrm{Im}\, \lambda_t \qquad (41)$$

With only $|\Delta \vec{I}| = \frac{1}{2}$ transitions we have

$$\langle I = 0 \mid O^{\mathrm{PENG}} \mid K^0 \rangle = \sqrt{\frac{3}{2}} \langle \pi^+ \pi^- \mid O^{\mathrm{PENG}} \mid K^0 \rangle \qquad (42)$$

For computing the matrix element of the penguin operator we use the vacuum saturation method. It is convenient to split the operator O^{PENG} into its lefthanded and righthanded parts $O^{PENG} \equiv O_R^P + O_L^P$ with

$$O_L^P{}_R = [\bar{d}\gamma^\mu (1 + \gamma_5) t^A s]\left[\sum_q \bar{q}\gamma_\mu (1 \pm \gamma_5) t^A q\right]$$

After Fierz rearrangement it turns out that the dominant contribution to the matrix element comes from O_R and one gets

$$\langle \pi^+ \pi^- \mid O_R^P \mid K^0 \rangle = \frac{32}{9}\Big\{ \langle \pi^+ \mid \bar{u}_R d_L \mid 0 \rangle \langle \pi^- \mid \bar{s}_L u_R \mid K^0 \rangle$$
$$+ \langle \pi^+ \pi^- \mid \bar{d}_R d_L \mid 0 \rangle \langle 0 \mid \bar{s}_L d_R \mid K^0 \rangle \Big\} \qquad (43)$$

The various terms occuring in equation (62) are now evaluated by using current algebra technics and PCAC. The results are

$$\langle \pi^+ \mid \bar{u}_R d_L \mid 0 \rangle = \frac{f_\pi m_\pi^2}{2(m_u + m_d)}$$

$$\langle 0 \mid \bar{s}_L d_R \mid K^0 \rangle = \frac{f_K m_K^2}{2(m_s + m_d)}$$

$$\langle \pi^- \mid \bar{s}_L u_R \mid K^0 \rangle = [f_+(m_K^2 - m_\pi^2) + f_- m_\pi^2]\frac{1}{2(m_s - m_u)}$$

$$\langle \pi^+ \pi^- \mid \bar{d}_R d_L \mid 0 \rangle = \frac{m_\pi^2}{2(m_u + m_d)}\Big[1 + \frac{(q_+ + q_-)^2}{m_\sigma^2}\Big]$$

where we have introduced a single scalar isoscalar resonance σ of effective mass m_σ. We then obtain

$$\langle \pi^+ \pi^- \mid O_R^P \mid K^0 \rangle \simeq \frac{8}{9}\frac{f_\pi m_K^2 m_\pi^2}{m_s(m_u + m_d)}\Big\{\frac{f_K}{f_\pi}\big(1 + \frac{m_K^2}{m_\sigma^2}\big) - 1\Big\} \qquad (44)$$

A numerical estimate of the matrix element is made with the scale $\mu = 1\,\mathrm{GeV}$

$$m_u + m_d \simeq (15,5 \pm 0,02)\,\mathrm{MeV} \qquad m_s = (200 \pm 33)\,\mathrm{MeV}$$

We choose m_σ in the range $m_\sigma \simeq (700 - 1000)\,\mathrm{MeV}$, we take f_π and f_K from experiment

$$f_\pi = 130\,\mathrm{MeV} \qquad f_K = 169\,\mathrm{MeV}$$

and we arrive to

$$\langle \pi^+\pi^- \mid O_R^P \mid K^0 \rangle = (0,125\,{}^{+0,069}_{-0,036})\,\mathrm{GeV}^3 \tag{45}$$

Using now equations (40), (41) and the value $\mathrm{Re}\,A_0 = 334\,\mathrm{eV}$ coming from a phenomenological analysis of the $K_S^0 \Rightarrow 2\pi$ experimental data we get

$$\sqrt{2}G_F \frac{\langle I = 0 \mid O^{\mathrm{PENG}} \mid K^0 \rangle}{\mathrm{Re}\,A_0} = 7,56\,{}^{+4,18}_{-2,18} \tag{46}$$

6) We are now in the position to have a numerical estimate of ξ_0 as written in equation (41). For the discussion it is convenient to introduce errors factors as follows
   i) for the QCD corrections (equation 39)

$$C_I(30\,\mathrm{GeV}) = 0,075\,\mathcal{E}_{\mathrm{QCD}} \qquad 0,59 < \mathcal{E}_{\mathrm{QCD}} < 1,41$$

   ii) for the penguin matrix element (equation 45)

$$\langle \pi^+\pi^- \mid O_R^P \mid K^0 \rangle = 0,125\,\mathrm{GeV}^3\,\mathcal{E}_{\mathrm{ME}} \qquad 0,71 < \mathcal{E}_{\mathrm{ME}} < 1,55$$

   iii) for the Cabibbo-Kobayashi-Maskawa term (equation 33) by using

$$A = 1,05 \pm 0,22 \qquad \lambda = 0,221 \pm 0,002$$

we get

$$A^2 \lambda^5 = 5,81\ 10^{-4}\mathcal{E}_{\mathrm{CKM}} \qquad 0,63 < \mathcal{E}_{\mathrm{CKM}} < 1,41$$

The final result is

$$\xi_0 = -3,29\ 10^{-4}\mathcal{E}_\xi\ g(m_t)[a\sin\delta] \tag{47}$$

where the error factor $\mathcal{E}_\xi = \mathcal{E}_{\mathrm{QCD}} \times \mathcal{E}_{\mathrm{ME}} \times \mathcal{E}_{\mathrm{CKM}}$ is given by

$$0,26 < \mathcal{E}_\xi < 3,08 \tag{48}$$

It is clear from equation (48) that the uncertainty for the computation of ξ_0 is larger than one order of magnitude.

7) In the usual phase convention chosen the amplitude A_2 is real and the imaginary part of A_0 is computed from the penguin diagram. However because of the expression of ϵ'

$$\epsilon' = i\omega\Big(\frac{\mathrm{Im}\,A_2}{\mathrm{Re}\,A_2} - \frac{\mathrm{Im}\,A_0}{\mathrm{Re}\,A_0}\Big)$$

a non zero value of $\mathrm{Im}\,A_2$ enters in ϵ' with an enhancement factor $\frac{\mathrm{Re}\,A_0}{\mathrm{Re}\,A_2}$ as compared to $\mathrm{Im}\,A_0$ and therefore it has to be carefully considered. The usual way to describe such effects is to define the quantity Ω by

$$\Omega = \frac{\xi_2}{\xi_0} \tag{49}$$

so that

$$\epsilon' = -i\omega(1 - \Omega)\xi_0 \tag{50}$$

Let us consider the isotopic spin breaking effects due to $m_u \neq m_d$ and which are directly related, in the SU(3) flavour language, to $\pi^0 - \eta - \eta'$ mixing. The decay amplitude $A(K^0 \Rightarrow \pi^0\pi^0)$ is affected as shown on Figure 8. Of course $A(K^0 \Rightarrow \pi^+\pi^-)$ is not modified.

In the framework of SU(3) flavour we have a configuration mixing between the singlet η_0 and the neutral isoscalar weight of the octet η_8. Such a mixing is described by the mixing angle θ_P and the physical states $\eta(548)$ and $\eta'(958)$ are given, in an obvious notation, by

$$\eta = \eta_8 \cos\theta_P - \eta_0 \sin\theta_P$$
$$\eta' = \eta_8 \sin\theta_P + \eta_0 \cos\theta_P$$

Using the Gell-Mann Okubo formula for squared masses we get $\theta_P = -10°$. Adding chiral loop corrections the value of θ_P becomes $= -20° \pm 4°$ in agreement with an estimate based on the $\frac{1}{N}$ expansion method which gives $\theta_P = -22°$. We shall use this last value in our numerical estimate.

According to Buras and Gerard the factor due to $\eta^0 - \eta - \eta'$ mixing has the following expression

$$\Omega_{\eta+\eta'} = \frac{1}{3\sqrt{2}}\frac{\mathrm{Re}\,A_0}{\mathrm{Re}\,A_2}\left\{(\cos\theta_P - \sqrt{2}\sin\theta_P)^2 + (\sin\theta_P + \sqrt{2}\cos\theta_P)^2\frac{m_\eta^2 - m_{\pi_0}^2}{m_{\eta'}^2 - m_{\pi_0}^2}\right\}\frac{m_d - m_u}{m_s} \tag{51}$$

Using $\theta_P = -22°$, $m_{\pi_0} = 134,96\,\mathrm{MeV}$ $m_\eta = 548,8\,\mathrm{MeV}$ $m_{\eta'} = 957,6\,\mathrm{MeV}$ and $\frac{m_u - m_d}{m_s} = 0,022$ we obtain

$$\Omega_{\eta+\eta'} = \frac{0,8797\ 10^{-2}}{|\omega|} \tag{52}$$

where

$$|\omega| = \frac{1}{\sqrt{2}}\frac{\mathrm{Re}\,A_2}{\mathrm{Re}\,A_0} \tag{53}$$

The quantity $|\omega|$ being few per cent the correction $\Omega_{\eta+\eta'}$ is important.

We are now interested in the electromagnetic correction associated to the penguin electromagnetic diagram of Figure 9 where the gluon has been replaced by a photon. In this case the amplitude $A(K^0 \Rightarrow \pi^+\pi^-)$ is affected and we get a non zero value for Ω_{em}

$$\Omega_{em} = \frac{\sqrt{2}}{3}\frac{\mathrm{Re}\,A_0}{\mathrm{Re}\,A_2}\alpha_{\mathrm{QED}}\frac{\langle\pi^+\pi^-\mid O_{em}^{\mathrm{PENG}}\mid K^0\rangle}{\langle\pi^+\pi^-\mid O^{\mathrm{PENG}}\mid K^0\rangle} \tag{54}$$

The numerical estimate of Buras and Gerard taking into account QCD corrections turns out to be very small and negative

$$\Omega_{em} \simeq -(2-3)10^{-3} \tag{55}$$

As a consequence these corrections can be neglected with respect to those coming from $\pi° - \eta - \eta'$ mixing and we write equation (52) in the form

$$|\omega|\xi_2 = 0,88\ 10^{-2}\xi_0 \tag{56}$$

IV - COMPARISON WITH EXPERIMENT

Let us first remark that on theoretical grounds we expect the ratio ϵ'/ϵ_0 to be quasi real and from experiment it has to be positive. Such a situation occurs when

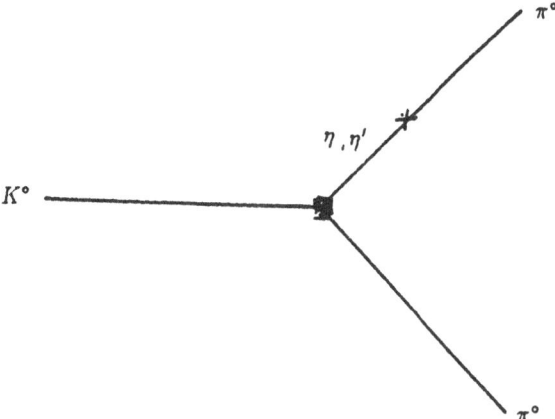

Figure 9 The electromagnetic penguin diagram

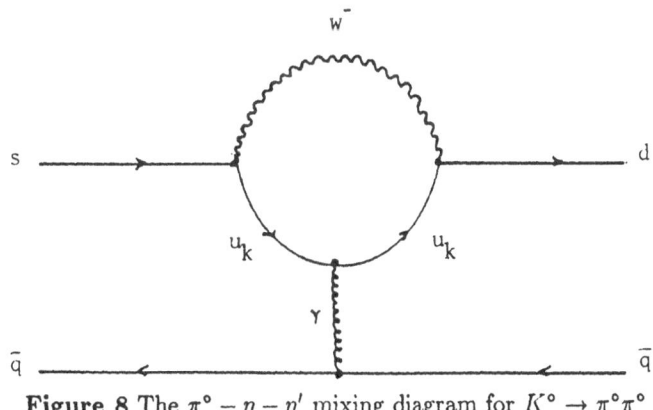

Figure 8 The $\pi^\circ - \eta - \eta'$ mixing diagram for $K^\circ \to \pi^\circ \pi^\circ$

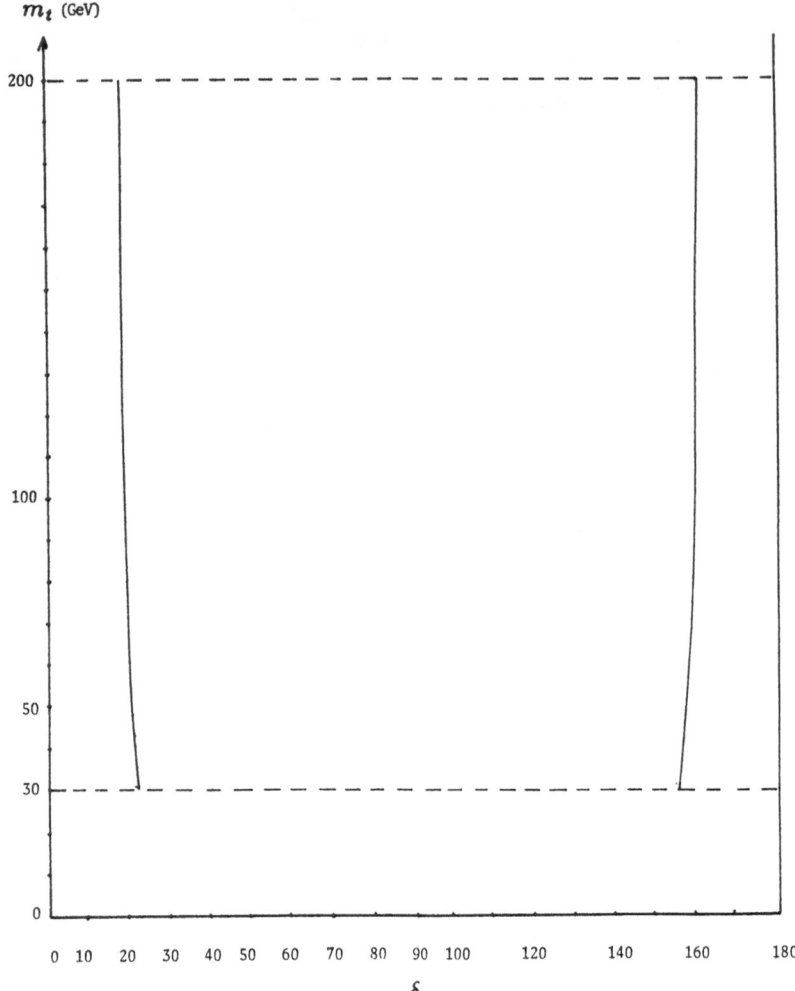

Figure 10 Correlation in the $m_t - \delta$ plane due to $|\epsilon'|$

$\xi_0 < 0$ or $\sin \delta > 0$. Such a result is hopefully in agreement with the one obtained for Re ϵ in the previous section.

In addition the sign of ξ_0 is relevant for the real part of $\mathcal{H}_{\text{eff}}^{\text{PENG}}$ which is supposed to partly explain the non leptonic $|\Delta \vec{I}| = \frac{1}{2}$ rule observed in $K_S^0 \Rightarrow 2\pi$ decay.

1) Parameter ϵ'

The parameter $|\epsilon'|$ is given by

$$|\epsilon'| = |\omega|[\xi_2 - \xi_0] \tag{57}$$

and using equation (75) we get

$$|\epsilon'| = [|\omega| - 0,88 \ 10^{-2}](-\xi_O) \tag{58}$$

We now take for $|\omega|$ the result of the phenomenological analysis of $K_S^0 \Rightarrow 2\pi$ decay

$$|\omega| = \left(2,30 \, {}^{+0,66}_{-0,47}\right) 10^{-2} \tag{59}$$

and we obtain

$$|\omega| - 0,88 \ 10^{-2} = 1,42 \ 10^{-2} \ \mathcal{E}_\omega \tag{60}$$

where

$$0,67 < \mathcal{E}_\omega < 1,46 \tag{61}$$

With equations (47) and (48) for ξ_0 the result is

$$|\epsilon'| = 4,67 \ 10^{-6} \ \mathcal{E} \ g(m_t)[a \sin \delta] \tag{62}$$

where now

$$0,17 < \mathcal{E} < 4,50 \tag{63}$$

Choosing now for $|\omega|$ the value $|\tilde{\omega}|$ implying the vanishing of the $|\Delta \vec{I}| = \frac{5}{2}$ transitions in $K \Rightarrow 2\pi$ decay

$$|\tilde{\omega}| = 3,21 \ 10^{-2} \tag{64}$$

the prediction for $|\epsilon'|$ becomes

$$|\tilde{\epsilon}'| = 7,67 \ 10^{-6} \ \mathcal{E}_\xi \ g(m_t)[a \sin \delta] \tag{65}$$

Experimentally we have

$$|\epsilon'|_{exp} = (7,44 \pm 2,42)10^{-6} \tag{66}$$

In the m_t, δ plane a lower bound curve can be obtained by using the maximal value of $\mathcal{E}$ and $a = 0,6$. From equations (62) and (63) we get the relation corresponding to the one standard lower limit of $|\epsilon'|_{exp}$

$$\sin \delta \ g(m_t) > 0,398 \tag{67}$$

The corresponding domain is represented on Figure 10. The largest range for δ is obtained for $m_t = 200 \,\text{GeV}$

$$17° < \delta < 163° \tag{68}$$

but due to the fact that $g(m_t)$ is a smooth function of m_t as shown on Figure 9 the range of allowed values of δ is poorly sensible to m_t.

If, instead of the value $|\omega|$ of equation (59) we take $|\tilde{\omega}|$ given in equation (64) the lower bound (67) reduces to $0,354$ and the range of allowed of δ turns out to be slightly larger

$$8° < \delta < 172° \tag{69}$$

213

2) Parameter Re ϵ

Using for simplicity $\alpha_K = 45°$ in the formula (64) of Part A we have

$$\mathrm{Re}\,\epsilon \simeq \tfrac{1}{2}\Big[\frac{\mathrm{Im}\,m_{12}}{\Delta_m} + \xi_0\Big] \tag{70}$$

The first term of the bracket has been computed in the box diagram approximation and using the results of Section II we have

$$\frac{\mathrm{Im}\,m_{12}^{box}}{\Delta_m} = 20,868\,B_K\,A^2\lambda^5 a\sin\delta\{\} \tag{71}$$

where the bracket $\{\}$ is given in equation (25). The second term ξ_0 has been computed in the penguin diagram approximation and using now the results of Section III we have

$$\xi_0 = -0,567\,\mathcal{E}_0\,A^2\lambda^5 a\sin\delta g(m_t) \tag{72}$$

with the error factor $\mathcal{E}_0$ given by

$$0,42 < \mathcal{E}_0 < 2,19 \tag{73}$$

Inserting the value of Re ϵ obtained from a phenomenological analysis of the experimental data

$$2\mathrm{Re}\,\epsilon_{exp} = 3,23\,10^{-3} \tag{74}$$

we obtain the relation

$$1,21 = \mathcal{E}_{\mathrm{CKM}} a\sin\delta[B_K\{\} - 0,123\,\mathcal{E}_0\,g(m_t)] \tag{75}$$

with as given in equation (27)

$$0,63 < \mathcal{E}_{\mathrm{CKM}} < 1,43 \tag{76}$$

The equation (75) gives a correlation between the unknown value of the top mass m_t and the PC violating CKM phase δ. Such a correlation is represented as a lower bound curve in the $m_t - \delta$ plane using $\mathcal{E}_{\mathrm{CKM}} = 1,43$, $\mathcal{E}_0 = 0,42$ and $a = 0,6$ for three values of B_K, $B_K = \frac{1}{3}$ in Figure 11, $B_K = \frac{2}{3}$ in Figure 12 and $B_K = 1$ on Figure 13.

For $B_K = \frac{1}{3}$ we obtain a lower bound on m_t at $\delta = 105°$

$$m_t \geq 60\,\mathrm{GeV} \tag{77}$$

and a restricted range of values of δ for $m_t = 200\,\mathrm{GeV}$

$$37° < 5 < 166° \tag{78}$$

For $B_K = \frac{2}{3}$ the lower bound on m_t is $25\,\mathrm{GeV}$ and therefore below our threshold of $30\,\mathrm{GeV}$ and the range of allowed values of δ increases

$$20° < 5 < 172° \tag{79}$$

For $B_K = 1$ again the lower bound on m_t is below $30\,\mathrm{GeV}$ and the range of allowed values of δ is slightly larger

$$14° < \delta < 175° \tag{80}$$

V - CONCLUSIONS

1) Improve the theoretical estimate of the parameter B_K
2) Improve the semi-phenomenological determination of the CKM parameters A and a
3) Make a reliable estimate of long range contributions to Δm
4) Improve the theoretical calculation of the matrix element of the penguin operator

MEASURE THE TOP QUARK MASS

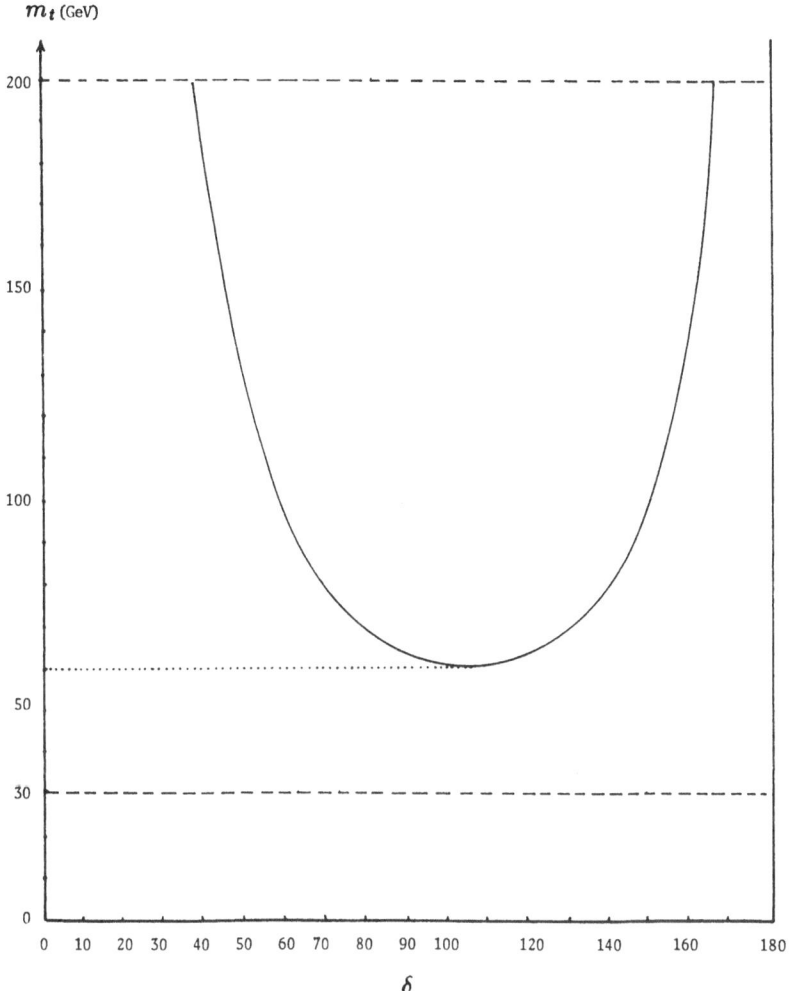

Figure 11 Correlation in the $m_t - \delta$ plane due to $\mathrm{Re}\,\epsilon$ for $B_K = \frac{1}{3}$

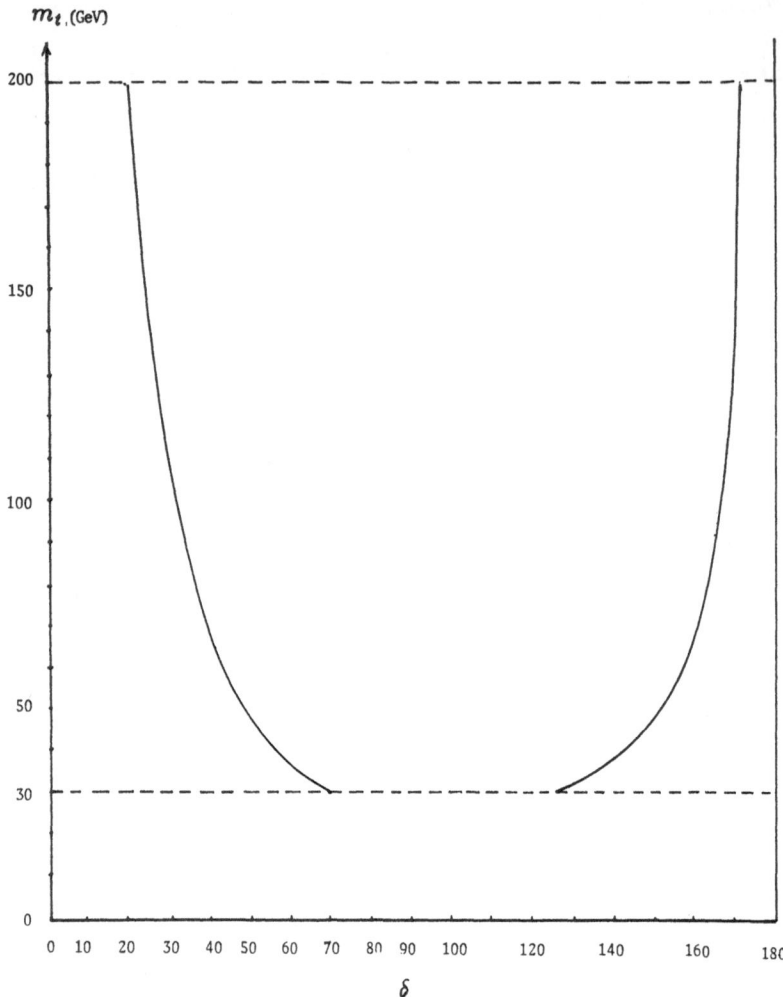

Figure 12 Correlation in the $m_t - \delta$ plane due to $\mathrm{Re}\,\epsilon$ for $B_K = \frac{2}{3}$

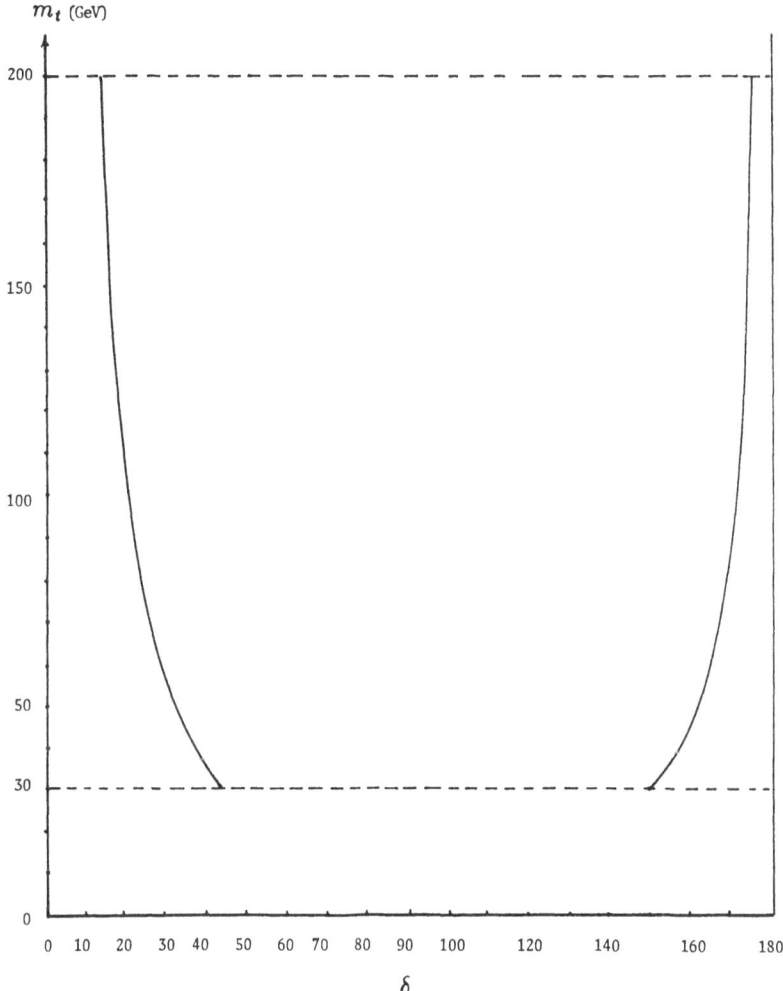

Figure 13 Correlation in the $m_t - \delta$ plane due to Re ϵ for $B_K = 1$

REFERENCES FOR BOX DIAGRAM CALCULATIONS IN THE STANDARD MODEL

M.K. Gaillard and B.W. Lee   Phys. Rev. **D10** 897 (1974)

J. Ellis, M.K. Gaillard and D.V. Nanopoulos   Nucl. Phys. **B109** 213 (1976)

S. Peksava and S. Sugawara   Phys. Rev. **D14** 305(1976)

L. Maiani   Phys. Lett. **62B** 183 (1976)

L. Wolfenstein   Nucl. Phys. **B160** 501 (1979)

C.T. Hill   Phys. Lett. **97B** 275 (1980)

F.J. Gilman and M.B. Wise
   Phys. Lett. **93B** 129 (1980), Phys. Rev. **D27** 1128 (1983)

J.S. Hagelin   Phys. Rev. **D23** 119 (1981)

T. Inami and C.S. Lim   Prog. Theor. Phys. **65** 297 (1981)

L.L. Chau   Phys. Reports **95** 1 (1983)

J.F. Donoghue, E. Golowich and B. Holstein   Phys. Rev. **131** 319 (1986)

I. Picek   Preprint DESY 86-140 (November 1986)

REFERENCES FOR THE CABIBBO-KOBAYASHI-MASKAWA MATRIX

M. Kobayashi and T. Maskawa   Prog. Theor. Phys. **49** 652 (1973)

L. Maiani   Proceedings of the International Symposium on Leptons and Photon
   Interactions at High Energies Hamburg (1977)

L. Wolfenstein   Phys. Rev. Lett. **51** 1945 (1983)

REFERENCES FOR B_K

[1] J.F. Donoghue, E. Golowich and B.R. Holstein   Phys. Lett. **119B** 412 (1982)

[2] A. Pich and E. de Rafael   Phys. Lett. **158B** 477 (1985)

[3] K.G. Chetyzkin, A.L. Kataev, A.B. Krasulin, H.A. Pivoravov
   Phys. Lett. **174B** 104 (1986)

[4] R. Decker   Nucl. Phys. **B277** 661 (1986)

[5] L.J. Reinders and S. Yasaki   Nucl. Phys. **B288** 789 (1987)

[6] N. Bilic, C.A. Dominguez, B. Guberina   Preprint Desy 87/162 (Dec. 1987)

[7] W.A. Bardeen, A.J. Buras and J.M. Gerard   Preprint MPI-PAE/PTh 22/87

M.B. Gavela, L. Maiani, S. Petrarca, F. Rapuano, G. Martinelli, O Pene,
   C.T. Sachrajda   Preprint CERN TH 4905/87 (Nov. 1987)

REFERENCES FOR THE QCD COEFFICIENTS

M.I. Vysotskii   Sov. J. Nucl. Phys. **31** 797 (1980)

F.J. Gilman and M.B. Wise   Phys. Rev.**D27** 1128 (1983)

J. Ellis and J.S. Hagelin   Nucl. Phys. **B217** 189 (1983)

A.J. Buras, W. Slominski and H. Steger
Nucl. Phys. **B238** 529 (1984) **245** 369 (1984)

W.A. Kaufman, H. Steger and Y.P. Yao
University of Michigan preprint UM-TH-87-13 (July 1987)

REFERENCES FOR THE PENGUIN DIAGRAM

M.A. Shifman, A.J. Vainshten and V.I. Zakharov   JETP Lett. **22** 55 (1975), Nucl.
Phys. **B120** 316 (1977), JETP Sov. Phys. **45** 670 (1977)

F.J. Gilman and M.B. Wise
Phys. Lett. **83B** 83 (1979), Phys. Rev. **D20** 2392 (1979), **D27** 1128 (1983)

B. Guberina and R. Peccei   Nucl. Phys. **B163** 289 (1980)

C.T. Hill   Phys. Lett. **97B** 275 (1980)

J.S. Hagelin   Nucl. Phys. **B193** 123 (1981)

R.D.C. Miller and B.H.J. Mc Kellar   Aust. J. Phys. **35** 235 (1982), J. Phys. G7
L247 (1981), Phys. Rep. 106C (1984)

F.J. Gilman and J.S. Hagelin   Phys. Lett. **126B** 111 (1983)

F. Bijnens and M.B. Wise   Phys. Lett. **137B** 245 (1984)

J.F. Donoghue, E. Golowich, B.R. Holstein and J. Trampetic
Phys. Lett. **179B** 361 (1986), **188B** 511 (1987)

A.J. Buras and J.M. Gérard   Phys. Lett. **192B** 156 (1987), **203B** 272 (1988)

Xuequio-Li, D. Wu, Z. Zhang   Preprint CERN, CERN TH 4793/87 (July 1987)

G. Altarelli and P. Franzini   Preprint CERN, CERN TH 4914/87 (Dec. 1987)

PART C - THE $B^0\overline{B}^0$ SYSTEMS

EXPERIMENTS AND THE MINIMAL STANDARD MODEL

I - GENERAL FORMALISM FOR $B^0\overline{B}^0$ MIXING

1) Basic equations

The $B^0\overline{B}^0$ system is described by an effective hamiltonian H which is written as usual as

$$H = M - i\frac{\Gamma}{2}$$

where the two hermitian matrices M and Γ are respectively the mass matrix and the decay matrix.

Assuming TCP invariance we have the following form for M and Γ in the $|B^0\rangle$, $|\overline{B}^0\rangle$ basis

$$M = \begin{vmatrix} m & m_{12} \\ m_{12}^* & m \end{vmatrix} \qquad \Gamma = \begin{vmatrix} \gamma & \gamma_{12} \\ \gamma_{12}^* & \gamma \end{vmatrix}$$

We now diagonalize the matrix H

$$H|B_\pm^0\rangle = (m_\pm - i\frac{\gamma_\pm}{2})|B_\pm^0\rangle \tag{1}$$

The eigenstates $|B_\pm^0\rangle$ are defined by one complex parameter ϵ

$$|B_\pm^0\rangle = \frac{1}{\sqrt{2(1+|\epsilon|^2)}}[(1+\epsilon)|B^0\rangle \pm (1-\epsilon)|\overline{B}^0\rangle] \tag{2}$$

they are not orthogonal and their hermitian product is real because of TCP invariance

$$\langle B_+^0 \mid B_-^0 \rangle = \frac{2\mathrm{Re}\,\epsilon}{1+|\epsilon|^2,} \tag{3}$$

The eigenvalues $m_\pm$, $\gamma_\pm$ are measurable quantities it is convenient to write in the form

$$m_\pm = m \mp \frac{\Delta m}{2}$$
$$\gamma_\pm = \gamma \mp \frac{\Delta\gamma}{2} \tag{4}$$

The constraints arising from the eigenvalue equation (1) are given in the form of two complex equations

$$\left(\frac{1+\epsilon}{1-\epsilon}\right)^2 = \frac{m_{12} - \frac{i}{2}\gamma_{12}}{m_{12}^* - \frac{i}{2}\gamma_{12}^*} \tag{5}$$

$$(\Delta m - i\frac{\Delta\gamma}{2})^2 = 4(m_{12} - \frac{i}{2}\gamma_{12})(m_{12}^* - \frac{i}{2}\gamma_{12}^*) \tag{6}$$

2) Solution for Δm and $\Delta\gamma$

In the $K^0\overline{K}^0$ case the width difference $\gamma_S - \gamma_L$ and the mass difference $m_L - m_S$ have similar values. Moreover the two lifetimes τ_S and τ_L differ by three orders of magnitude. The situation is different in the $B^0\overline{B}^0$ case where we have a large available phase space. The two lifetimes $\tau_{B_+^0}$ and $\tau_{B_-^0}$ have no reason to be drastically different and we must have $|\Delta\gamma| \ll \gamma$. On the other hand the mass difference Δm is probably larger than $m_L - m_S$ and we expect to be in the situation where

$$|\frac{\Delta\gamma}{\Delta m}| \ll 1 \tag{7}$$

We shall discuss later the a posteriori validity of this assumption in the standard model.

Let us first consider the relation (6) which is equivalently writen as

$$(\Delta m - i\frac{\Delta\gamma}{2})^2 = 4|m_{12}|^2 \left\{ 1 - \frac{1}{4}\left|\frac{\gamma_{12}}{m_{12}}\right|^2 - i\,\mathrm{Re}\,\frac{\gamma_{12}}{m_{12}} \right\}$$

In the approximation (7) we also have $\left|\frac{\gamma_{12}}{m_{12}}\right|^2 \ll 1$ and we immediately obtain

$$\Delta m - i\frac{\Delta\gamma}{2} \simeq 2|m_{12}| \left\{ 1 - \frac{i}{2}\,\mathrm{Re}\,\frac{\gamma_{12}}{m_{12}} \right\}$$

corresponding to the two relations

$$\Delta m = 2|m_{12}| \tag{8}$$

$$\frac{\Delta\gamma}{\Delta m} = \mathrm{Re}\left(\frac{\gamma_{12}}{m_{12}}\right) \tag{9}$$

Let us notice that we have chosen the eigenstates $|B_+^0\rangle$ and $|B_-^0\rangle$ such that $\Delta m = m_- - m_+ > 0$. Of course $\Delta\gamma = \gamma_- - \gamma_+$ is either positive or negative.

3) Solution for $\mathrm{Re}\,\epsilon$

The complex parameter ϵ introduced in the equation (1) depends on the choice of phases done for the basic vectors $|B^0\rangle$ and $|\overline{B}^0\rangle$. Therefore ϵ is not an observable quantity. However the hermitian product $\langle B_+^0 \mid B_-^0 \rangle$ as given in equation (3) can be experimentally determined. Such a quantity is precisely the one appearing in the modulus of the left hand side of equation (5)

$$\left|\frac{1+\epsilon}{1-\epsilon}\right|^2 = \frac{1 + \frac{2\mathrm{Re}\,\epsilon}{1+|\epsilon|^2}}{1 - \frac{2\mathrm{Re}\,\epsilon}{1+|\epsilon|^2}}$$

and, from equation (5) we get

$$\frac{2\mathrm{Re}\,\epsilon}{1+|\epsilon|^2} = \frac{|m_{12} - \frac{i}{2}\gamma_{12}| - |m_{12}^* - \frac{i}{2}\gamma_{12}^*|}{|m_{12} - \frac{i}{2}\gamma_{12}| + |m_{12}^* - \frac{i}{2}\gamma_{12}^*|} \tag{10}$$

The calculation is analogous to the previous one and using again the assumption $|\frac{\gamma_{12}}{m_{12}}|^2 \ll 1$ we obtain

$$\frac{\mathrm{Re}\,\epsilon}{1+|\epsilon|^2} = \frac{1}{4}\mathrm{Im}\left(\frac{\gamma_{12}}{m_{12}}\right) \tag{11}$$

and to lowest order in ϵ

$$\mathrm{Re}\,\epsilon \simeq \frac{1}{4}\mathrm{Im}\left(\frac{\gamma_{12}}{m_{12}}\right) \tag{12}$$

221

II - TIME EVOLUTION OF B^0 AND $\overline{B}^0$ BEAMS

1) Let us produce by a strong interaction mechanism at $t = 0$ a B^0 meson or a $\overline{B}^0$ meson. We are now interested to what is happening at a time $t > 0$.

The time evolution is simple in terms of eigenstates $B^0_\pm$

$$|B^0_\pm(t)\rangle = e^{-i(m_\pm - \frac{i}{2}\gamma_\pm)t}|B_\pm(0)\rangle$$

The two $B^0\overline{B}^0$ and $B^0_+ B^0_-$ basis are related by the PC violating parameter ϵ and inverting equations (2) we get

$$|B^0\rangle = \sqrt{\frac{1 + |\epsilon|^2}{2}} \frac{1}{1 + \epsilon}\left[|B^0_+\rangle + |B^0_-\rangle\right]$$

$$|\overline{B}^0\rangle = \sqrt{\frac{1 + |\epsilon|^2}{2}} \frac{1}{1 - \epsilon}\left[|B^0_+\rangle + |B^0_-\rangle\right]$$

Therefore the time evolution of a beam which is purely B^0 or $\overline{B}^0$ at $t = 0$ becomes

$$|B^0(t)\rangle = f_+(t)|B^0\rangle + \frac{1 - \epsilon}{1 + \epsilon}f_-(t)|\overline{B}^0\rangle$$

$$|\overline{B}^0(t)\rangle = f_+(t)|\overline{B}^0\rangle + \frac{1 + \epsilon}{1 - \epsilon}f_-(t)|B^0\rangle \tag{13}$$

where the function $f_\pm(t)$ are given by

$$f_\pm(t) = \tfrac{1}{2}\left[e^{-i(m_+ - \frac{i}{2}\gamma_+)t} \pm e^{-i(m_- - \frac{i}{2}\gamma_-)t}\right] \tag{14}$$

By using now the decomposition (4) for the eigenvalues $m_\pm$ and $\gamma_\pm$ we get

$$f_\pm(t) = \tfrac{1}{2}e^{-imt}e^{-\frac{1}{2}\gamma t}\left[e^{+\frac{i}{2}\Delta mt}e^{+\frac{1}{2}\Delta\gamma t} \pm e^{-\frac{i}{2}\Delta mt}e^{-\frac{1}{2}\Delta\gamma t}\right] \tag{15}$$

In the approximation where $\Delta\gamma$ can be neglected we simply have

$$f_+(t) = e^{-imt}e^{-\frac{1}{2}\gamma t}\cos\tfrac{1}{2}\Delta mt$$

$$f_-(t) = ie^{-imt}e^{-\frac{1}{2}\gamma t}\sin\tfrac{1}{2}\Delta mt \tag{16}$$

The various effects due to the $B^0\overline{B}^0$ oscillations will then appear through 2 real and 1 complex quantity

$$|f_+(t)|^2 = e^{-\gamma t}\cos^2\tfrac{1}{2}\Delta mt$$

$$|f_-(t)|^2 = e^{-\gamma t}\sin^2\tfrac{1}{2}\Delta mt$$

$$f_+(t)f_-^*(t) = \frac{1}{2i}e^{-\gamma t}\sin\Delta mt \tag{17}$$

2) Because of the smallness of the $B^0_\pm$ lifetimes $O(10^{-13}s)$ most of the observations will take place at a time $t \gg \frac{1}{\gamma}$. In this case what will be measured are time integrated quantities. We simply have in the approximation where $\Delta\gamma$ is neglected

$$\int_0^\infty |f_+(t)|^2 dt = \frac{1}{2\gamma}\frac{2 + x^2}{1 + x^2}$$

$$\int_0^\infty |f_-(t)|^2 dt = \frac{1}{2\gamma}\frac{x^2}{1 + x^2} \tag{18}$$

$$\int_0^\infty f_+(t)f_-^*(t)dt = \frac{1}{2\gamma}\frac{-ix}{1 + x^2}$$

222

where the dimensionless ratio x is defined by

$$x = \frac{\Delta m}{\gamma} \tag{19}$$

III - EXPERIMENTAL SITUATION FOR LEPTONIC ASYMMETRIES

1) Let us consider the semi-leptonic decay of B^0 and $\overline{B}^0$ mesons

$$B^0, \overline{B}^0 \Rightarrow l^{\pm} \nu_l X^{\mp}$$

and let us assume the validity of the $\Delta B = \Delta Q$ rule. We have only two allowed amplitudes

$$B^0 \Rightarrow l^- \overline{\nu}_l X^+ \qquad \overline{B}^0 \Rightarrow l^+ \nu_l X^-$$

However because of the mixing of B^0 and $\overline{B}^0$ it is possible starting at $t = 0$ with a pure B^0 or $\overline{B}^0$ state to observe, at a time $t > 0$, a wrong sign lepton. We then define two ratios

$$r = \frac{\Gamma(B^0(t) \to l^+ \nu_l X^-)}{\Gamma(B^0(t) \to l^- \overline{\nu}_l X^+)} \qquad \overline{r} = \frac{\Gamma(\overline{B}^0(t) \to l^- \overline{\nu}_l X^+)}{\Gamma(\overline{B}^0(t) \to l^+ \nu_l X^-)} \tag{20}$$

or, equivalently

$$\chi = \frac{\Gamma(B^0(t) \to l^+ \nu_l X^-)}{\Gamma(B^0(t) \to l^{\pm} \nu_l X^{\mp})} \qquad \overline{\chi} = \frac{\Gamma(\overline{B}^0(t) \to l^- \overline{\nu}_l X^+)}{\Gamma(\overline{B}^0(t) \to l^{\pm} \nu_l X^{\mp})} \tag{21}$$

Of course

$$\chi = \frac{r}{1+r} \qquad \overline{\chi} = \frac{\overline{r}}{1+\overline{r}} \tag{22}$$

2) Using now the results of the time integration given in equation (18) we get

$$r = \left|\frac{1-\epsilon}{1+\epsilon}\right|^2 \frac{x^2}{2+x^2} \qquad \overline{r} = \left|\frac{1+\epsilon}{1-\epsilon}\right|^2 \frac{x^2}{2+x^2} \tag{23}$$

Therefore r and $\overline{r}$ are different if and only if $\epsilon \neq 0$ The quantity $\left|\frac{1-\epsilon}{1+\epsilon}\right|^2$ is clearly governed by $\frac{\text{Re}\,\epsilon}{1+|\epsilon|^2}$ which is expected to be very small for the $B^0 \overline{B}^0$ systems. In the present status of experiments it can be safely neglected. We shall come back on this point in part D. In this approximation we have

$$r = \overline{r} = \frac{x^2}{2+x^2}$$

$$\chi = \overline{\chi} = \frac{x^2}{2(1+x^2)} \tag{24}$$

with the positivity bounds

$$0 \leq r \leq 1 \qquad 0 \leq \chi \leq 0,5 \tag{25}$$

3) Double oscillations

Producing a $B^0 \overline{B}^0$ state at the time $t = 0$ we can observe at the time $t > 0$ dileptons, in the final state, of opposite charges or of same charge because of the possible transformations $B^0 \Rightarrow \overline{B}^0$ and $\overline{B}^0 \Rightarrow B^0$ due to CP violation. We then define the parameter R

$$R = \frac{N(l^+l^) + N(l^-l^-)}{N(l^+l^-)} \tag{26}$$

and we have two possible situations

    i) the production of B^0 and $\overline{B}^0$ is coherent, as coming, for instance, from the decay of the Υ (4S) resonance which is a $J^P = 1^-$ state, then we get

$$R = \frac{\chi}{1-\chi} = r = \frac{x^2}{2+x^2} \tag{27}$$

    ii) the B^0 and the $\overline{B}^0$ are not created in a state of definite angular momentum and the transformations $B^0 \Rightarrow \overline{B}^0$ and $\overline{B}^0 \Rightarrow B^0$ are independent. Then the relevant relations are different

$$R = \frac{2\chi(1-\chi)}{\chi^2 + (1-\chi)^2} = \frac{2r}{1+r^2} = \frac{x^2(2+x^2)}{2+2x^2+x^4} \tag{28}$$

4) Experimental results coming from the Υ (4S) decay
    The data correspond to the $B_d^0 \overline{B}_d^0$ mixing
      i) ARGUS

$$r_d = 0,21 \pm 0,08 \tag{29}$$

      ii) CLEO

$$r_d = 0,17 \pm 0,05 \tag{30}$$

Combining both results we obtain in the 90% C.L.

$$\begin{aligned} 0,09 &< r_d < 0,24 \\ 0,44 &< x_d < 0,79 \end{aligned} \tag{31}$$

5) Experimental data coming from the continuum
    In this case the $B_d^0 \overline{B}_d^0$ and the $B_s^0 \overline{B}_s^0$ systems have not get been separated and only the effective parameter χ_{eff} has been measured

$$\chi_{\text{eff}} = \frac{P_s BR_s \chi_s + P_d BR_d \chi_d}{\langle BR \rangle} \tag{32}$$

where $\begin{array}{l} P_q \text{ is the fraction of } b \text{ jets containing } B_q^0 \\ BR_q \text{ is the fraction of semi-leptonic decays for } B_q^0 \end{array}$ $q = d, s.$

    i) UA1 $p\overline{p}$ annihilation at $\sqrt{s} = 650 GeV$

$$\chi_{\text{eff}} = 0,121 \pm 0,047 \tag{33}$$

    and in the 90% confidence level $\chi_{\text{eff}} > 0,065$

    ii) MARK 2 e^+e_- annihilation at $\sqrt{s} = 29$ GeV No effect has been seen and in the 90% confidence level $\chi_{\text{eff}} < 0,12$

Combining both results —if justified— we obtain in the 90% C.L.

$$0,065 < \chi_{\text{eff}} < 0,12 \tag{34}$$

III - BOX DIAGRAMS FOR m_{12}

1) The box diagram for the $\Delta B = 2$, $B^0 - \overline{B}^0$, transition are very similar to those of the $K^0 - \overline{K}^0$ system and they are represented on the Figure 1. The light quark q is either d or s and the virtual quarks u_k and u_l have the electric charge $Q = +\frac{2}{3}$: u, c, t.

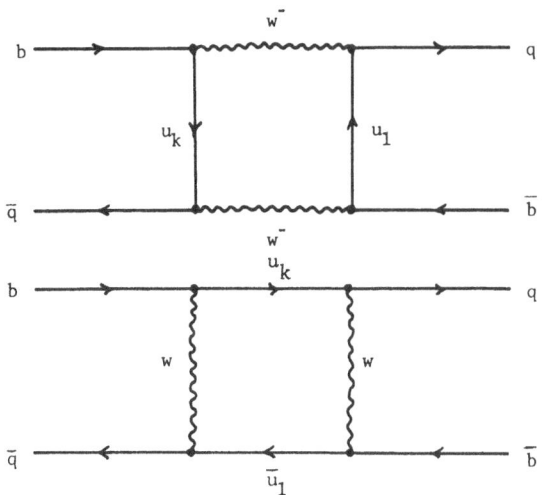

Figure 1 The box diagrams for the transition $\overline{B}^0 \Rightarrow B^0$ (m_{12})

TABLE 1

METHOD	f_{B_d}(MeV)	f_{B_s}(MeV)	REFERENCE
Potential Model	125	175	(1)
	160	200	(2)
Bag Model	100		(3)
	60–148	148–198	(4)
	86	96	(5)
Mock-meson	106–191	140–236	(6)
Harmonic Oscillator	84	106	(6)
Non-Relativistic	58	68	(6)
Hyperfine Mass-Splitting	75–126	86–146	(7)
QCD Sum Rule	140		(8)
	90		(9)
	130		(10)
	230		(11)
	190 ± 30	210 ± 30	(12)
	100–110		(13)
	182 ± 18		(14)
	147–212		(15)
	129 ± 13		(16)
Lattice QCD	120	150	(17)
	90 ± 32	162 ± 42	(18)

2) Case $B_d^0 \overline{B}_d^0$

We define the parameters $\lambda_k = V_{u_k d}^* V_{u_k b}$ and we use the parametrization of Wolfenstein for the CKM matrix elements

$$\left| \begin{array}{l} \lambda_u = A\,a\,\lambda^3 e^{i\delta} \\ \lambda_c = -A\lambda^3 \\ \lambda_t = A\lambda^3(1 - ae^{-i\delta}) \end{array} \right. \tag{35}$$

Among the six quantities $\operatorname{Re}\lambda_k$ and $\operatorname{Im}\lambda_k$ five are of the order $\mathcal{O}(\lambda^3)$ and only $\operatorname{Im}\lambda_c$ is of the order $\mathcal{O}(\lambda^7)$. Therefore we have the dominance of the tt exchange contributions for large masses of the top quark $m_t > 30\,\mathrm{GeV}$

3) Case $B_s^0 \overline{B}_s^0$

Now the parameters $\lambda_k = V_{u_k s}^* V_{u_k b}$ are given in the parametrization of Wolfenstein by

$$\left| \begin{array}{l} \lambda_u = A\lambda^4 a e^{-i\delta} \\ \lambda_c = A\lambda^2(1 - \dfrac{\lambda^2}{2}) \\ \lambda_t = -A\lambda^2(1 - \dfrac{\lambda^2}{2}) - A\lambda^4 a e^{-i\delta} \end{array} \right. \tag{36}$$

Again we have the dominance of the tt exchange contribution.

4) The computation of m_{12}^{box} is very similar to the one performed for neutral K mesons. Retaining only the tt exchange contribution we get

$$m_{12}^{\mathrm{box}} = \frac{G_F^2}{12\pi^2} f_B^2 m_B B_B \eta_{tt} m_t^2 F(z_t) \lambda_t^2 \tag{37}$$

where the definition of the various terms is as follows
  - f_B is the coupling constant for the leptonic decay mode $B^+ \to l^+ + \nu_l$
  - B_B is the dimensionless parameter describing the departure from the vacuum insertion approximation of the matrix element of the quark field operators between the B^0 and $\overline{B}^0$ states
  - η_{tt} is a QCD correction factor slightly dependent of the top mass as shown on Part B (Figure 4)
  - $F(z_t)$ is a function of the top mass through the dimensionless parameter $z_t = \left(\frac{m_t}{m_W}\right)^2$ already introduced in Part B.

5) Theoretical estimates for f_B and B_B

The table 1 shows the main results concerning the theoretical estimates of the parameter f_B for the B_d^0 and B_s^0 mesons in the absence of experimental data. We clearly have very large uncertainties.

On the other hand the vacuum insertion approximation is expected to work better for the B meson than for the K meson and B_B is probably close to unity.

For numerical computations we shall use the following values

$$\begin{aligned} f_{B_d} B_{B_d}^{\frac{1}{2}} &= (140 \pm 40)\,\mathrm{MeV} \\ f_{B_s} B_{B_s}^{\frac{1}{2}} &= (170 \pm 50)\,\mathrm{MeV} \end{aligned} \tag{38}$$

IV - PARAMETER x FOR THE $B_d^0 \overline{B}_d^0$ SYSTEM

By using equations (8), (35) and (37) we obtain the expression of the parameter x_d

$$x_d = \frac{\tau_B}{\hbar} \frac{G_F^2}{6\pi^2} f_B^2 B_B m_B \eta_{tt} m_t^2 F(z_t) A^2 \lambda^6 [1 - 2a \cos \delta + a^2] \tag{39}$$

where all indices B refer to the B_d^0 meson.

1) Numerical estimate of x_d

We use the following input values

$$\tau_B = (1,18 \pm 0,14)10^{-12}\text{s} \qquad m_B = 5,275\,\text{GeV}$$
$$A = 1,05 \pm 0,17 \qquad \lambda = 0,221 \pm 0,002$$

$$f_B B_B^{\frac{1}{2}} = (140 \pm 40)\,\text{MeV}$$

and the result is

$$x_d = 0,024257\, \mathcal{E}_d\, f(m_t)[1 - 2a \cos \delta + a^2] \tag{40}$$

where the function $f(m_t)$ represented on Fig. 2 is nothing but the product $\eta_{tt} m_t^2 F(z_t)$ normalized to unity at $m_t = 30\,\text{GeV}$. The error factor $\mathcal{E}_d$ is computed by taking into account one standard deviation limits for the input quantities. The numerical bounds on $\mathcal{E}_d$ are

$$0,30 < \mathcal{E}_d < 2,64 \tag{41}$$

and we unfortunately have one order of magnitude uncertainty primarily due to the errors on f_{B_d} and on A.

2) Comparison with experiment

Let us use 90% confidence limits (31)

$$0,44 < x_d < 0,79$$

From the equation (40) we can obtain in the $m_t - \delta$ plane two correlation curves between these parameters

i) a lower bound curve associated to $x_d > 0,44$ and computed with $\mathcal{E}_d = 2,64$ and $a = 0,6$

ii) an upper bound curve associated to $x_d < 0,79$ and computed with $\mathcal{E}_d = 0,30$ and $a = 0,3$

It turns out that the upper bound curve is entirely above $m_t = 200\,\text{GeV}$ and it has not been represented on Figure 3 where only the lower bound curve appears. As a result we obtain a significant lower bound on m_t

$$m_t > 55\,\text{GeV} \tag{42}$$

and a restriction on δ given by the upper limit $m_t = 200$ GeV

$$\delta > 45°$$

V - PARAMETER x FOR THE $B_s^0 \overline{B}_s^0$ SYSTEM

The expression of x_s is given by

$$x_s = \frac{\tau_B}{\hbar} \frac{G_F^2}{6\pi^2} f_B^2 B_B m_B \eta_{tt} m_t^2 F(z_t) A^2 \lambda^4 \tag{43}$$

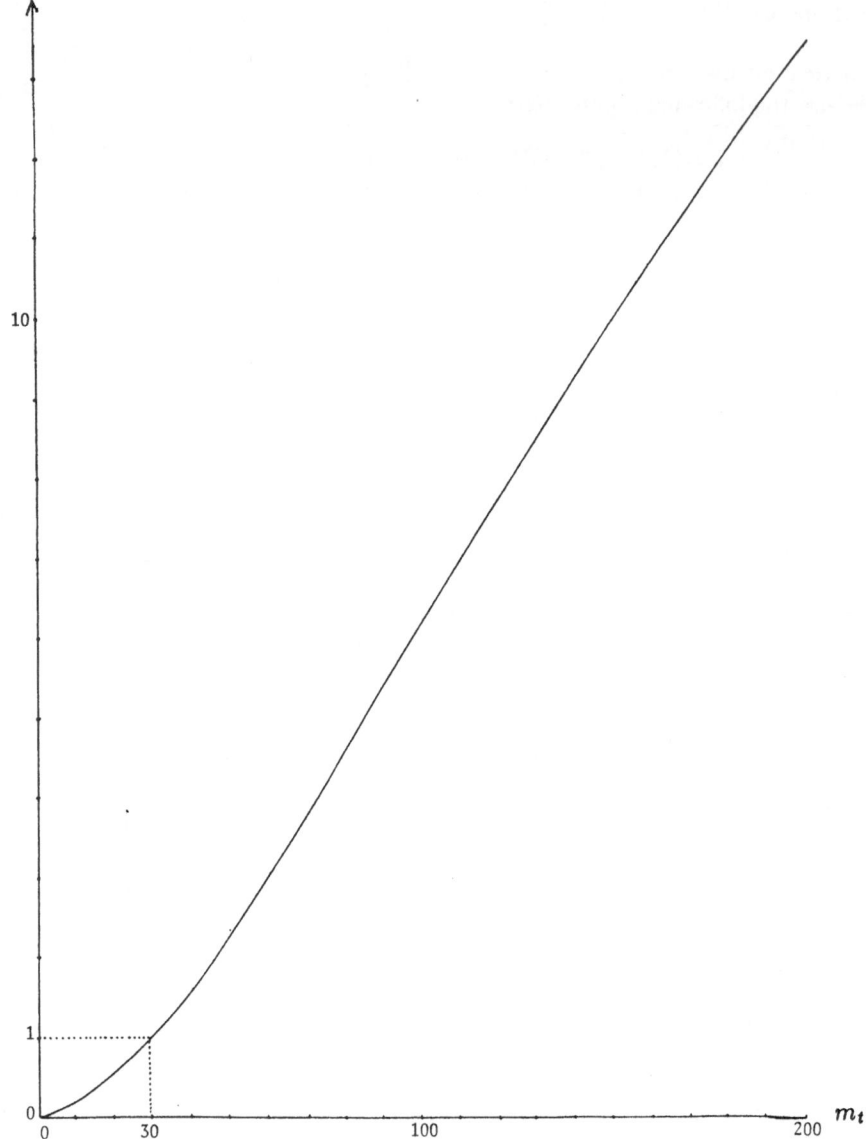

Figure 2 The function $f(m_t)$ for $0 \leq m_t \leq 200$ GeV e.g. $\eta_{tt} m_t^2 F(z_t)$ normalized to one at $m_t = 30$ GeV

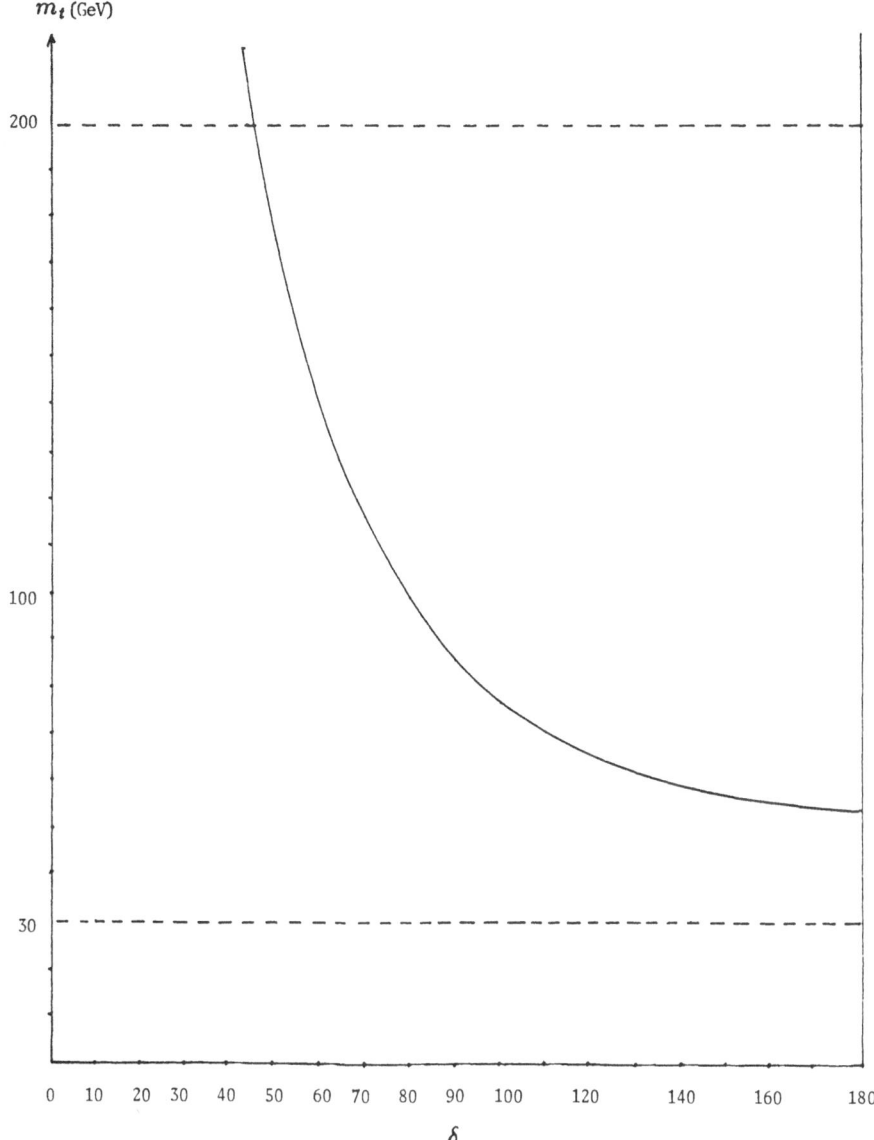

Figure 3 Correlation in the $m_t - \delta$ plane due to x_d

where now the indices B refer to the B_s^0 meson.

1) Numerical estimate of x_s

We use the following input values

$$\tau_B = (1,18 \pm 0,14)10^{-12}\text{s} \quad m_B = 5,354\,\text{GeV}$$
$$A = 1,05 \pm 0,17 \quad \lambda = 0,221 \pm 0,002$$

$$f_{B_s} B_{B_s}^{\frac{1}{2}} = (170 \pm 50)\,\text{MeV}$$

and the result is

$$x_s = 0,743\,\mathcal{E}_s\,f(m_t) \tag{44}$$

where again the error factor is dominated by the uncertainties on f_B and A

$$0,30 < \mathcal{E}_s < 2,62 \tag{45}$$

2) Using the result coming from the previous section, $m_t > 55$ GeV and $\mathcal{E}_s = 0,3$ we obtain a lower bound for x_s

$$x_s > 0,60 \tag{46}$$

An upper bound is obtained with $m_t = 200$ GeV and $\mathcal{E}_s = 2,62$

$$x_s < 26,2 \tag{47}$$

3) Ratio x_s/x_d

Most of the uncertainties disappear if we consider the ratio x_s/x_d which is independent of the top mass. We have

$$\frac{x_s}{x_d} = F \left| \frac{V_{ts}}{V_{td}} \right|^2 \tag{48}$$

with

$$F = \frac{\tau_{B_s}}{\tau_{B_d}} \left(\frac{f_{B_s}}{f_{B_d}} \right)^2 \frac{B_s}{B_d} \frac{m_{B_s}}{m_{B_d}} \tag{49}$$

and

$$\left| \frac{V_{ts}}{V_{td}} \right|^{-2} = \lambda^2 [1 - 2a \cos \delta + a^2] \tag{50}$$

The factor F differs from unity because of flavour $SU(3)$ breaking a d quark being replaced by a s quark. All the reasonable estimates expect the breaking corrections to the positive definite and $F > 1$. Making the following guesses

$$\frac{\tau_{B_s}}{\tau_{B_d}} \simeq 1 \quad \frac{B_s}{B_d} \simeq 1$$

$$\frac{m_{B_s}}{m_{B_d}} = 1,015 \quad \text{from potential models}$$

$$\frac{f_{B_s}}{f_{B_d}} = 1,22 \pm 0,22 \quad \text{from potential models}$$

we obtain

$$1 < F < 2 \tag{51}$$

Bounds on the ratio of the two CKM matrix elements $\left|\frac{V_{ts}}{V_{td}}\right|$ are obtained by using the upper limit on a, $a = 0,6$ and the range of values of δ allowed by the curve of Figure 3, $45° < \delta < 180°$. The result is

$$7,86 < \left|\frac{V_{ts}}{V_{td}}\right|^2 < 41 \tag{52}$$

and for the ratio x_s/x_d we get

$$7,86 < \frac{x_s}{x_d} < 82 \tag{53}$$

4) Comparison with experiment

We use the two 90% confidence limit results

$$0,083 < \chi_d < 0,194$$
$$0,065 < \chi_{\text{eff}} < 0,120 \tag{54}$$

Unfortunately we do not have a clean determination of χ_s and we now try to extract some information coming from χ_{eff} which is a linear combination of χ_d and χ_s as given by the equation (32). For the fractions P_s and P_d only guesses can be done and reasonable values are

$$P_s = 0,1\text{--}0,2 \qquad P_d = 0,375\text{--}0,400$$

For computations we use $P_s = 0,150$, $P_d = 0,375$ and we assume the equality of the two semi-leptonic branching ratios $BR_s = BR_d$. The effective χ_{eff} is then simply

$$\chi_{\text{eff}} = 0,150\chi_s + 0,375\chi_d$$

The experimental bounds (54) are represented by straight lines in the $\chi_s \chi_d$ plane restricted to the square $0 \le \chi_s, \chi_d \le 0,5$ and clearly the various sets of data are compatible with each other as shown on Figure 4.

We now use the theoretical information $\frac{x_s}{x_d} > 7,86$ which is represented in the χ_s, χ_d plane by a branch of hyperbola

$$\left(\frac{x_s}{x_d}\right)^2 = \frac{\chi_s(1 - 2\chi_d)}{\chi_d(1 - 2\chi_s)} \tag{55}$$

The Figure 4 shows the experimental bounds and the theoretical constraint coming from the box diagram calculation. The dashed area is the allowed region for χ_s and χ_d and the points A and B give interesting informations

A a lower limit for χ_s $\chi_s > 0,46$

B an upper limit for χ_d $\chi_d < 0,13$

For theoretical considerations the variables x_s and x_d are more convenient and the Figure 5 is simply the translation of the informations of Figure 4 in the x language. Now the positivity bounds are simply $x_s, x_d \ge 0$. The straight lines $\chi_d = $ const are straight lines $x_d = $ const but the straight lines $\chi_{\text{eff}} = $ const become quartic curves. The branch of the theoretical hyperbola (55) is a simple straight line $\frac{x_s}{x_d} = $ const. Finally for the points A and B we get

A a lower limit for x_s $x_s > 3,46$

B an upper limit for x_d $x_d < 0,56$

and the dashed area of Figure 5 corresponds to a part of the dashed area of Figure 4.

An upper limit for the ratio $\frac{x_s}{x_d}$ has been previously obtained in equation (53) and it would correspond to a second straight line starting from the origin on Figure 5. Because of the poor relevance of this upper limit it has not been represented on Figure 5 but obviously it would make finite the dashed region of the x_s, x_d plane.

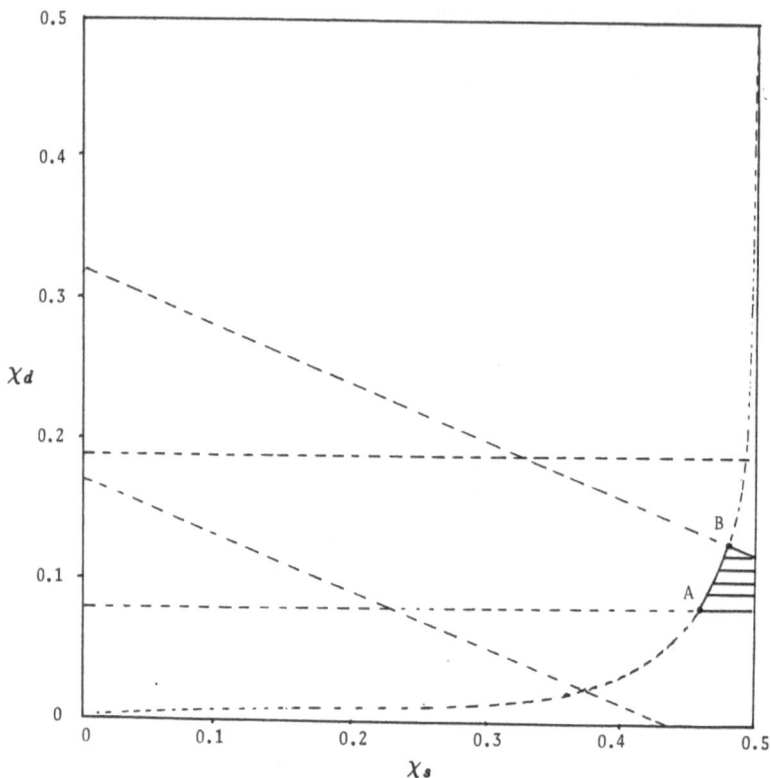

Figure 4 Allowed region in the χ_s, χ_d plane

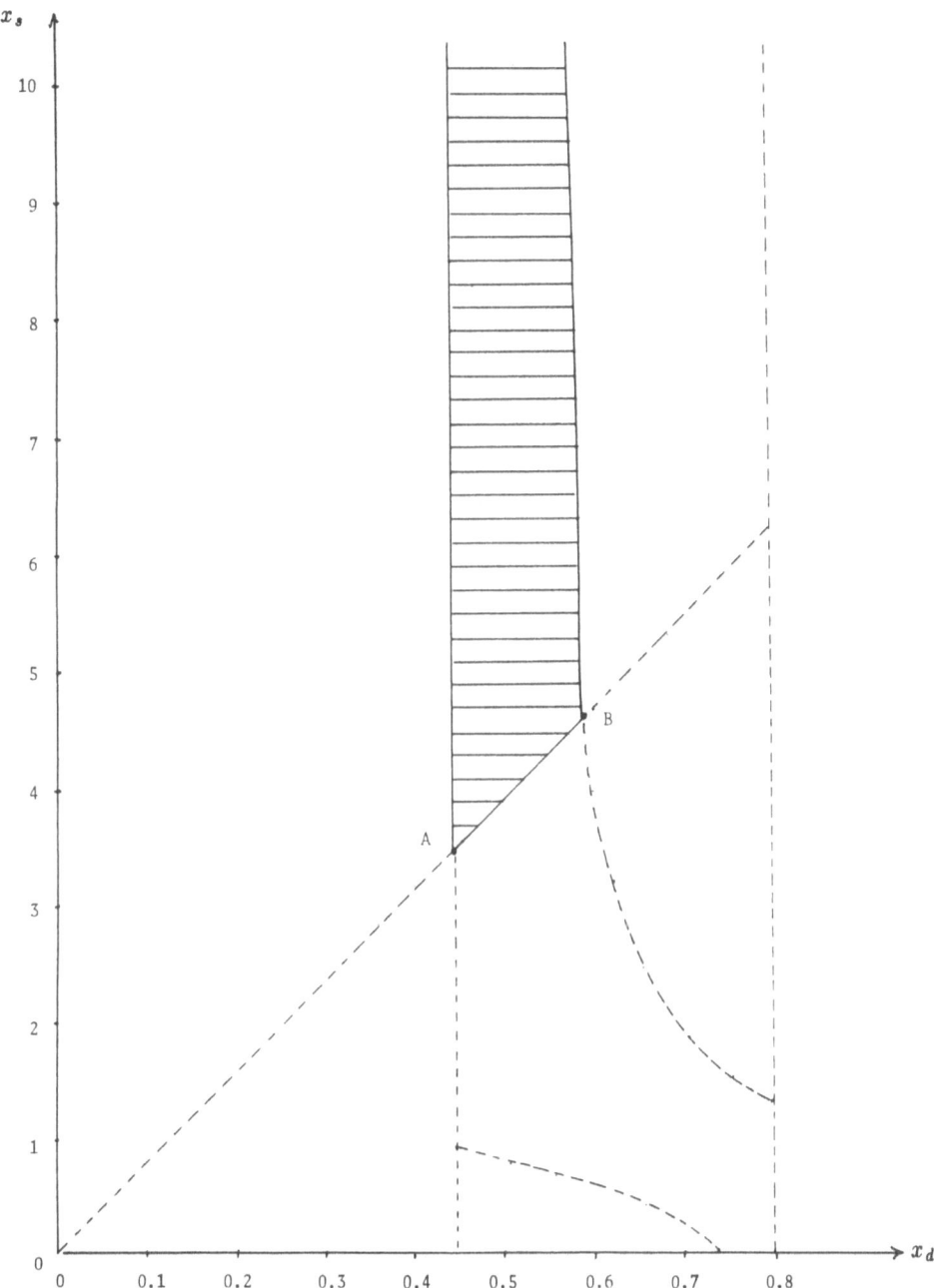

Figure 5 Allowed region in the x_s, x_d plane

VI CONCLUSIONS

1) Improve the theoretical estimates of $f_{B_d}, f_{B_s}, B_{B_d}, B_{B_s}$
2) Improve the semi-phenomenological determination of the CKM parameters A and a
3) For experimentalists

    i) measure separately the masses and the lifetimes for B_d^0 and B_s^0

    ii) observe the purely leptonic decays

$$B_u^+ \rightarrow l^+ + \nu_l \quad B_c^+ \rightarrow l^+ + \nu_l$$

    iii) measure χ_s and if possible x_s with a reasonable accuracy

MEASURE THE TOP QUARK MASS

REFERENCES FOR EXPERIMENTAL DATA

ARGUS	H. Albrecht et al.	Phys. Lett. **192B** 245 (1987).
CLEO	A. Bean et al.	Phys. Rev. Lett. **58** 183 (1987).
UA1	C. Albajar et al.	Phys. Lett. **186B** 247 (1987).
MARK II	T. Schaad et al.	Phys. Lett. **160B** 188 (1985).
JADE	W. Bartel et al.	Phys. Lett. **146B** 437 (1984).
MAC	R.B. Hurst	Proceedings of the 22nd Rencontre de Moriond Les Arcs (March 1987)

REFERENCES FOR BOX DIAGRAM COMPUTATIONS

M.K. Gaillard and B.W. Lee   Phys. Rev. **D10** 897 (1974).

J. Ellis, M.K. Gaillard and D.V. Nanopoulos   Nucl. Phys. **B109** 213 (1976).

J. Ellis, M.K. Gaillard, D.V. Nanopoulos and S. Rudaz
   Nucl. Phys. **B131** 285 (1977).

A. Ali and Z.Z. Aydin   Nucl. Phys. **B148** 165 (1979).

T. Inami and C.S. Lim Prog. Theor. Phys. **65** 297 (1981).

J.S. Hagelin   Nucl. Phys. **B193** 123 (1981).

E. Franco, M. Lusignole and A. Pugliese   Nucl. Phys. **B194** 403 (1982).

H.Y. Cheng   Phys. Rev. **D26** 143 (1982).

Ling-Lie Chau   Physics Reports **95** 1 (1983).

A.J. Buras, W. Slominski and H. Steger
   Nucl. Phys. **B238** 529 (1984), **B245** 369 (1984).

E.A. Paschos and U. Türke   Nucl. Phys. **B243** 29 (1984).

REFERENCES FOR THE PARAMETERS F_{B_Q} AND F_{B_s}

1 H. Kraseman   Phys. Lett. **96B** 397 (1980).

2 A. Ali and C. Jarskog   Phys. Lett. **144B** 266 (1984).

3 E. Golowich   Phys. Lett. **91B** 271 (1980).

4 M. Claudson   Harvard University Report 81-546 (1981) unpublished.

5 I.I. Bigi and A.I. Sanda   Phys. Rev. **D29** 1393 (1984).

6 S. Godfrey   Phys. Rev. **D33** 1391 (1986).

7 M. Suzuki   Phys. Lett. **142B** 207 (1984) **162B** 392 (1985).

8 E.Y. Shuryak   Nucl. Phys. **B198** 23 (1982).

9 A.R. Zhitnitskii, I.R. Zhinitskii and V.L. Chernyak
   Sov. J. Nucl. Phys. **38** 773 (1983).

10 T.M. Aliev and V.L. Eletskii   Sov. J. Nucl. Phys. **38** 936 (1983).

11 V.S. Mathur and M.T. Yamawaki   Phys. Rev. **D29** 2057 (1984).

12 L.J. Reinders, H. Rubinstein and S. Yazaki   Phys. Lett. **104B** 305 (1981),
   Phys. Rep. **C127** 1 (1985).

13 M.A. Shifman and M.B. Voloshin   Sov. J. Nucl. Phys. **45** 292 (1987).

14 S. Narison   Phys. Lett. **198B** 104 (1987).

15 C.A. Dominguez and N. Paver   Phys. Lett. **197B** 423 (1987).

16 A. Pich   Cern preprint TH 4951/88 (1988).

17 M.B. Gavela, L. Maiani, S. Petrarca, G. Martinelli, O. Pene
   Phys. Lett. **B206** 113 (1988).

18 L. Bernard, T. Drapper, G.Hockney and A. Soni
   Indiana preprint UCLA/87/TEP/40 (1987).

REFERENCES FOR THE QCD COEFFICIENTS

M.I. Vysotskii   Sov. J. Nucl. Phys. **31** 797 (1980).

F.J. Gilman and M.B. Wise   Phys. Lett. **93B** 129 (1980)
Phys. Rev. **D27** 1128 (1983).

J. Ellis and J.S. Hagelin   Nucl. Phys. **B217** 189 (1983).

A.J. Buras, W. Slominski and H. Steger   Nucl. Phys. **B238** 529 (1984).

W. Kaufman, H. Steger and Y.P. Yao
University of Michigan preprint UM-TH-87-13 (July 1987).

M. Lusignoli   Cern preprint CERN TH-5043/88 (May 1988).

PROSPECTS

I - BOX DIAGRAMS FOR γ_{12}

1) In the computation of γ_{12} we restrict to intermediate physical states which can be reached from both B^0 and $\overline{B}^0$ mesons. Because of the $\Delta B = \Delta Q$ rule these states are dominantly hadronic and we shall consider essentially two types of contributions completed with QCD corrections.

    i) those coming from the spectator diagram where the heavy b ($\overline{b}$) quark decays;

    ii) those coming from the annihilation diagram where the W boson is exchanged between the quark and the antiquark of the B^0 or $\overline{B}^0$ mesons

    The contributions to γ_{12} coming from these two types of decay are given by the absorptive part of the $\Delta B = 2$ matrix element evaluated from the box diagrams and the cuts made for computing such absorptive parts are indicated on Figure 1. Let us notice that the intermediate states being physical the quarks u_k and u_l can only be the u and c quarks. As previously q is either d or s.

2) The computation of γ_{12}^{box} gives the following result

$$\gamma_{12}^{\text{box}} = \frac{G_F^2}{8\pi} f_B^2 \, B_B \, m_B \, m_b^2 \{ \eta_1 [\lambda_u^2 + (1+2x)\sqrt{1-4x}\lambda_c^2 + 2(1+2x)(1-x)^2 \lambda_c \lambda_u]$$
$$- \frac{8}{3}\eta_2 x[\sqrt{1-4x}\lambda_c^2 + (1-x)^2 \lambda_c \lambda_u]\}$$

(1)

where η_1 and η_2 are QCD correction factors and x the dimensionless ratio of quark masses $x = \left(\frac{m_c}{m_b}\right)^2$.

    It is convenient to eliminate λ_u by using the unitarity relation

$$\lambda_u + \lambda_c + \lambda_t = 0$$

The result is

$$\gamma_{12}^{\text{box}} = \frac{G_F^2}{8\pi} f_B^2 \, B_B \, m_B \, m_b^2 \, \eta_1 [\lambda_t^2 + A\lambda_c\lambda_t + B\lambda_c^2]$$

(2)

where the constant A and B are given by

$$A = \frac{8}{3}\frac{\eta_2}{\eta_1} x(1-x)^2 + 2x^2(3-2x)$$
$$B = \frac{8}{3}\frac{\eta_2}{\eta_1} x\left[(1-x)^2 - \sqrt{1-4x}\right] + \left[(1+2x)\sqrt{1-4x} - 1 + 2x^2(3-2x)\right]$$

(3)

3) The QCD coefficients η_1 and η_2 are related to gluonic corrections of the spectator and annihilation diagrams

$$\eta_1 = \eta_S \qquad \eta_2 = \frac{1}{4}(3\eta_S + \eta_A)$$

It turns out that the spectator contribution is only slightly increased while the annihilation one is strongly suppressed

$$\eta_S \simeq 1,06 - 1,09$$
$$\eta_A \simeq 0,16 - 0,27$$

(4)

and we get

$$\frac{\eta_2}{\eta_1} \simeq 0,79 - 0,81$$

(5)

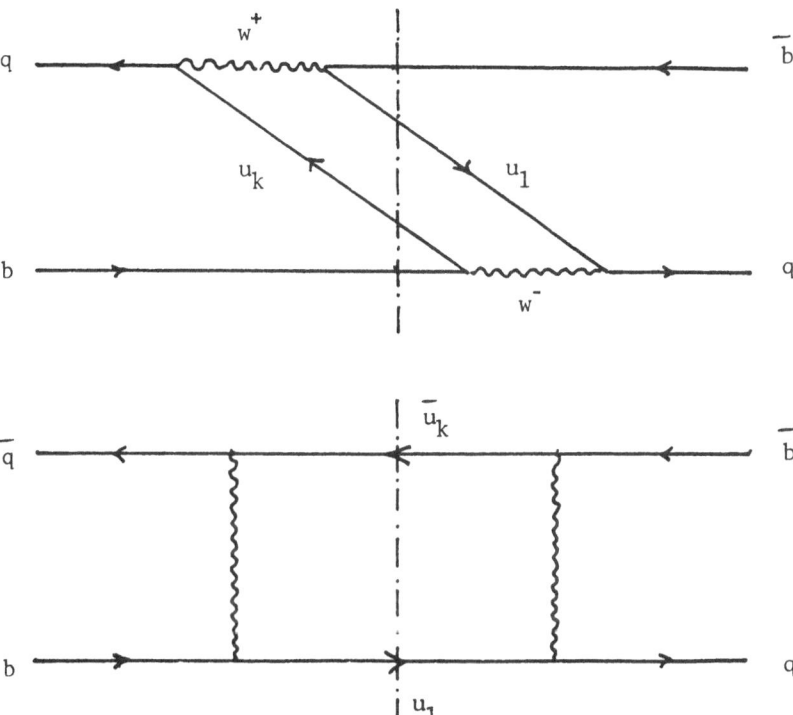

Figure 1   The box diagrams for γ_{12}

4) Using the expression found in Part C for the box diagram calculation of m_{12} and equation (2) for γ_{12} we obtain the ratio γ_{12}/m_{12} in the form

$$\frac{\gamma_{12}}{m_{12}} = \frac{3\pi}{2} \left(\frac{m_b}{m_t}\right)^2 \frac{1}{F(z_t)} \frac{\eta_S}{\eta_{tt}} \left\{ 1 + A\frac{\lambda_c}{\lambda_t} + B\left(\frac{\lambda_c}{\lambda_t}\right)^2 \right\} \tag{6}$$

The coefficients A and B are computed by using $m_c = 1,5$ GeV, $m_b = 5,0$ GeV and $\eta_2/\eta_1 = 0,8$ as given in equation (5). The results are

$$A = 0,205 \qquad B = -0,005 \tag{7}$$

As a consequence it is legitimate to neglect, in equation (6), the term $\left(\frac{\lambda_c}{\lambda_t}\right)^2$.

Of course the value of m_{12} depends on the unknown top mass m_t and it is convenient to use the function $f(m_t)$ defined in Part C and normalized to unity at $m_t = 30$ GeV. With $\eta_S = 1,075$ we obtain

$$\frac{3\pi}{2} \left(\frac{m_b}{m_t}\right)^2 \frac{1}{F(z_t)} \frac{\eta_S}{\eta_{tt}} = \frac{0,2856}{f(m_t)}$$

and the equation (6) takes the form

$$\frac{\gamma_{12}}{m_{12}} = \frac{0,2856}{f(m_t)} [1 + 0,205\frac{\lambda_c}{\lambda_t}] \tag{8}$$

It is also interesting to introduce the relative phase $\phi(\Delta B = 2)$ of γ_{12} and m_{12}. From equation (8) it is given by

$$\tan \phi(\Delta B = 2) = \frac{0,205 \text{Im} \frac{\lambda_c}{\lambda_t}}{1 + 0,205 \text{Re} \frac{\lambda_c}{\lambda_t}} \tag{9}$$

5) $B_d^0 \overline{B}_d^0$ system

Both real and imaginary parts of λ_c and λ_t are of the order $0(\lambda^3)$ and for the ratio λ_c/λ_t

$$\frac{\lambda_c}{\lambda_t} = -\frac{1}{1 - a\,e^{-i\delta}} \tag{10}$$

both $\text{Re}\,\lambda_c/\lambda_t$ and $\text{Im}\,\lambda_c/\lambda_t$ are independent of λ. Their values depend on
  i) the CKM parameter a with $0,3 < a < 0,6$
  ii) the PC violating phase δ with $45° < \delta < 170°$ as a result of the analysis of Parts B and C.

The phase $\phi_d(\Delta B = 2)$ defined in equation (9) is restricted as follows

$$0,5° < \phi_d(\Delta B = 2) < 12°5 \tag{11}$$

Using now the equation (9) and (12) of the Part C we can derive bounds for the physical quantities $\Delta\gamma/\Delta m$ and $\text{Re}\,\epsilon$

$$0,22 < f(m_t)\frac{\Delta\gamma}{\Delta m} < 0,25$$
$$6\ 10^{-4} < f(m_t)\text{Re}\,\epsilon < 1,2\ 10^{-2} \tag{12}$$

If we now take into account the bound $m_t > 55$ GeV derived from x_d in the Part C we can use

$$2,7 < f(m_t) < 12,28 \tag{13}$$

and we get

$$1,5\ 10^{-2} < \frac{\Delta\gamma}{\Delta m} < 9,2\ 10^{-2}$$
$$1,7\ 10^{-4} < \text{Re}\,\epsilon < 4,5\ 10^{-3} \tag{14}$$

6) $B_s^0 \overline{B}_s^0$ system

Now both Re λ_c and Re λ_t are of the order $O(\lambda^2)$ but Im λ_t is of the order $O(\lambda^4)$ and Im λ_c is of the order $O(\lambda^8)$. For the ratio

$$\frac{\lambda_c}{\lambda_t} \simeq -\frac{1}{1 + a\lambda^2 e^{-i\delta}} \tag{15}$$

we get Re $\frac{\lambda_c}{\lambda_t}$ independent of λ and Im $\frac{\lambda_c}{\lambda_t}$ proportionnal to λ^2. Bounds on these quantities can be derived as previously and we obtain

$$0°07 < -\phi_s(\Delta B = 2) < 0°43 \tag{16}$$

and

$$f(m_t)\frac{\Delta\gamma}{\Delta m} = 0,227 \pm 0,001$$
$$7\ 10^{-5} < -f(m_t)\mathrm{Re}\,\epsilon < 4,3\ 10^{-4} \tag{17}$$

Using now the top mass constraint (18) we get

$$1,6\ 10^{-2} < \frac{\Delta\gamma}{\Delta m} < 8,5\ 10^{-2}$$
$$4,9\ 10^{-6} < -\mathrm{Re}\,\epsilon < 1,6\ 10^{-4} \tag{18}$$

II - LEPTONIC ASYMMETRIES

1) We have already discuss in Part C a first class of leptonic asymmetries measuring the ratio of the same sign to the opposite sign dileptons. These asymmetries are related to the mass difference Δm or equivalently to the parameter $x = \frac{\Delta m}{\gamma}$

2) A second class of leptonic asymmetries will bring a direct information on the parameter ϵ. We simply have to measure the charge asymmetry in the same sign dilepton production.

$$\alpha = \frac{N(l^+l^+) - N(l^-l^-)}{N(l^+l^+) + N(l^-l^-)} \tag{19}$$

The parameter α is related to r and $\overline{r}$ introduced in part C

$$\alpha = \frac{r - \overline{r}}{r + \overline{r}} \tag{20}$$

In the ratio α the x dependence disappears and only remain the ϵ dependence. Using a first order approximation in ϵ for the ratio

$$\frac{\overline{r}}{r} = \left|\frac{1 + \epsilon}{1 - \epsilon}\right|^4 \tag{21}$$

we get

$$\alpha \simeq -4\frac{\mathrm{Re}\,\epsilon}{1 + |\epsilon|^2} \simeq -\mathrm{Im}\,\frac{\gamma_{12}}{m_{12}} \tag{22}$$

3) In electron-positron annihilation into quark-antiquark pairs we have a forward-backward charge asymmetry due to the interference between the one photon and the one Z° exchange. Such a asymmetry is calculable in the framework of the standard model. Consider the $b\overline{b}$ production followed by the leptonic decay of the b and an hadronic jet coming from the $\overline{b}$

$$e^+ + e^- \Rightarrow \gamma, Z^0 \Rightarrow b \quad \overline{b} \tag{23}$$

$$\overset{\displaystyle\llcorner\!\!\to\, \text{Jet}}{\underset{\displaystyle\llcorner\!\!\to\, l^\pm X}{}}$$

Because of the $B^0\overline{B}^0$ mixing we shall have a flip of the charge of the observed lepton and therefore we expect a reduction of the charge asymmetry by a factor $1-2\chi$. Such a method is called forward-backward lepton jet asymmetry.

III - HADRONIC ASYMMETRIES

We know describe qualitatively two types of hadronic asymmetries where the PC violation can be detected without going into details of specific examples.

1) The first class of hadronic symmetries is due to the existence of $B^0\overline{B}^0$ oscillations already introduced in Part C. Let us call as f an hadronic final state and as $\overline{f}$ its PC transformed. We introduce four possible decay amplitudes

$$A(B^0 \to f) = A(f) \quad A(B^0 \to \overline{f}) = A(\overline{f})$$
$$A(\overline{B}^0 \to f) = \overline{A}(f) \quad A(\overline{B}^0 \to \overline{f}) = \overline{A}(\overline{f})$$

Starting at $t = 0$ with a pure B^0 or $\overline{B}^0$ state we have, at $t > 0$ a mixture of B^0 and $\overline{B}^0$ and we study the following time dependent amplitudes

$$A(B^0(t) \to f) = A(f)\left[f_+(t) + f_-(t)\frac{1-\epsilon}{1+\epsilon}\frac{\overline{A}(f)}{A(f)}\right]$$

$$A(\overline{B}^0(t) \to \overline{f}) = \overline{A}(\overline{f})\left[f_+(t) + f_-(t)\frac{1+\epsilon}{1-\epsilon}\frac{A(\overline{f})}{\overline{A}(\overline{f})}\right] \tag{24}$$

where the functions $f_\pm(t)$ have been introduced in Part C. It is convenient to define the following complex ratios

$$\rho_f = \frac{\overline{A}(f)}{A(f)} \quad \lambda_f = \frac{1-\epsilon}{1+\epsilon}\rho_f$$

$$\overline{\rho}_{\overline{f}} = \frac{A(\overline{f})}{\overline{A}(\overline{f})} \quad \overline{\lambda}_{\overline{f}} = \frac{1+\epsilon}{1-\epsilon}\overline{\rho}_f \tag{25}$$

It is interesting to notice that the parameters λ_f and $\overline{\lambda}_{\overline{f}}$ are independent of the phase convention used for the basic states B^0 and $\overline{B}^0$. Therefore they are measurable complex numbers as their phases are associated to the violation of PC. The time dependent widths are given by

$$\Gamma(B^0(t) \to f) \propto |A(f)|^2 |f_+(t) + f_-(t)\lambda_f|^2$$

$$\Gamma(\overline{B}^0(t) \to \overline{f}) \propto |\overline{A}(\overline{f})|^2 |f_+(t) + f_-(t)\overline{\lambda}_{\overline{f}}|^2 \tag{26}$$

and we define the hadronic asymmetry C_f^{00} in the following way

$$C_f^{00} = \frac{\Gamma(B^0(t) \to f) - \Gamma(\overline{B}^0(t) \to \overline{f})}{\Gamma(B^0(t) \to f) + \Gamma(\overline{B}^0(t) \to \overline{f})} \tag{27}$$

We now assume that the violation of PC does not occur in magnitude but in phase

$$|\overline{A}(\overline{f})| = |A(f)| \tag{28}$$

In this case we get

$$C_f^{00} = \frac{2\mathrm{Re}\,[f_+(t)f_-^*(t)(\lambda_f^* - \overline{\lambda}_{\overline{f}}^*)] + |f_-(t)|^2[|\lambda_f|^2 - |\overline{\lambda}_{\overline{f}}|^2]}{2|f_+(t)|^2 + 2\mathrm{Re}\,[f_+(t)f_-^*(t)(\lambda_f^* + \overline{\lambda}_{\overline{f}}^*)] + |f_-(t)|^2[|\lambda_f|^2 + |\overline{\lambda}_{\overline{f}}|^2]} \tag{29}$$

The previous asumption (28) is realized for instance when only one CKM parameter occurs in the decay amplitude. If, in addition, we have only one strong interaction channel with one phase δ_f due to final state interaction we can define the reduced amplitudes free of these phases

$$\begin{aligned} A(f) = A_f\,e^{i\delta_f} & \quad A(\overline{f}) = A_{\overline{f}}\,e^{i\delta_f} \\ \overline{A}(f) = \overline{A}_f\,e^{i\delta_f} & \quad \overline{A}(\overline{f}) = \overline{A}_{\overline{f}}\,e^{i\delta_f} \end{aligned} \tag{30}$$

As a consequence of TCP invariance we have the relations

$$\overline{A}_f = A_{\overline{f}}^* \quad \overline{A}_{\overline{f}} = A_f^* \tag{31}$$

Therefore

$$\overline{\rho}_{\overline{f}} = \frac{A_{\overline{f}}}{A_f^*} = \rho_f^* \tag{32}$$

The two ratios ρ_f and $\overline{\rho}_{\overline{f}}$ are complex conjugate numbers. On the other hand the two quantities $\frac{1-\epsilon}{1+\epsilon}$ and $\frac{1+\epsilon}{1-\epsilon}$ have opposite phases and the same property extends to the physical ratios λ_f and $\overline{\lambda}_{\overline{f}}$. Only the phase ϕ_f of λ_f is a PC violating observable phase. The moduli of the ratios $\frac{1-\epsilon}{1+\epsilon}$ and $\frac{1+\epsilon}{1-\epsilon}$ are simply expressed in terms of the observable parameter

$$\frac{\mathrm{Re}\,\epsilon}{1 + |\epsilon|^2}$$

As discussed previously such a quantity is expected to be small for both $B_d^0 \overline{B}_d^0$ and $B_s^0 \overline{B}_s^0$ systems. If we neglect it we obtain the simple relation

$$\overline{\lambda}_{\overline{f}} = \lambda_f^* \tag{33}$$

and the hadronic asymmetry C_f^{00} becomes

$$C_f^{00} = \frac{2\,\mathrm{Im}\,[f_+(t)f_-^*(t)]\mathrm{Im}\,\lambda_f}{|f_+(t)|^2 + |f_-(t)|^2|\rho_f|^2 + 2\,\mathrm{Re}\,[f_+(t)f_-^*(t)]\mathrm{Re}\,\lambda_f} \tag{34}$$

In practice the first generation of measurements will concern time integrated asymmetries. Using the results of the Part C

$$\int |f_+(t)|^2 dt \propto 2 + x^2$$

$$\int |f_-(t)|^2 dt \propto x^2 \tag{35}$$

$$\int f_+(t) f_-^*(t) \, dt \propto -i\,x$$

we obtain

$$C_f^{00} = \frac{-2x \, \mathrm{Im} \, \lambda_f}{2 + x^2 + x^2 |\rho_f|^2} \tag{36}$$

A particularly interesting case is when f is an eigenstate of PC. Then $|\rho_f| = 1$ and the asymmetry C_f^{00} reduces to

$$C_f^{00} = \frac{-x}{1 + x^2} \, \mathrm{Im} \, \lambda_f \tag{37}$$

For x non negative the function $\frac{x}{1+x^2}$ vanishes at $x = 0$ and when $x \to \infty$ having its maximal value at $x = 1$. As a consequence the situation looks more promising for the $B_d^0 \overline{B}_d^0$ system than for the $B_s^0 \overline{B}_s^0$ one where x_s might be large. The second term in equation (37) , $\mathrm{Im} \, \lambda_f$, can be computed from the CKM matrix elements and it is proportional to $\sin \delta$. Unfortunately because of the small decay rates for the relevant hadronic states f these asymmetries are difficult to measure experimentally.

2) We now consider a second class of hadronic asymmetries asociated to charged $B^\pm$ mesons and we call as $f^\pm$ two PC conjugate hadronic states occuring is the decay of $B^\pm$.

We assume here that there exists two different strong interaction channels 1 and 2 with two different phases δ_1 and δ_2 due to final state interaction.

$$A(B^+ \to f^+) = A_1 \, e^{i\delta_1} + A_2 \, e^{i\delta_2}$$

$$A(B^- \to f^-) = A_1^* \, e^{i\delta_1} + A_2^* \, e^{i\delta_2} \tag{38}$$

Here TCP invariance has been used for the reduced amplitudes A_1 and A_2.

The hadronic asymmetry C_f^{+-} is defined by

$$C_f^{+-} = \frac{\Gamma(B^+ \to f^+) - \Gamma(B^- \to f^-)}{\Gamma(B^+ \to f^+) + \Gamma(B^- \to f^-)} \tag{39}$$

Using the previous decomposition of the decay amplitudes we obtain

$$C_f^{+-} = -\frac{\mathrm{Im}\,(A_1 \, A_2^*) \, \sin(\delta_1 - \delta_2)}{|A_1|^2 + |A_2|^2 + 2\,\mathrm{Re}\,(A_1 \, A_2^*)\,\cos(\delta_1 - \delta_2)} \tag{40}$$

Therefore the asymmetry C_f^{+-} is different from zero if and only if the relative phase of A_1 and A_2 and the difference of phases $\delta_1 - \delta_2$ are both non zero modulo π. In that case the asymmetry C_f^{+-} is easy to measure experimentally but it is difficult to produce reliable theoretical calculations.

242

REFERENCES

M. Bander, D. Silverman and A. Soni   Phys. Rev. Lett. **43** 242 (1974).

L.B. Okun, V.I. Zakharov and B.M. Pontecorvo
Lett. Nuovo Cim. **13** 218 (1975).

A. Pais and S.B. Treiman   Phys. Rev. **D12** 244 (1975).

A.B. Carter and A.I. Sanda   Phys. Rev. Lett. **45** 952 (1980),
Phys. Rev. **D23** 1567 (1981).

J. Barnabeu and C. Jarlskog   Z. Phys. **C8** 233 (1981).

I.I. Bigi and A.I. Sanda   Nucl. Phys. **B193** 85 (1981),
Phys. Rev. **D19** 1393 (1984),   Nucl. Phys. **B281** 41 (1987).

J.S. Hagelin   Nucl. Phys. **B193** 123 (1981).

A.J. Buras, W. Slominski and H. Steger
Nucl. Phys. **B238** 529 (1984), **B245** 369 (1984).

L.L. Chau and H.Y. Cheng   Phys. Rev. Lett. **53** 1037 (1984), **59** 958 (1987).

A. Ali and T. Barreiro   Z. Phys. **C30** 635 (1986).

D. Du, I. Dunietz and D.D. Wu   Phys. Rev. **D34** 8414 (1986).

I. Dunietz and T. Nakada   Preprint SIN PR-87-06 (June 1987).

R.D. Peccei   Preprint DESY 87-134 (Oct. 1987).

Y. Azimov, V. Khoze and M. Uraltsev   Sov. J. Nucl. Phys. **45** 878 (1987).

J.F. Donoghue, T. Nakada, E.A. Paschos, D. Wyler
Phys. Lett. **B195** 285 (1987).

S. Stone   Preprint Cornell CLNS 87/110 (Nov. 1987).

P. Krawczyk, D. London, R.D. Peccei, H. Steger   Nucl. Phys. **B307** 19 (1988).

Y.L. Wu   Preprint Dortmund DO-TH-88/8)March 1988).

M. Lusignoli   CERN Preprints CERN-TH 4977/88 (Feb. 1988),
CERN TH 5045/88 (May 1988)

DISCUSSION

– *Volkas:*

Could you please briefly comment on the experimental status of Weinberg's model of spontaneous CP violation?

– *Gourdin:*

The answer will be in the written version of my talk, and it is essentially the following one: the minimal standard model accomodate all existing experiments and it is a success. This model provides a mechanism for CP violation. Therefore the first comparison between theory and experiment we must make is within the minimal standard model, before asking for more complicated and sophisticated explanations. If we cannot explain all the data on CP violation with the same phase δ we shall be obliged, by experiment, to go beyond the minimal standard model and to do that the Weinberg model is only one possible mechanism among others.

– *Gonzales:*

Why is there no attempt of measuring CP violation effects in $D^0 - \overline{D}^0$ system? Which is the difference between $D^0 - \overline{D}^0$ system and $K^0 - \overline{K}^0$ or $B^0 - \overline{B}^0$?

– *Gourdin:*

I never said that there was no attempt to measure $D^0 - \overline{D}^0$ mixing but only that no effect has been seen. Such a fact can be expected on theoretical basis. From experiment the best upper limit on the parameter x_D is coming from the Fermilab experiment E691.

$$x_D < 8.6 \times 10^{-2}$$

Such an upper limit is well above the estimate made by considering only the box diagram:

$$x_D^{box} < 3.4 \times 10^{-4}$$

But following the Wolfenstein argument the box diagram will probably represent only a part of the actual x_D and the rest will be due to long range effects which have not yet been properly evaluated.

– *Rahal:*

Do the inconsistencies you mentioned disappear if you perform a two standard deviation analysis?

– Gourdin:

The answer is given just by considering the graph I already showed during my talk (fig. 2). Now, looking at the experimental value of ϕ_{00}, if you accept 2 standard deviation you will obviously be compatible with $\phi_{00} \simeq \phi_{+-}$ and the inconsistency disappears.

– Rahal:

In contrast to P or CP violation, CPT conservation follows from relativistic invariance, which is a solid principle (i.e. fulfilled by observation). Is $m_{K_L} - m_{K_S} \simeq 10^{-12}$ MeV to be considered as a CPT test or not?

– Gourdin:

In quantum field theory the TCP theorem follows in particular from special Lorentz invariance and locality. I agree that the most accurate test is the $K^0 - \overline{K}^0$ mass difference bounded by 10^{-15} in relative value.

But if at the end we cannot solve the inconsistency problem, for instance if the phase difference $\phi_{+-} - \phi_{00}$ in absolute value is definitively larger than, let us say 2^0 with 5 s.d., we shall be obliged to conclude that TPC invariance is violated in some processes, even if such a violation is at a tiny level.

– Rahal:

Is the effective Hamiltonian hermitian?

– Gourdin:

The effective hamiltonian H does not have to be hermitian because the neutral K's are unstable particles. The hermitian part is the mass matrix M, the skew hermitian part is i times the decay matrix Γ. The formalism is due to Wigner and Weisskopf and it is perfectly in agreement with quantum mechanics assuming, in particular, the superposition principle and the conservation of probabilities.

– Rahal:

Why do you take $B_K > 1/3$?

– Gourdin:

Please, look at Table 1 where there are different theoretical predictions for B_K and you will realize that the statement:

$$1/3 < B_K < 1$$

is very conservative and pessimistic. But it proves the necessity of having more accurate estimate of B_K as soon as possible.

– *Rahal:*

Do you have any comment on the outcome of the measurement of the electric dipole moment of the neutron in Grenoble? (N. Ramsey).

– *Gourdin:*

The neutron electric dipole moment is interesting because is a direct test of time reversal invariance and for this reason it will give a complementary information to that coming from CP violation. If TCP is a good symmetry then the violation of CP implies the violation of T. Now you certainly know that the theoretical estimates of the neutron electric dipole moment are strongly model dependent and therefore poorly reliable if you wish to conclude into a contradiction between theory and experiment, assuming that the experiments have reached a reasonable level of accuracy.

– *Skarke:*

What are the chances of systematic error in the data supplied by the experimentalists?

– *Gourdin:*

The systematic errors are now given separately and for instance in the best NA31 CERN experiment we have

$$\left| \frac{\eta_{00}}{\eta_{+-}} \right|^2 = 0.980 \pm 0.004 \pm 0.003$$

$$\uparrow \qquad \uparrow$$

statistical systematic

error error

Systematic and statistical errors are of the same magnitude.

Probably there exists a problem for the phase ϕ_{00} and I personally suspect that the errors given by the experimentalists are underestimated and probably a 10^0 error is more realistic than a 5^0 one. But this is not my job and I shall ask the experimentalists to be very careful on this point.

– *Colas:*

Is there an experimental way of measuring f_B and B_b, outside the $B^0 - \overline{B}^0$ mixing?

– Gourdin:

Yes, the process $B^{\pm} \to \mu^{\pm} + \nu_{\mu}$. For the moment we have no experimental data.

Similarly, from the non observation of:

$$D^{\pm} \to \mu^{\pm} + \nu_{\mu}$$

we can deduce at a 90% confidence level that:

$$f_D < 290 \ MeV$$

which unfortunately is a little far from the theoretical expectation:

$$f_D \simeq (170 \pm 40) \ MeV$$

– Bobbink:

In the section of incoherent production from the continuum you considered $B_d^0 - \overline{B}_d^0$ and $B_s^0 - \overline{B}_s^0$ systems only. However, there are excited states and charged states. How are they taken into account?

– Gourdin:

Let me point out that experimentalists – not me – take only B_d^0 and B_s^0 into account. I agree that the vector mesons $B_d^{0^*}, B_s^{0^*}$ might be present and such a question is also related to the knowledge of the percentages P_s and P_d of B_s^0 and B_d^0 in the jet. For the moment we are only at the paleolitic age of this type of experiments ad before making reliable estimates we must have a correct knowledge of the content of the jets. I think that we are very far from such a situation.

GRAN SASSO PHYSICS

L. Votano

Istituto Nazionale di Fisica Nucleare
Laboratori Nazionali di Frascati
Via E. Fermi 40, 00044 Frascati, Rome, Italy

INTRODUCTION

Underground physics plays a leading role in current-day physics research and the experimentalists and theoreticians are showing a growing interest in the possibilities and availability of underground laboratories. The importance of this branch of physics and the relevant laboratories can be attributed to two main aspects:

1) Neutrino astronomy, i.e., the study of the universe using neutrinos as probes instead of the more traditional e.m. radiation in the form of visible radiation, X-rays, etc., is a new method of exploring the sky and represents a grand opportunity for physicists to further their knowledge of the universe, its origin and evolution, as well as the evolution of the stars.

2) The second aspect is connected to the study of elementary particles: physicists have long sought the unification of the forces of nature that are supposed generated by a unique force. The next objective is to unify electroweak with strong interaction, i.e., quarks with leptons. However, the energy scale for this to happen ($>10^{14}$ GeV) is not accessible to the particle accelerators either in operation or foreseeable in the near future. On the other hand, some of the predictions of the Grand Unification Theory regarding the stability of matter, the neutrino mass and the magnetic monopoles can be addressed in experiments carried out in underground laboratories which could, therefore, greatly contribute to the unification of the fundamental forces.

The Superworld III
Edited by A. Zichichi
Plenum Press, New York, 1990

Anticipating the interest in underground physics, at the beginning of the '80's Antonino Zichichi proposed[1] the Gran Sasso project. Its 200,000 m^3 of excavations and nine approved experiments make it the biggest international underground laboratory of the moment. The winning idea was to have a laboratory with the same characteristics and general facilities as the largest and best equipped connected to the particle accelerators; it had to be able to house complex and sophisticated apparatus equal to any of the biggest particle physics experiments. The Gran Sasso Laboratory has other extremely favourable conditions such as its geographic location and its proximity to the Gran Sasso Rome-Adriatic motorway which permits ease of access and heavy transport. In fact, before the Gran Sasso, underground experiments were installed in mines or in small "garages" near motorway tunnels, the only exception being Baksan which was conceived exclusively as a laboratory, but on a much reduced scale. From the physics point of view, the Gran Sasso can boast the following advantages. The laboratory depth (3,600 mwe) is a crucial parameter: in fact, for particles deriving from rare cosmic phenomena, there is an optimum depth at which these particles can be detected by the apparatus without being completely absorbed by the overlying rock when it is too thick, or being hidden by the background of low energy cosmic rays when it is too shallow. Another basic parameter is the natural radioactivity in the surrounding rock, which can constitute undesirable noise and submerge the signals from the rare phenomena under observation: the Gran Sasso is formed mainly of limestone which has a very low content of natural radioactivity[2] (for instance, the gamma and neutron fluxes are from 5-10 lower than those measured in Mont Blanc).

Table I outlines some of the principle characteristics of the Gran Sasso Laboratory.

TABLE I

Total volume	~200000 m^3
Depth	>1400 m of rock
	CaCO3 ~2.81g cm^{-3}, <Z>~9.4>3600 mwe
Location	47°27'09" Lat N
	13°34'28' Long E
	963 m a.s.l.
Rock activity	^{40}K<5.1 Bq/Kg
	^{114}Bi<4.2 Bq/Kg
	^{232}Th<0.25 Bq/Kg
	^{238}U<5.2 Bq/Kg
Muons	1 $m^{-2}h^{-1}$
Thermal neutrons	$(1.09\pm0.08)10^{-6}cm^{-2}s^{-1}$
Fast neutrons	$0.77\times10^{-6}<\Phi<1.25\times10^{-6}cm^{-2}s^{-1}$
Natural conditions	6°C,98% humidity
Operational conditions	20°C,50% humidity
Ventilation	~30000 m^3h^{-1} steady
	~200000 m^3h^{-1} in emergency

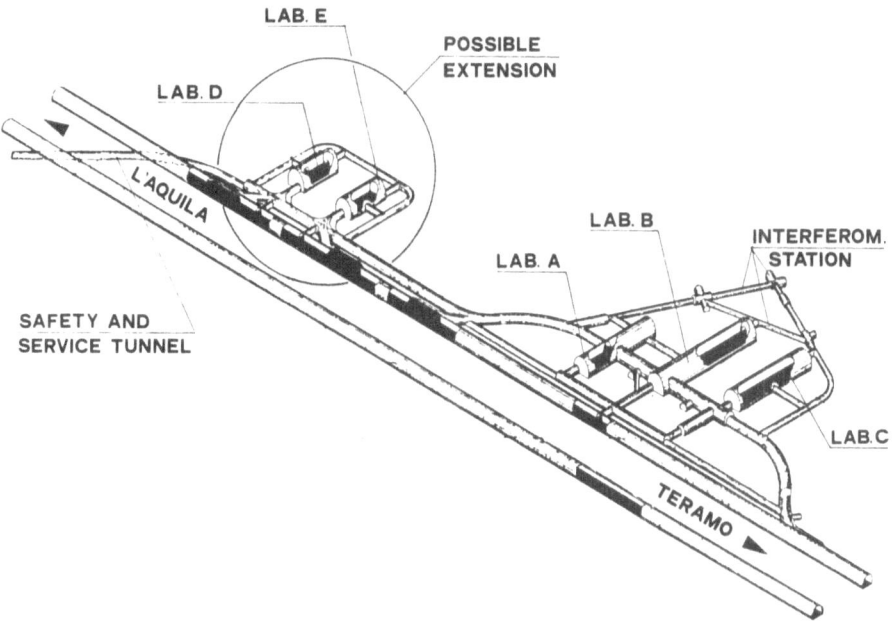

Fig.1. Present Gran Sasso Laboratory and probable extensions.

The scientific community has shown such an enormous interest in the laboratory that not only has the available space been completely allocated, but it has been felt necessary to increase it to make place for further experiments. In June 1988, the International Scientific Committee for the Gran Sasso unanimously approved a document requesting the Italian authorities to promote the excavation of more tunnels. A plan of the laboratory and its probable extensions is shown in Fig.1.

The Main Lines of Underground Physics

If we look at the main physics and astrophysical issues which will be researched at the Gran Sasso, two stand out, with neutrino physics playing a prominent role:

a) Neutrino astronomy comprising

- *the study of stellar collapse* (Eν~10 MeV), the main research object of the LVD experiment;

- *solar neutrino physics*, the primary objective of the GALLEX and ICARUS experiments and of the new one under study, BOREX;

- *the search for astrophysical point sources of very high energy neutrinos and gammas*, primarily under the LVD and MACRO experiments.

- *the search for supersymmetric particles*, good candidates for dark matter, through their annihilation in neutrinos.

b) <u>Problems related to the Grand Unification Theory</u>

- *neutrino masses* studied from two different aspects: the possible oscillations of neutrinos, using as source atmospheric neutrinos or solar neutrinos, and research on the neutrinoless double β *decay*;

- *the search for monopoles*, the main objective of the MACRO experiment and also feasible with the LVD;

- *proton decay in new channels*, one of the aims of ICARUS and LVD.

The study of the penetrating components of cosmic rays, tied to the problem of the composition and origin of primary cosmic rays, the research on gravitational waves from stellar collapse, as well as some geophysical experiments contribute to paint an extremely exciting picture of the Gran Sasso Laboratory.

STELLAR COLLAPSE AND TYPE II SUPERNOVAE

A Type II supernova explosion[3] is one of the most spectacular events in the sky and is the outer brilliant sign of a catastrophe marking the end of the evolution of massive stars: $M > 8M_\odot$..

These events are associated with neutron star or black hole birth and are copious sources of neutrinos detectable in underground experiments.

The final stage of the evolution of massive stars is an "onion skin" configuration: a central iron core surrounded by burning layers of Si, O, Ne, C, He, and H. When the fuel has been consumed, the star cannot support itself any longer against gravity and the core begins to collapse. As soon as the density of the inner region of the core exceeds the nuclear density, the collapse of the core stops and a shock wave begins to move outwards. If the wave reaches the outer envelope with enough energy, the supernova explosion can take place.

It has been calculated that about 3×10^{53} ergs (gravitational binding energy) must be released to form a neutron star: during the supernova explosion, $\sim 10^{49}$ ergs photons and $\sim 10^{51}$ ergs kinetic energy are emitted. The difference between the calculated and observed energies is in the form of neutrinos (99%) and gravitational waves (1%). Less than 10% of the neutrinos is radiated in the delta capture (neutronization) of the $\sim 10^{57}$ protons: $p + e^- \rightarrow n + \nu_e$, and the remainder in pair processes (deleptonization): $e^- + e^+ \rightarrow \nu_i + \overline{\nu}_i$ $i = e, \mu, \tau$.

The delta capture occurs in the initial collapse, the initial ν_e burst takes place in $< 10^{-2}$ s. When the density is $\geq 2 \times 10^{11}$ g cm^{-3}, the inner core is no longer transparent to the neutrinos and they are in equilibrium with matter. The pair neutrinos are thermally radiated in a time scale of the order of the diffusion process ($\sim$ seconds). Core implosion, bounce and shock wave take about 1 s during which the first

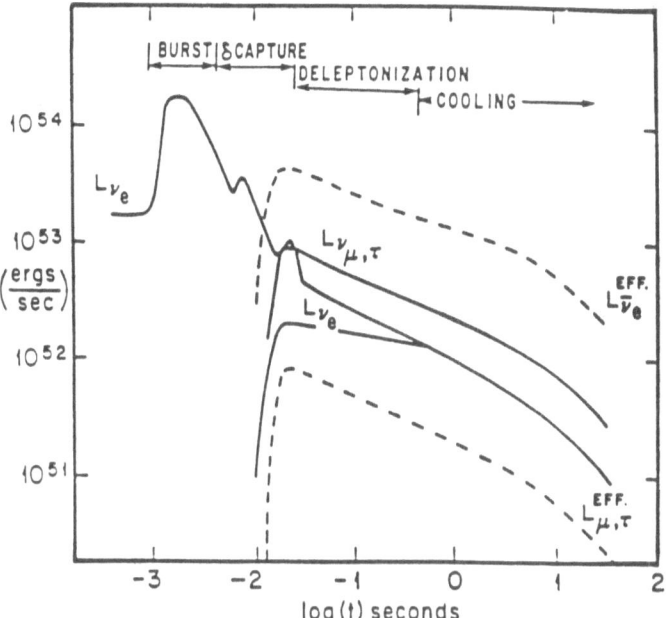

Fig.2. Neutrino emission time scale.

half of the neutrinos is emitted. The second half is emitted over the next few tens of seconds as the hot newborn neutron star cools down to become a standard cold neutron star (cooling phase). See Fig.2.

The general scenario of collapse is well understood, but the explosion mechanism and envelope ejection still need clarification. On the basis of the present models, the average neutrino luminosity, mean neutrino energy and total emitted energy depend only on the initial iron core mass and are independent of the explosion mechanism. As the time structure of the neutrino luminosity is strictly related to the explosion mechanism, it is crucial for an experiment to perform detailed studies on the time scale structure of neutrino emission.

THE LVD EXPERIMENT

The detector[4] consists of a large volume of liquid scintillator divided into modules surrounded by streamer chambers. Its design fulfills the following objectives:

The detection of neutrino interactions inside the detector: low energy neutrino interactions and measurements of neutrino energy; pattern identification of neutrino-induced events of higher energy.

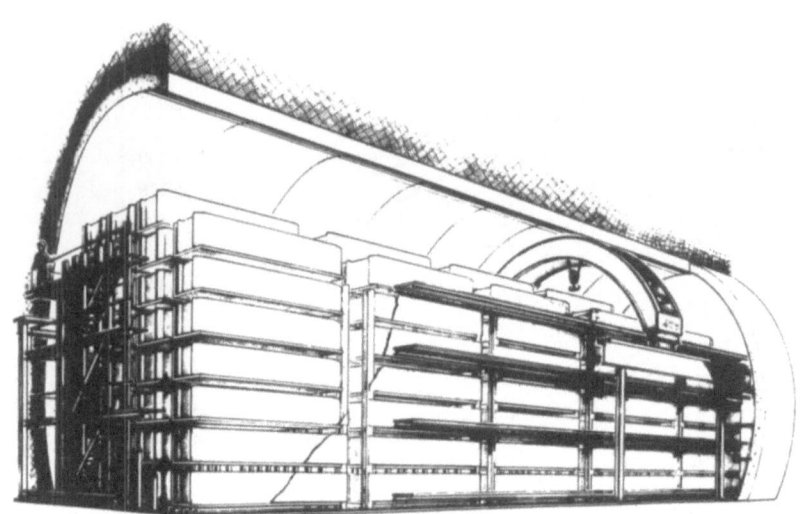

Fig.3. LVD - layout.

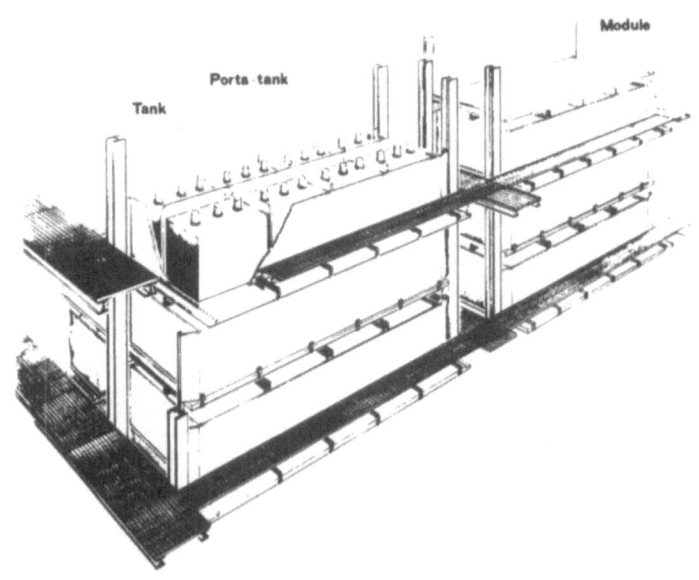

Fig.4. LVD - assembly.

The detection of mu mesons and measurements of their direction of flight: cosmic muons of very high energy ($E \geq 2$ TeV) crossing the apparatus; muons induced by neutrinos in the surrounding rock.

The LVD is basically formed of 190 identical modules, each one containing 9.6 tons of liquid scintillator, 6.7 tons of steel and surrounded on the bottom and one one side by a double layer of streamer tubes which makes up the tracking system (see Figs.3 and 4). The modules can be either 6.6 m × 2.1 m × 1.1 m or 6.6 m × 2.6 m × 1.1 m placed in an array of 40 m × 12 m × 13 m and are inserted in an iron support structure. The total weight including the support structure and tracking system is 3,600 tons.

The scintillator system. The basic element is a 1m × 1m × 1.5m stainless steel tank filled with well-tested Russian scintillator and housed eight-by-eight in the iron "portatank". Each tank is viewed by three 15 cm phototubes. The characteristics of the scintillator system are as follows: structure C_nH_{2n+2} with n=10; density 0.8 g cm^{-3}; attenuation length 20 m (λ=420 nm); decay time of a signal in the scintillator 5 ns; output light ~5 photoelectrons each PM for 1 MeV of energy loss; energy resolution 20%/$\sqrt{E}$.

The tracking system. The basic element of the tracking sytem is an L-shaped chamber containing ~80 6.3 m long standard PVC streamer tubes (LST) for a total of 15,000, with eight 1 cm × 1 cm cells assembled in a double layer. The chambers are arranged in eight (seven) double horizontal planes and five vertical double planes. The LST signals are picked up by 4-cm-wide Al+PVC strips. The characteristics of the tracking system are angular resolution 0.5 degrees, double layer efficiency ~100% and geometrical acceptance 7,700 m^{2}sr.

Detection of Stellar Collapse in LVD

The most convenient way to study stellar collapse is the detection of antineutrinos through the inverse β decay reaction (σ=10.3 10^{-42}cm^2 for E>4 MeV):

$$\bar{\nu}_e + p \rightarrow n + e^+$$
$$\downarrow$$
$$\rightarrow n + p \rightarrow D + \gamma \ (E_\gamma = 2.2 \text{ MeV})$$

Another source of information can be the ν-e scattering, essentially at the very first phase of collapse, even if it produces a lower number of interactions in the detector (σ=6.41 10^{-44}cm^2 for E>4 MeV).

When operating in a low background environment, the scintillator modular counters of the LVD are suitable for the detection of both the pulses that give the signature for an antineutrino:

- a prompt pulse from the positron with the energy above the high energy threshold (~6 MeV);

- a delayed pulse from the neutron during a gate width Δt~500 μs and energy threshold of about 0.8 MeV.

Antineutrino detection with this double signature in the same type of scintillator counter has already been checked in LSD at Mont Blanc[5]. The neutron moderation time plus deuterium fusion have been measured experimentally using a ^{252}Cf source and the efficiency for the detection of the neutron was found to be 70%. At Mont Blanc the trigger threshold is 6 MeV, limited by the natural radioactivity in the laboratory. The threshold could be lower in the Gran Sasso Laboratory.

In LVD we can expect about 1000 events (E$\geq$6 MeV) in 20 s for a collapse in the galactic centre (10 Kpc). On the basis of the background measured by LSD at Mont Blanc, and of the lower background presumed at the Gran Sasso, we can expect a noise rate of 0.1 counts/s, which means that the ratio signal/noise should be optimal. For an extragalactic source (i.e., Magellanic Clouds ~52 Kpc), the signal will obviously be lower but still detectable. Recent observations of neutrino signals from SN1987A have increased the interest in this field.

CRIOGENIC GRAVITATIONAL ANTENNA EXPERIMENT

A supernova explosion should also lead to a burst of gravitational waves as a consequence of the large and fast variation of the supernova quadrupole moment. Two different experimental methods can be used to observe gravitational waves: the laser interferometric system and resonating bars at a very low temperature.

A criogenic gravitational wave antenna[6] (3 m long, 2,300 Kg of Al 50/56) will be installed in the Gran Sasso laboratory to operate at about 50 m°K. The sensitivity to energy variation of the antenna fundamental vibration will be 10^{-6}-10^{-7} °K, sufficient to detect pulses coming from the Virgo Cluster ($\geq$20 Mpc). At this level, the antenna is also sensitive to cosmic rays; this can be avoided by veto shielding, but the effective "duty cycle" will be optimal only in an underground laboratory.

SOLAR NEUTRINOS

Solar neutrinos are produced in the hot interior of the sun (15×10^6 °K) by a series of nuclear fusion reactions which, according to the standard solar model, provide more than 98% of the energy required to explain the observed solar luminosity. Due to the small mean energy E_ν of neutrinos and their weak interaction with matter, they are the best probe for studying the interior of the sun. In fact, neutrinos can escape directly from a stellar interior, unlike photons which are emitted from the stellar surface - the photosphere - and give us conventional information about the stars; the mean free path for photons in stellar interiors where the nuclear fusion occurs is less than a centimetre. By letting us look inside a star, the neutrinos allow us to study and test

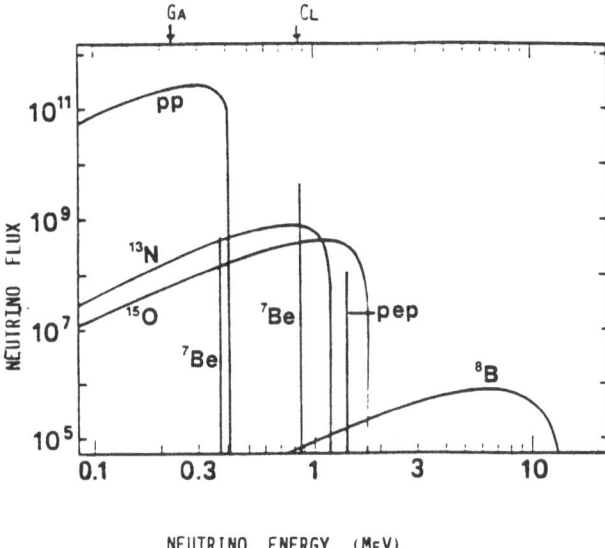

Fig.5. Flux of solar neutrinos reaching the earth vs neutrino energy.

the predictions of the Standard Solar Model. The measurement of the expected solar neutrino fluxes is thus the best test of the theory of stellar evolution.

We know more about the sun than any other star; we know all about its mass, luminosity, radius, temperature, surface composition, age, and its evolutionary phases in the context of the quiescent main sequence. Consequently, we can use the solar model to formulate theories on stellar evolution and their astronomical and cosmological applications.

Figure 5 shows the expected flux[7] at the earth of solar neutrinos vs neutrino energy. Most of the neutrinos come from the $pp \rightarrow de^+\nu_e$ reaction with neutrinos $0 < E_\nu < 0.47$ MeV, while the most energetic neutrinos come from the ^8B decay with $0 < E_\nu < 14.06$ MeV.

Experimental Situation and Solar Neutrino Puzzle

The first solar neutrino experiment was performed by Davis[8] et al. using ^{37}Cl and the capture reaction $\nu_e + {}^{37}Cl \rightarrow e^- + {}^{37}Ar$. The radioactive ^{37}Ar is extracted from a large volume of C_2Cl_4 and put in a small proportional counter where the Auger electron from the $^{37}Ar \rightarrow {}^{37}Cl$ decay is measured. The threshold of the reaction is 814 KeV. The Davis results are

- (1970-1984): A rate of 2.18 + 0.25 SNU corresponding to a ν_e flux at the earth $\Phi(\nu_e)(^8B) < 2 \times 10^6 cm^2 s^{-1}$.

-   (1986-1987): The results seem to agree better with SSM
    4.2±0.7 SNU.

Other preliminary and recent results come from the KAMIOKANDE[9] experiment which measures the $\nu_e e^-$ scattering with a threshold of 9.3 MeV in a fiducial volume of 680 tons of water Cerenkov detector:

-   (1987-1988): $\Phi \nu_e(^8B)=2.6 \ 10^6 cm^{-2}s^{-1}\pm30\%$

The origin of discrepancies between the experimental results and theoretical predictions can derive from a lack of knowledge of the nuclear reactions inside the sun (Standard Solar Model), or the behaviour of neutrinos (neutrino oscillations such as $\nu_e \to \nu_\mu$ on their path from the sun to the earth, or matter oscillations in the interior of the sun).

As a general comment we have to say that both Davis and Kamiokande are only sensitive to the most energetic neutrinos originating from the decay of ^8B, which are a strongly solar-model-dependent production and a rare side-branch of the fusion reaction chain. On the other hand, the neutrinos from the $pp \to D+\nu_e+e^+$ reaction are somewhat model independent and constitute the majority of the neutrinos emitted.

Gran Sasso Laboratory - A Challenge for the Solar Neutrino Puzzle

Many experiments are foreseen at the Gran Sasso regarding such a crucial problem:

- GALLEX -   a radiochemical experiment able to measure the pp neutrinos.

- ICARUS -   a direct counting experiment of the ^8B neutrinos with additional information on the direction of the incoming neutrinos.

The LVD experiment may well give additional measurements of the flux of ^8B neutrinos provided the noise is kept very low.

A new experiment - BOREX - is also foreseen for measuring both neutral and charge current interaction in ^{11}B. This could allow the simultaneous measurement of the total flux of neutrinos of all flavours and of the ν_e flux and is particularly important to distinguish between problems related to neutrino oscillations or to the Standard Solar Model.

THE GALLEX EXPERIMENT

The GALLEX[10] experiment (Fig.6) will provide measurements of solar neutrino fluxes using a 30-ton Gallium detector to study the reaction: $\nu_e+^{71}Ga \to ^{71}Ge+e^-$. As the energy threshold is 233.2 keV, the experiment is sensitive to pp neutrinos. The expected capture rates on the basis of SSM are given in Table II.

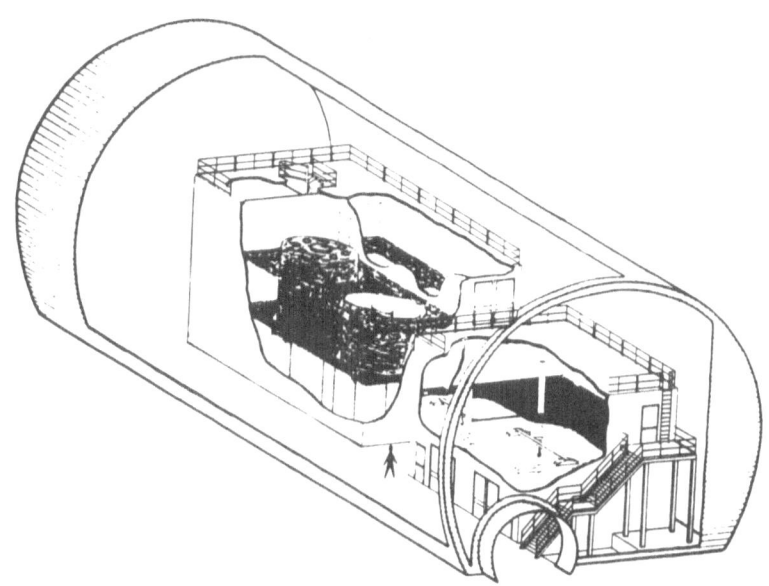

Fig.6. Sketch of the GALLEX experimental setup.

TABLE II

ν_e Source	Capture Rate
pp (0÷0.42 MeV)	71.0 SNU
pep (1.44 MeV)	2.5 SNU
^{7}Be (0.38÷0.86 MeV)	31.3 SNU
^{8}B (0÷14.06 MeV)	1.4 SNU
^{13}N (0÷1.20 MeV)	2.9 SNU
^{15}O	4.0 SNU

Detection Technique

The target material is a solution of $GaCl_3$ and the expected ^{71}Ge production rate is ~1 atom/day. The resulting $GeCl_4$ compound produced is highly volatile and has a half-life of 11.4 days; thus, every two weeks, the $GeCl_4$ will be swept away by a circulating stream of He. The $GeCl_4$ extracted will then be reduced to gaseous GeH_4 and after chromotographic purification put into the proportional counter where the decay ^{71}Ge+e$^-$→^{71}Ga+ν_e is measured by X-ray and Auger electrons. With the one neutrino capture/day expected in the 30-ton detector, it is clear that the background in the proportional counter (β particles from natural radioactivity, Compton electrons caused by external gamma rays, and electronic noise) must be kept as low as possible by using ultra pure materials and passive and anticoincidence shieldings. The aim of the experiment is to have a counting background of less than one count/week that will ensure a ±10% measurement with a three years' run.

Further problems could arise from concurrent reactions such as $^{71}Ge(p,n)^{71}Ge$ where the protons provoking this reaction are produced as secondaries of (α,p), (n,p) or cosmic ray reactions. This contribution has been estimated as less than a few percent. It will also be possible to calibrate the experimental setup using a 1-MC artificial ν source of ^{51}Cr yielding ν_e of 746 keV, which can populate both the 175 and 500 keV states in ^{71}Ge.

ICARUS I EXPERIMENT

ICARUS I (Imaging Cosmic and Rare Underground Signals) is the first step to a more ambitious programme, ICARUS,[11] which foresees a multikiloton liquid argon detector with an analysing magnetic field and which is mainly devoted to proton decay and solar neutrino physics. While most of the approved experiments are based on well-tested techniques, ICARUS represents a large scale innovation in technology; however, as pointed out by its authors, it must grow step-by-step.

The ICARUS I detector consists of a cryostat filled with 300 tons of ultra pure liquid argon. Due to its drift time technique and nondestructive and continuous read-out, this cryogenic image chamber allows a recording of the complete electron image produced by the ionizing event. The chamber has an excellent spatial resolution (±1 mm) and a colorimetric energy resolution of ~3% for 1 MeV electron. The direct observation of solar ν_e from ^8B can be done by measuring the neutrino-electron elastic scattering $\nu+e^-\rightarrow\nu+e^-$ and the neutrino absorption $\nu_e+^{40}Ar\rightarrow^{40}K*+e^-$. Assuming a measured flux of ν $2\times10^6cm^2s^{-1}$ in 200 tons of fiducial volume, the expected rates per year with an energy cut-off of 5 MeV are 80 scattering events and 90 absorbtion events.

We have to remember that the signature of solar neutrino elastic scattering events is the forward peaked angular distribution of the recoil electrons, so the possibility of measuring the electron direction in the image chamber will be a powerful tool for discriminating events from background. Figure 7 shows the neutrino-electron scattering as it will be recorded in the chamber.

MAGNETIC MONOPOLES

While the concept of magnetic monopoles had already been presented as a possibility before the theory of electromagnetism was formulated, today their existence is required by the Grand Unification and Supersymmetric Theories as a consequence of the breaking of a Grand or

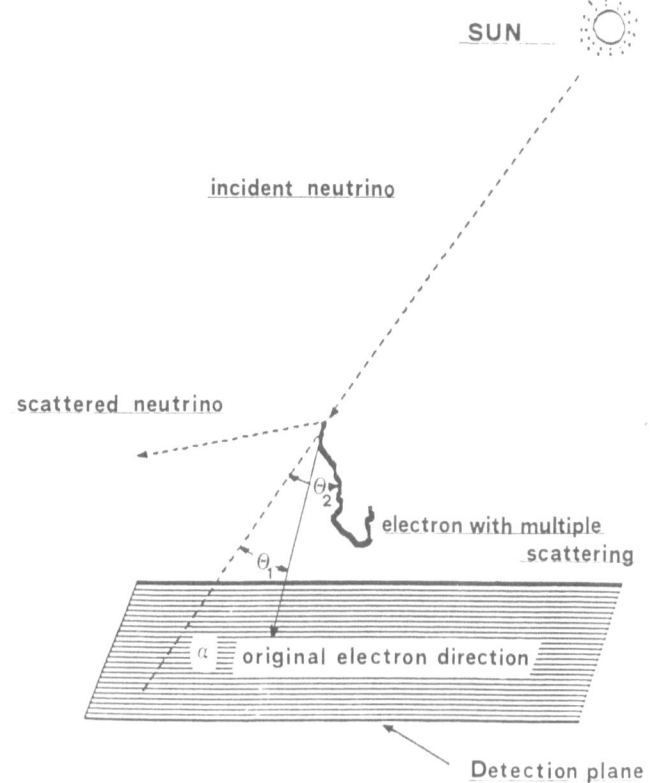

Fig.7. Schematic of the neutrino-electron scattering in ICARUS.

Supersymmetric Group G. Monopole properties can also be predicted by theory: the mass of the monopole is assumed $m_m = m_x/G \sim 10^{16}$ GeV, where m_x (X boson mass) is about 10^{14}-10^{15} GeV and the coupling constant G=0.03.

If unification also includes gravity, the monopole mass should be bigger, with the situation being complicated further if the Supersymmetric Theory is also considered. We can conclude[12] that the Grand Unification and Supersymmetric Theories predict the existence of monopoles of large masses, but that their exact value depends on the theory. Consequently, it is necessary for the experiments regarding monopoles to be sensitive to a large range of masses.

Due to their large mass, monopoles cannot be produced either in the accelerators of today and the near future, or in the cosmic ray collisions of high energy; the only time when such a high energy was available was during the first instants of the Big Bang when the temperature of the universe was comparable to the monopole mass. In the Standard Universe Model, GUT monopoles are thought to be produced at 10^{-35}s after the Big Bang during a phase transition when the unification between strong and electroweak interactions was broken down. It has been shown that due to the fast expansion of the universe, the pairs of stable monopoles produced had no chance to annihilate, so have had the possibility of surviving up to the present. In this case, nevertheless, their number density should be comparable to the baryon number density and monopoles should dominate the present density of the universe. The necessity of explaining the apparent suppression of monopoles by a factor $\geq 10^{15}$ has suggested several hypotheses: in the Inflationary Model of the universe, for instance, an exponential expansion of the universe is thought to have reduced the number of monopoles, provided that the inflation took place between the generation of the monopoles and that of the baryons. This model can also reproduce other important physical properties of the universe, while the monopole density, although delimited by inflation, is, in any case, measurable.

Astrophysical limits on the flux and velocity of monopoles

An upper limit on the monopole density and flux can be derived as a consequence of the existence of the galactic magnetic field, based on the consideration that too large a number of monopoles can extract energy at a rate which is unacceptable compared to the value of the galactic magnetic field. This limit - known as the Parker limit - which can be calculated as a function of mass and β of monopoles, $\Phi \sim 10^{-15}$ cm^{-2} sr^{-1}s^{-1}, is not fully accepted, but is a reference point in monopole research.

The velocity of monopoles depends on their origin. After their production, the monopoles may have also lost kinetic energy, may have clustered in galaxies, or been accelerated by the galactic magnetic field. For instance, if the monopoles are of extragalactic origin, their velocity would be comparable to the velocity of our galaxy in respect to the rest of the universe $\beta \sim 2 \times 10^{-3}$. On the other hand, if the monopoles are concentrated in our galaxy, their velocity would be comparable to the velocity of the sun in respect to the galactic centre $\beta \sim 8 \times 10^{-3}$, or to the galactic escape velocity $\beta \sim 10^{-3}$; if they are concentrated in the solar system, they would have a velocity of the order of 10^{-4}. The expected β-distribution could therefore present a minimum speed of 3×10^{-5} with a peak around 10^{-3} and some poles with higher energy.

It is clear that the Grand Unification Theory and cosmological considerations need more experimental verification in this field.

Detection Techniques

Several different techniques are thought suitable for detecting monopoles.

Induction technique (direct). The very small persistent currents induced by a monopole passing through a superconducting coil can be measured by a Josephson junction device (SQUID).

<u>Ionization/excitation technique.</u> The fundamental parameter is the monopole velocity. The calculation and measurement of the proton ionizing power at low β confirm that the slow monopoles are detectable[13] by a scintillator down to a velocity β of 6×10^{-4} (or 3×10^{-4} according to some hypotheses). For a gaseous detector the conventional threshold is 10^{-3}; through excitation by the Drell[14] mechanism and consequent ionization by Penning effect, the threshold can be as low as 10^{-4} in He.

<u>Track-etch technique.</u> The passage of charged particles creates submicroscopic damage trails in the lattice of the detector and these can be enlarged by chemical etching. Based on the well-defined response function for the development of the etch pit from Z/β, a threshold as low as 2×10^{-5} can be reached.

MACRO EXPERIMENT

The Monopole Astrophysics and Cosmic Rays Observatory experiment is a large area detector[15] with a planar structure and geometrical acceptance of 10,000 m²sr. The experiment has mainly been dedicated to monopole research, neutrino astrophysics and cosmic ray physics and consists of three types of detector:

- 2(3) horizontal planes of liquid scintillator counters $(0.75\times0.25\times 12)m^3$, each one viewed by two photomultipliers;

- 9(18) horizontal layers of streamer tubes; the eight-wire PVC chambers $(0.25\times0.03\times12)m^3$ are filled with a gas mixture of $He:CO_2$ npentane;

- a track-etch detector consisting of three layers of CR39 and five layers of Lexan.

Slabs of concrete absorbers are placed between the various layers of the detector.

One MACRO supermodule is $(12\times2\times4.5)m^3$ and the final dimensions are $(72\times12\times9)m^3$ (see Fig.8).

The Detection of Monopoles in MACRO

The use of three different kinds of detector allows a good redundancy. In the scintillators the monopole is univocally identified by the long light pulse emitted at the passage of the pole. A waveform digitizer records the pulse with a time granularity of 1% of total pulse duration. The time-of-flight of a particle between two layers is also recorded. In the streamer tubes the monopole appears as a track in space of uniformly delayed hits. The streamer pulse height is recorded too as tests on relativistic ion beams have shown that large ionization losses can be measured. The response from the track-etch detector will be analysed in the case of a monopole candidate.

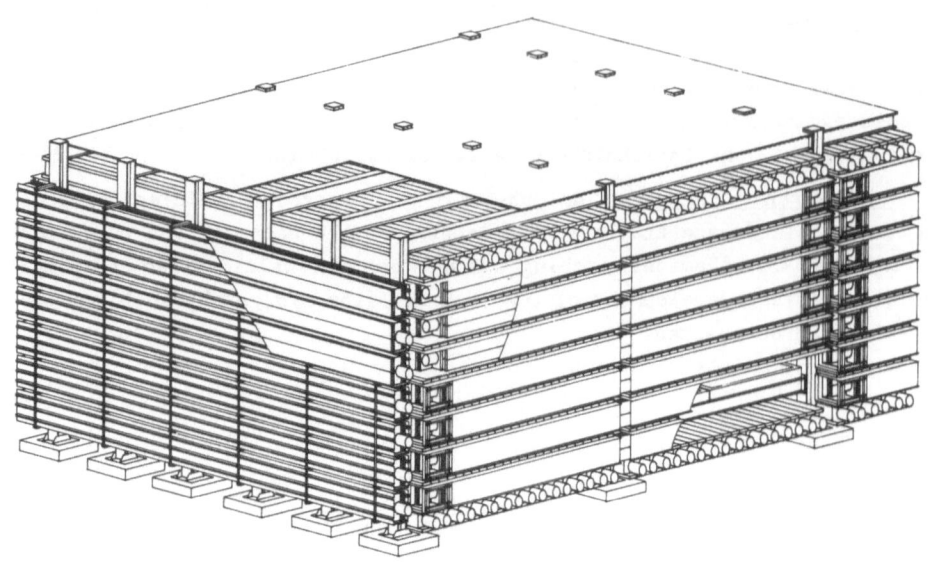

Fig.8.   A MACRO supermodule.

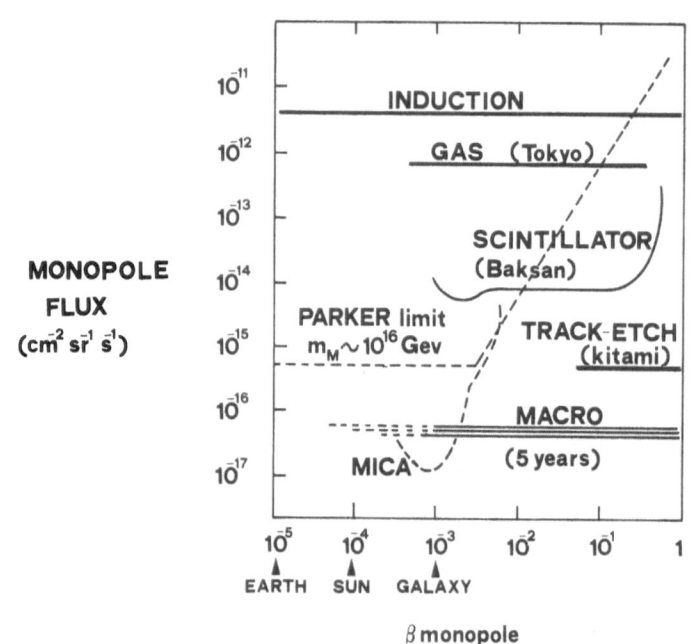

Fig.9.   Summary of the monopole research experiment limits.

Figure 9 gives a summary of the results from the main monopole research experiments using the various techniques and shows the expected MACRO sensitivities with respect to the Parker limit.

NEUTRINO MASS

The question of whether the neutrino has a nonzero mass is most important for both particle physics and astrophysics. Masses of the order of (1-10) eV could solve the problem of the dark matter in the universe, while masses of $<10^{-2}$ eV could explain the solar neutrino puzzle.

In the Standard Model, neutrinos are massless and both the individual lepton flavour numbers and the total lepton number are automatically conserved. However, most extensions of the Standard $SU_2 \times U_1$ Electroweak Model predict nonzero neutrino masses.

Direct kinematic limits on the masses of ν_e can be measured from tritium β-decay; for instance, in the Zurich-SIN[16] experiment, $m_{\nu_e} < 18$ eV. In addition, the measured flux and arrival times of the neutrinos from Supernova 1987A[17,18] also place a limit on the 20 eV range, while the limits on the m_{ν_μ} for $\pi-\mu$ decay, $m_{\nu_\mu} < 0.25$ MeV (SIN),[19] and on the m_{ν_τ} (from $\tau \rightarrow \nu_\tau + 5\pi$) $m_{\nu_\tau} < 50$ MeV (Argus) are relatively weak.

A less direct, but more sensitive approach to the problem of detecting the neutrino mass is to look for neutrino oscillations[20]. Two parameters are fundamental: $\Delta m^2 |m_1^2 - m_2^2| eV^2$, the difference in mass between the two-mass eigenstates, and the lepton mixing angle, $\sin^2\theta$. Oscillations can be looked for in

-   appearance experiments - search at a certain distance from the source for the appearance of a different neutrino type;

-   disappearance experiments - search for a reduction in the expected flux of a definite neutrino type.

There are presently many limits from accelerator experiments and reactors and some preliminary results from underground experiments (see Fig.10)[21].

The Gran Sasso Approach to the Neutrino Mass

The approach will be twofold:

-   a search for neutrino oscillations using atmospheric or solar neutrinos as source;

-   a neutrinoless double β decay experiment, $(A,Z) \rightarrow (A,Z+2)+2e^-$, that violates the lepton number by two units and that can take place only if the neutrinos have Majorana masses.

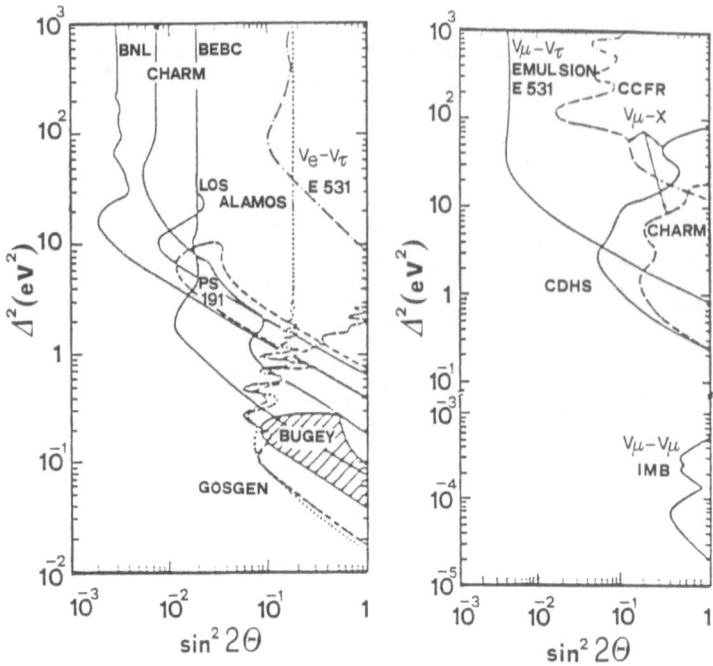

Fig.10.   90% CL limits for neutrino oscillations (from P. Langacker).

Neutrino Oscillations with Atmospheric Neutrinos

The idea of using underground detectors for neutrino flavour oscillations by employing the neutrinos produced in the atmosphere by cosmic rays[22] is very attractive because of the small value of the neutrino mass difference that can be explored. The neutrinos are originated in the (10-20)Km atmospheric shell surrounding the earth: the ones from the nearest zenith are called down-going, while the those which have travelled over a large fraction of the earth's diameter are known as up-going. The distance a neutrino has travelled is a function of its zenith angle $L = [r^2 - (r-d)^2 \sin^2\theta]^{1/2} - (r-d)\cos\theta$, where r is the earth's radius and d is the height of the atmosphere. If neutrino oscillations occur, it is possible to see differences in the interactions inside the detector of the neutrinos in the up-going solid angle and those in the down-going angle that have travelled only about 10^4 m. Thus, for a full-wavelength oscillation,

$$\Delta m^2 (eV^2) \equiv |m_1^2 - m_2^2| = 2.5 \, <E_\nu \, (MeV)>/L(m)$$

$$\Delta m^2 \approx 2.5 \times 600/1.3 \times 10^7 \ eV^2 \approx 10^{-4} \ eV^2$$

a small value of the neutrino mass difference may be explored. In the LVD

266

experiment with its large sensitive volume, it will be possible to search for neutrino flavour oscillations and a 3σ effect could be given after several years of operation.

Due to the low flux of atmospheric neutrinos, another attempt is looking at up-going muons produced by ν_μ in the rock surrounding the detector and their angular distribution. However, in this method the relevant neutrino energy (1-100 GeV) is much higher, which means less sensitivity in Δm^2. Furthermore, there are more uncertainties on the expected muon rate calculation.

DOUBLE B DECAY EXPERIMENT

This experiment[23] is devoted to the search for neutrinoless double β decay, which is forbidden by lepton number conservation. The detector is a multiwire proportional chamber of 61 cells, 80 cm long and with a cross section of 2.5 cm side. Each cell is filled with xenon at high pressure (10 bar) and the decay studied will be $^{136}Xe \rightarrow ^{136}Ba + 2e^-$.

A year of effective running without any detection will yield a 10^{23}-year limit on the lifetime corresponding to $m_\nu < 1.5$ eV.

In order to increase the performance of the detector, enriched xenon can be used in a low activity environment.

THE ORIGIN OF COSMIC RAYS

One of the most debated problems in the field of astrophysics concerns the origin of cosmic rays. One of their main characteristics is the fact that the chemical composition, at least up to a few GeV/nucleon, exactly reproduces that of the solar system. The power radiated in our galaxy is very high; it emits 10^{40}-10^{41} ergs/s of cosmic rays; 10^{44} ergs/s of visible light; 10^{39} ergs/s of radiowaves; 2×10^{39} ergs/s of X-rays

The most credible models today hold that cosmic rays must be generated either by an object of particular composition or by many different exotic sources. Several different sources have been proposed which would be able to act as extremely powerful particle accelerators and thus explain the energy of cosmic rays: pulsars, binary systems, supernova explosions, quasars, etc. A few discrete sources of cosmic radiation have been discovered in recent years,[24] in particular gammas of ultra high energy such as Cygnus X3, Vela X1, LMC X4, with the gammas having a spectral power of E^{-2} and a characteristic modulation time of the signal emitted. In order to study these point-form-sources of cosmic rays, it is necessary to use neutral particles only as charged particles have completely lost the memory of their initial direction due to the magnetic field of the galaxy. The obvious candidates are gamma-rays and neutrinos.

Research regarding discrete sources can be successfully carried out in underground detectors by examining either gammas in their interaction with the atmosphere, or neutrinos in their interaction with the rock surrounding the apparatus.

The large area and good spatial resolution of the MACRO and LVD will allow the study of the composition and origin of cosmic rays and the search for discrete sources of gammas and neutrinos of ultra high energy.

CONCLUSIONS

The Gran Sasso Laboratory with its great "observer" experiments provides an unequaled occasion to tackle the vast array of problems connected to both particle physics and astrophysics. This logically explains the enormous interest and participation in the experiments by physicists of different extractions and from all over the world. In the next few years we will no doubt be witness to extremely exciting results and new prospects in frontier physics.

ACKNOWLEDGEMENTS

I would like to thank R. Bernabei, L. Paoluzi and M. Spinetti for useful information and discussions.

REFERENCES

1. A. Zichichi, Invited Paper at the Int. Workshop, Rome, 29-30 October 1981, preprint INFN/AE-82/1; Proceedings of the Workshop on Science Underground, Los Alamos, 27 September - 1 October, 1982, p.52; Proceedings of the ICOMAN 83, Frascati, 17-21 January, 1983, p.8 - preprint INFN/AE-83/8.

2. G. Campos Venuti et al., INFN/LNF 82/78 (1982); E. Bellotti et al., INFN/TC 85/19 (1985).

3. R. Mayle et al., Ap. J. 318:288 (1987).

4. C. Alberini et al., Nuovo Cimento 9C:237 (1986); G. Bari et al., N.I.M. A264:5 (1988).

5. G. Badino et al., Nuovo Cimento C7:573 (1984).

6. G. Pizzella et al., "Proposal for a Gravitational Wave Detector at the Gran Sasso Lab." (1985).

7. J.N. Bachal, Rev. of Mod. Phys. 54:767 (1982).

8. R. Davis, Phys. Rev. 79:749 (1964).

9. K. Hirata et al., Contribution to the XVI INS Int. Symp. on Neutrino Mass and Related Topics, Tokyo (16-18) March 1988.

10. T. Kirsten et al., "GALLEX, a Proposal for the Gran Sasso Lab." (1985).

11. C. Rubbia et al., "ICARUS, a Proposal for the Gran Sasso Lab". (1985).

12. P. Musset, Nuovo Cimento 9C:559 (1986).

13. S. Ahlen et al., Phys. Rev. D27:688 (1983).

14. S.D. Drell et al., Phys. Rev. Lett. 50:644 (1983).

15. M. Calicchio et al., Nuovo Cimento 9C:281 (1986).

16. M. Fritschi et al., Phys. Lett., 173B:485 (1986).

17. K. Hirakata et al., Phys. Rev. Lett. 58:1490 (1987).

18. R.M. Bionta et al., Phys. Rev. Lett. 58:1494 (1987).

19. R. Abela et al., Phys. Lett. B146:431 (1984).

20. B. Pontecorvo, JETP 3:247 (1958); JETP 7:171 (1958).

21. P. Langacker, DESY 88-023 February 1988.

22. D.S. Ayres et al., Phys. Rev. D29:902 (1984).

23. E. Bellotti et al.,"A Proposal for an Experiment on Double Beta Decay in the Gran Sasso Lab." (1985).

24. M. Samorsky et al., Ap. J. Lett. L27:268 (1983); R.J. Protheroe et al., Ap. J. Lett., L47:280 (1984).

DISCUSSION

– *Mincer:*

Is there a surface air-shower array planned at Gran-Sasso, that would operate in conjunction with the underground detectors?

– *Votano:*

At the moment it is a planned experiment. It is in a certain sense small: 28 elements of 10 m^2 liquid scintillator separated by 25 m in the central part and 100 m in the external part.

– *Zichichi:*

The top laboratory is an intrinsic part of the original project of the Gran Sasso. It still has to be done.

– *Mincer:*

Will that information be tied into the other experiments, for instance will MACRO have timing information?

– *Votano:*

Yes, as this is very important for many experiments.

– *Colas:*

As the absolute time measurement is needed to compare experiments and to get precise absolute directions of stellar sources, is there a general clock facility foreseen in the Gran Sasso Laboratory?

– *Votano:*

A facility to have the absolute time with the best accuracy is foreseen at Gran Sasso.

– *Bobbink:*

How much energy does a slowly moving monopole loose in going thru the rock above Gran Sasso laboratory?

– *Votano:*

A monopole has a large energy loss in matter due to excitation and ionization. These losses are proportional to $g^2\beta^2$ and for an ultra-relativistic monopole are of the order of 10 GeV/g.cm^{-2}. For low β there exists a maximum energy transfer which fixes also the threshold for various detectors. As I showed in my lecture, the energy loss in the rock, at least for the standard model, is not a problem, as the monopoles we consider are so massive that they have an enormous energy.

– *Volkas:*

Has the monopole-like event observed by Cabrera a few years ago been explained as a more conventional phenomenon?

– *Zichichi:*

This experiment has been repeated with sensitivity increased by two or three orders of magnitude, and nothing more has been observed. However, this event should still be included in the data sample.

– *Fordham:*

I believe he has written that a physical shock produces the same signal.

– *Marchioni:*

How much can the present limits on proton decay be improved by Gran Sasso?

– *Votano:*

In the LVD, it is possible to make a search for the channel predicted by supersymmetry, $p \to K^+ + \overline{\nu}$. LVD can have a good signal since it can, in principle, detect the signal from K^+ and then from μ^+ decay. In ten years of running, we should reach a limit greater than 10^{32} years.

– *Calafiura:*

Will the Gran Sasso Laboratory ever have a Kamiokande-like experiment with a water volume two to three orders of magnitude larger than Kamiokande?

– *Zichichi:*

The whole thing was originally planned to be Kamiokande–like technology, but the results from Kamiokande were not considered worth investing an immense amount of money in it. Kamiokande is relevant only if you increase the water volume by three orders of magnitude, which is at present science fiction.

– *Colas:*

What is the fiducial mass of the large volume experiment?

– *Votano:*

It is about 1.2 kilotons.

– *Zichichi:*

The 1.2 kilotons is what we considered the first approach, in order to get experience, but the real size of the experiment is to be 10 kilotons. The originality of the experiment is that we should have the same efficiency in all the decay channels. This is the aim of the experiment.

– *Sarid:*

Will Icarus be able to say anything about neutrinos from supernovae?

– *Votano:*

In the first phase of the detector, Icarus I, it seems to be quite impossible, but it will be possible if the giant chamber is built and filled with methane instead of liquid argon.

– *Zichichi:*

The first Icarus is called "Baby Icarus".

– *Votano:*

Icarus I is 300 tons instead of 4.5 kilotons for the full-size project.

– *Brandt:*

There were about a hundred papers that came out of the 1987 supernova. Have any modifications been made to the detectors as a result of all this information?

– *Votano:*

The results from SN 1987 A are hotly debated because if both Kamio- kande and IMB on one side and Mont-Blanc on the other side are true measurements of the supernova explosion, a special model needs to be invoked. Anyway, no reasons appear from this debate in favor of changes in the detector.

– *Polychronakos:*

I am curious to know what veto shielding is.

– *Votano:*

Cosmic rays can be a source of noise in the gravitational wave detector because of its level of sensitivity. The veto shielding can be used to silence the detector during the time of passage of cosmic rays, but on the surface of the earth the density of cosmic particles is so high that the antenna would be silenced continuously. That is why the detector is underground.

272

– Servoli:

Are there plans to measure the very high energy tail of the cosmic ray spectrum at Gran Sasso?

– Votano:

LVD and MACRO have very large tracking detectors, so the physics of multi-muon events, anisotropies, etc., can be studied with very great sensitivity.

– Mannel:

Is there any estimate of the rate of stellar collapses?

– Votano:

The rate should be one about every thirty years in our galaxy. I would like to underline that due to their association to massive stars, Type-II supernovae are seen only in spiral arms of spiral galaxies, regions which are normally optically obscured. However, neutrinos are not obscured, and supernovae can be detected everywhere in our galaxy.

– Bobbink:

How many electronics channels are their on the LVD detector?

– Votano:

The tracking system is something like 10^5, with 15,000 streamer tubes. The scintillator system has about the same number of channels.

– Sarid:

What results do you expect to have one year from now, in summer 1989?

– Votano:

LVD and MACRO should have part of their detectors ready and some results. The GALLEX detector, the solar neutrino experiment, should have part of its gallium ready also.

TUTORIAL GUIDE TO THE TAU LEPTON AND

CLOSE–MASS LEPTON PAIRS*

Martin L. Perl

Stanford Linear Accelerator Center
Stanford University
Stanford, California 94309

ABSTRACT

This is a tutorial guide to present knowledge of the tau lepton, to the tau decay mode puzzle, and to present searches for close-mass lepton pairs. The text is minimal; the emphasis is on figures, tables and literature references. It is based on a lecture given at the 1988 International School of Subnuclear Physics: The Super World III.

1. HISTORY OF CHARGED LEPTON DISCOVERIES

Each of the known charged leptons, e, μ, and τ, was discovered through a different technique. The electron was discovered in the 1890's by Thomson[1] using a cathode ray tube. The muon was discovered in the 1930's by Neddermeyer and Anderson.[2]

The modern history of the search for heavier leptons using the signature

$$e^+ + e^- \rightarrow L^+ + L^- \ ,$$
$$L^+ \rightarrow e^+ + \nu_e + \bar{\nu}_L \ ,$$
$$L^- \rightarrow \mu^- + \bar{\nu}_\mu + \nu_L \ , \tag{1}$$

began at the Adone e^+e^- storage ring with the work of Bernardini et al.[3], Fig. 1, and of S. Orito et al.[4]

The tau was discovered in 1974–1975 at the SPEAR e^+e^- storage ring by Perl et al.[5] using the $e\ \mu$ signature, Fig. 2. In the period 1975–1978 the basic properties of the τ were established by numerous experiments at the SPEAR and DORIS rings. Since then the detailed properties of the τ have been measured by many experiments at the DORIS, PEP, PETRA, SPEAR, and TRISTAN rings.

No other charged leptons have been found,[6] Secs. 8, 9.

2. BASIC PROPERTIES OF THE τ

Almost all data on the τ comes from

$$e^+ + e^- \rightarrow \tau^+ + \tau^- \ , \tag{2}$$

through both γ and Z^0 s–channel exchange, Fig. 3. Up to the highest energy at which the τ has been detected, 56 GeV at TRISTAN, γ–exchange is the main amplitude.

---

*Work supported by the Department of Energy, contract DE–AC03–76SF00515.

The Superworld III
Edited by A. Zichichi
Plenum Press, New York, 1990

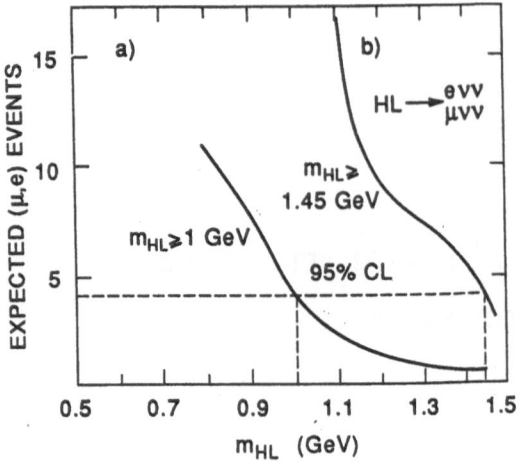

Fig. 1. Results of the search for a heavy lepton, called HL, by M. Bernardini *et al.*[3]

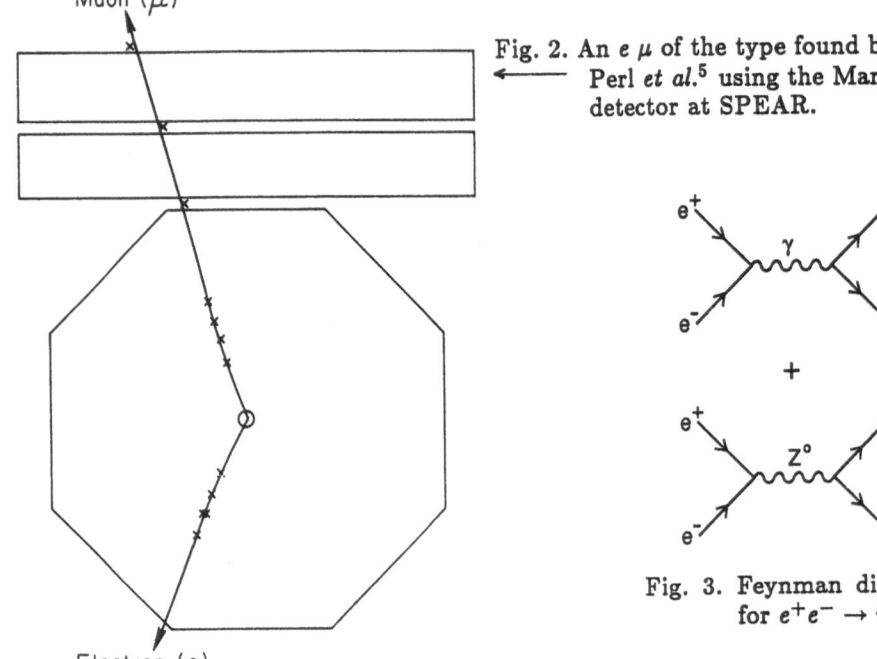

Muon (μ)

Fig. 2. An $e\,\mu$ of the type found by Perl *et al.*[5] using the Mark I detector at SPEAR.

Electron (e)

Fig. 3. Feynman diagrams for $e^+e^- \to \tau^+\tau^-$.

All existing data agrees with the τ being a spin 1/2 point particle of unit charge with the V–A weak interaction, and with no strong interaction. The mass based mostly on an old measurement[7] is

$$m_\tau = 1784 \pm 3 \text{ MeV}/c^2 \; . \tag{3}$$

The lifetime,[8] Sec. 6.2, is

$$\tau_\tau = (3.03 \pm 0.09) \times 10^{-13} \; s \; . \tag{4}$$

All known decay modes of the τ are consistent with τ lepton number conservation

$$\tau^- \to \nu_\tau + \text{ other particles} \tag{5}$$

No violations have been found, Table 1.

Table 1. Upper Limits on Branching Ratios for τ Decay Modes that would Violate τ Lepton Number Conservation. Limits at 90% Confidence Level. ℓ^- Means e^- or μ^-.

Decay Mode	Upper Limit	Experimental Group	Reference
$\tau^- \to e^- e^+ e^-$	3.8×10^{-5}	ARGUS	H. Albrecht *et al.*,
$e^- \mu^+ \mu^-$	3.3×10^{-5}		Phys. Lett. **185B**, 228 (1987)
$\mu^- e^+ e^-$	3.3×10^{-5}		
$\mu^- \mu^+ \mu^-$	2.9×10^{-5}		
$\ell^- \ell^\mp \ell^\pm$	3.8×10^{-5}		
$e^- \pi^+ \pi^-$	4.2×10^{-5}		
$\mu^- \pi^+ \pi^-$	4.0×10^{-5}		
$e^- \rho^0$	3.9×10^{-5}		
$\mu^- \rho^0$	3.8×10^{-5}		
$\ell^\mp \pi^\pm \pi^-$	6.3×10^{-5}		
$e^- \pi^+ K^-$	4.2×10^{-5}		
$\mu^- \pi^+ K^-$	1.2×10^{-4}		
$e^- K^{*0}$	5.4×10^{-5}		
$\mu^- K^{*0}$	5.9×10^{-5}		
$\ell^\mp \pi^\pm K^-$	1.2×10^{-4}		
$e^- \gamma$	2.0×10^{-4}	CRYSTAL	S. Keh *et al.*, (1988)
$e^- \pi^0$	1.4×10^{-4}	BALL	DESY 88-065
$e^- \eta$	2.4×10^{-4}		SLAC-PUB 4634
			HEN-25
$e^- K^0$	1.3×10^{-3}	MARK II	K. G. Hayes *et al.*,
$\mu^- K^0$	1.0×10^{-3}		Phys. Rev. **D25**, 2829 (1982)
$\mu^- \gamma$	5.5×10^{-4}		
$\mu^- \pi^0$	8.2×10^{-4}		
$e^- \pi^0$	2.1×10^{-4}		

The tau neutrino, ν_τ has never been directly detected. All its properties are deduced from τ decays, Eq. 5. The deductions are consistent with the ν_τ being a spin 1/2 point particle with the V–A weak interaction, and with no strong interaction. The 95% C.L. upper limit[9] on the mass is

$$m_{\nu_\tau} < 35 \text{ MeV/c}^2 . \tag{6}$$

3. τ DECAYS: THEORETICAL CONCEPTS AND BRANCHING FRACTION MEASUREMENTS

The decay of the τ takes place through W–exchange, Fig. 4. If the three fermion pairs, $(e^-, \bar{\nu}_e), (u^-, \bar{\nu}_\mu), (d, \bar{u})$ are treated equally the following branching fractions are predicted:

$$
\begin{aligned}
B_e &= B(\tau^- \to \nu_\tau e^- \bar{\nu}_e) = 20\% , \\
B_\mu &= B(\tau^- \to \nu_\tau u^- \bar{\nu}_\mu) = 20\% , \\
B_{had} &= B(\tau^- \to \nu_\tau \text{ hadrons}) = 60\%
\end{aligned}
\tag{7}
$$

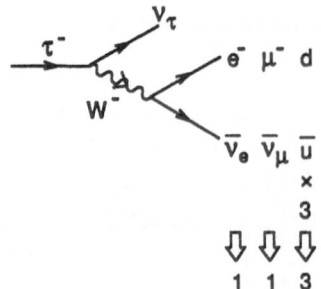

Fig. 4.   Feynman diagram for τ decay.

Measurement gives:[8]

$$B_e = (17.6 \pm 0.4)\% ,$$
$$B_\mu = (17.7 \pm 0.4)\% \qquad\qquad (8)$$

and by subtraction from 100%,

$$B_{had} = (64.7 \pm 0.6)\% .$$

The difference between the B's in Eqs. 7 and 8 is mostly caused by final state strong interactions in

$$\tau^- \to \nu_\tau + \text{hadrons} . \qquad\qquad (9)$$

The branching fraction B_i for mode i is given by

$$B_i = \Gamma_i/\Gamma , \qquad\qquad (10)$$

where Γ_i and Γ are the decay widths for mode i and for the sum of all modes. The Γ_i's for the e and μ modes are exactly calculated[10] from weak interaction theory. The theory predicts

$$B_e/B_\mu = 0.973 , \qquad\qquad (11)$$

and measurement, Eq. 8, agrees.

Some Γ's for modes containing hadrons can be calculated[11-13] from non-τ data. These modes include

$$\tau^- \to \nu_\tau + \pi^- ,$$
$$\tau^- \to \nu_\tau + K^- ,$$
$$\qquad\qquad (12)$$
$$\tau^- \to \nu_\tau + \rho^- ,$$
$$\tau^- \to \nu_\tau + (4\pi)^- .$$

At present there is no way to calculate the Γ_i for some hadron-containing modes such as

$$\tau^- \to \nu_\tau + (3\pi)^- . \qquad\qquad (13)$$

The calculation of the total width for all hadron-containing modes, Γ_{had}, is difficult[14] and uncertain by 5 to 10%. Therefore at present calculations of all branching fractions

$$B_i = \frac{\Gamma_i}{\Gamma_e + \Gamma_\mu + \Gamma_{had}} \qquad\qquad (14)$$

are uncertain by 5 to 10%.

4. τ DECAYS: WELL–MEASURED BRANCHING FRACTIONS

4.1 Topological Branching Fractions

The average measured values of the inclusive or topological, branching fractions into 1, 3, 5, or 7–charged particles are[8,15,16]

$$B_1 = (86.6 \pm 0.3)\% , \qquad B_5 = (0.10 \pm 0.03)\% ,$$
$$B_3 = (13.3 \pm 0.3)\% , \qquad B_7 \leq 0.019\% , \qquad 90\% \text{ CL} . \qquad (15)$$

Thus, most decays have 1–charged particle, almost all the rest have 3–charged particles.

4.2 Well–Measured One–Charged Particle Branching Fractions

The well-measured 1–charged particle branching fractions are given in Table 2. The sum of these branching fractions is $(77.9 \pm 1.5)\%$. Comparing this sum to B_1 in Eq. 15, there must exist another 8 or 9% in poorly measured or unmeasured 1–charged particle modes, such as

$$\tau^- \to \nu_\tau + \pi^- + n\pi^0 \quad , \qquad n > 2 ,$$
$$\tau^- \to \nu_\tau + \pi^- + n\eta \quad , \qquad n > 0 . \qquad (16)$$

4.3 Three–Charged Particle Branching Fractions

The value $B_3 = (13.3 \pm 0.3)\%$ is better understood,[6] Table 3.

Table 2. Well–Measured One–Charged Particle Branching Fractions.

Symbol	Decay Mode	Branching Fraction (%)	Ref.
B_e	$\nu_\tau + e^- + \bar{\nu}_e$	17.6 ± 0.4	8
B_μ	$\nu_\tau + \mu^- + \bar{\nu}_\mu$	17.7 ± 0.4	8
B_π	$\nu_\tau + \pi^-$	10.8 ± 0.6	8
B_ρ	$\nu_\tau + \rho^-$	22.5 ± 0.9	8
$B_{\pi 2\pi^0}$	$\nu_\tau + \pi^- + 2\pi^0$	7.6 ± 0.8	17–19
B_{K1}	$\nu_\tau + mK + n\pi^0$ $\longrightarrow$ 1–charged particle $m \geq 1, n \geq 0, K = K^0$ or K^-	1.7 ± 0.3	20
	Sum of above	77.9 ± 1.5	

Table 3. Three–Charged Particle Branching Fractions.

Symbol	Decay Mode	Branching Fraction (%)
$B_{2\pi^- \pi^+}$	$\nu_\tau + \pi^- + \pi^+ + \pi^-$	6.7 ± 0.4
$B_{2\pi^- \pi^+ n\pi^0}$	$\nu_\tau + \pi^- + \pi^+ + \pi^- + n\pi^0 , n > 0$	5.0 ± 0.5
B_{K3}	$\nu_\tau + mK + n\pi^0$ $\longrightarrow$ 3–charged particles $m \geq 1, n \geq 0, K = K^0$ or K^-	0.9 ± 0.4
	Sum of above	12.6 ± 0.7

5. THE ONE–CHARGED PARTICLE DECAY MODE PROBLEM

5.1 Use of Only Direct Branching Fraction Measurements

Table 4 gives the sum of direct measurements compared with B_1. There is no problem with this restricted information.

Table 4. Summary of Direct Measurements of Branching Fractions of One–Charged Particle Modes Using Only One–Charged Particle Decays.

Type of Information	Row	Decay Mode	Branching Fraction (%)
Sum of well measured modes in Table 3	A		77.9 ± 1.5
Upper limit deduced or estimated in 1–charged particle decays	B	$\nu_\tau \pi^- 3\pi^0$	< 2.5
	C	$\nu_\tau \pi^- 4\pi^0 + \nu_\tau \pi^- 5\pi^0$	$\lesssim 4.$
	D	$\nu_\tau \eta$	< 0.3
	E	$\nu_\tau \eta n\pi^0$	< 2.1
	F	$\nu_\tau 2\eta$	< 1.4
Sum of rows B–F	G		$\lesssim 10.3$
Sum of A + G			$\lesssim 88.2 \pm 1.5$
1–charged particle topological B_1			86.6 ± 0.3

5.2 Use of Theory and Other Data

The 1–charged particle decay mode problem appears when theory[11-13] and other data are used to evaluate or set upper limits on the branching fractions in Rows B–F of Table 4. I paraphrase Sec. III of Ref. 21 to explain the use of theory. There are four methods

In method (a), a directly measured 3–particle or 5–charged particle branching fraction is used to set an upper limit on a 1–charged particle branching fraction by invoking strong isospin conservation. For example, direct measurement gives

$$B(3\pi^- 2\pi^+ \nu_\tau) = (0.051 \pm 0.020)\% ,$$

and strong isospin conservation requires

$$B(\pi^- 4\pi^0 \nu_\tau) \leq \frac{3}{4} B(3\pi^- 2\pi^+ \nu_\tau) ;$$

hence,

$$B(\pi^- 4\pi^0 \nu_\tau) \leq 0.06\% , \qquad 95\% \text{C.L.}$$

In method (b) the η decay mode

$$\eta \to \pi^+ + \pi^- + \pi^0 ,$$

is used in the direct measurement of an η containing mode.

In method (c) we calculate a 1–charged particle branching fraction using the conserved vector current rule and a corresponding $e^+ e^-$ cross section.

In method (d) the rule against a second class current forbids the decay mode

$$\tau^- \to \pi^- + \eta + \nu_\tau .$$

The results of these considerations are given in Table 5. The 10.3% upper limit in Row G of Table 4 is replaced by 2.7%.

Table 5. Values and Upper Limits of Branching Fractions for One–Charged Particle Modes Deduced from Theory and Other Measurements. The Sum Does Not Include Modes with $\nu_\tau \pi \eta n \pi^0$, $n > 2$.

Mode	Method	Value (%)	Upper Limit (%) 95% C.L.
$\nu_\tau \pi^- 3\pi^0$	c	1.0 ± 0.15	1.25
$\nu_\tau \pi^- 4\pi^0$	a		0.06
$\nu_\tau \pi^- 5\pi^0$	a		0.11
$\nu_\tau \pi^- \eta$	d		0.00
$\nu_\tau \pi^- \eta \pi^0$	c		0.24
$\nu_\tau \pi^- \eta 2\pi^0$	a		0.40
$\nu_\tau \pi^- \eta \eta n \pi^0$, $n \geq 0$	b		0.60
Sum			2.7

5.3 The τ Decay Mode Problem

The τ 1–charged particle decay mode problem appears when the upper limits from Table 5 are added to the well-measured branching fractions of Table 2. As shown in Table 6 about 6% of the 86.6% in B_1 is not explained.

Table 6. Branching Fractions for One–Charged Particle Decays.

Source of Information	Branching Fraction (%)
Sum of well-measured modes from Table 2	77.9 ± 1.5
Sum of 95% C.L. upper limits from Table 5	≤ 2.7
Sum of above	$\leq 80.6 \pm 1.5$
Topological branching fraction B_1	86.6 ± 0.3

6. DISCUSSION OF τ DECAY MODE PROBLEM

6.1 Error Analysis

The significance of the τ decay problem depends upon the validity of the error analysis. The validity has been examined in two recent paper: Hayes and Perl[8] and Hayes, Perl, and Efron.[22] The former paper uses Gaussian error analyses, the latter uses the much more general bootstrap analysis method, applied to the branching fractions:

B_1 based on 11 measurements ,

B_e based on 10 measurements ,

B_μ based on 16 measurements , $\qquad\qquad$ (17)

B_π based on 7 measurements ,

B_ρ based on 6 measurements .

The Gaussian error analysis shows:

(a) The errors associated with an individual measurement by the experimenters who made the measurement are either about right or too large. Therefore the decay problem cannot be explained away by arbitrarily enlarging these errors.

(b) There is evidence for bias in the B_ρ measurements and hints of bias in other measurements in the sense that the individual measurements cluster more about their central value than their individual errors would predict. We cannot tell if this bias has shifted the central value from the true value.

(c) The Gaussian error analysis does not resolve the decay mode problem.

The bootstrap analysis method finds:

(1) The mean values of the branching fractions in Eq. 17 are similar to, but not identical to, the means found by the Gaussian analysis.

(2) The bootstrap method still shows the decay mode problem, but with smaller statistical significance compared to the Gaussian error analysis.

6.2   Comparison of B_e and B_μ with τ Lifetime

The τ lifeline, τ_τ, calculated[8] from B_e and B_μ, is

$$\tau_\tau \ (\text{predicted}) = (2.87 \pm 0.04) \times 10^{-13} \ \text{s} \ ,$$

compared to

$$\tau_\tau \ (\text{measured}) = (3.03 \pm 0.09) \times 10^{-13} \ \text{s} \ .$$

The difference

$$\tau_\tau(\text{measured}) - \tau_\tau \ (\text{predicted}) \ = (0.15 \pm 0.10) \times 10^{-13} \ \text{s} \ ,$$

is 1.5 standard deviations. This does not have enough significance to require B_e and B_μ to be larger than the values in Table 2.

6.3   Search for an Unconventional Explanation of the Decay Mode Problem

I don't know if unconventional physics in tau decay is the explanation of the decay problem, no satisfactory unconventional explanation has been found. Experiments have ruled out[22] the possibility that the missing 6% could come from η–containing modes. A recent idea of mine has failed,[23] the hypothesized existence of a second tau neutrino with mass close to m_τ.

7.   FUTURE RESEARCH ON THE τ

There is much experimental research to be carried out on the τ:

(a) resolution of the 1–charged particle decay mode problem;

(b) modern measurements of m_τ and tests of V-A;

(c) more sensitive study of m_{ν_τ};

(d) detection and properties of ν_τ;

(e) devise a method to measure $g_\tau - 2$;

(f) precise studies of the known decay modes with respect to branching fractions and decay dynamics;

(g) study of strong interaction physics in the 1 GeV region.

Some of these goals have been discussed by Burchat,[24] Stroynowski[25] and Perl.[26]

8.   CLOSE–MASS LEPTON PAIRS: CONCEPT

About two years ago I pointed out[27] that the standard e^+e^- search methods for heavy charged leptons using

$$e^+ + e^- \to L^+ + L^- \ ,$$
$$L^+ \to \bar{L}^0 + \text{other particles} \ , \tag{18}$$
$$L^- \to L^0 + \text{other particles} \ ,$$

assume that the L^0 mass, m_0, is much less than the L^- mass, m_-. Indeed most searches set

$$m_0 = 0 .$$

If m_0 is close to m_-, still with

$$m_0 < m_- ,$$

the detected energy, usually called visible energy, will be relatively small in the events described by Eq. 18. Defining the mass difference

$$\delta = m_- - m_0 , \tag{19}$$

the standard search methods fail[27] when $\delta \lesssim 4$ GeV/c^2. Stoker and I[28,29] have devised methods to search the $m_- - m_0$ region with δ values as small as 0.3 GeV/c^2.

Riles[30] has developed a different small-δ search method using the radiative process

$$e^+ + e^- \to L^+ + L^- + \gamma . \tag{20}$$

This suppresses the backgrounds from the two-virtual-photon processes.

$$e^+ + e^- \to e^+ + e^- + e^+ + e^-, \ e^+ + e^- + \mu^+ + \mu^- . \tag{21}$$

The small-δ problem also limits[27] the significance of searches for heavy charged leptons at $\bar{p}p$ colliders. These searches[31] use

$$\bar{p} + p \to W^- + \text{other particles} ,$$
$$W^- \to L^- + \bar{L}^0 , \tag{22}$$
$$L^- \to L^0 + \text{other particles} ,$$

and depend on a relatively large missing transverse momentum[32] in these events.

9. CLOSE–MASS LEPTON PAIRS AND LIMITS ON THE EXISTENCE OF NEW HEAVY CHARGED LEPTONS

Table 7 lists the published experiments on the existence of new heavy charged leptons where $m_0 > 0$ has specifically been considered in the publication. In the case of the experiments at TRISTAN, AMY[34] and VENUS,[35] and the UA1 result,[31]

Table 7. Publications on limits on new heavy lepton masses, m_- and m_0, when $m_0 \geq 0$.

Method	Lower limit on m_- (GeV/c^2) when $m_0 = 0$	Experiment	Figure	Reference
$e^+e^- \to L^+ + L^-$ at 29 GeV		Mark II	5	28, 29
$e^+e^- \to L^+ + L^-$ at 29 GeV		TPC	6	33
$e^+e^- \to L^+ + L^-$ at 56 GeV	27.6 , 95% C.L.	AMY	7	34
$e^+e^- \to L^+ + L^-$ at 56 GeV	27.6 , 95% C.L.	VENUS	8	35
$\bar{p}p \to W^- + \ldots$ $W^- \to L^- + \bar{L}^0$	41. , 90% C.L.	UA1	9	31, 32

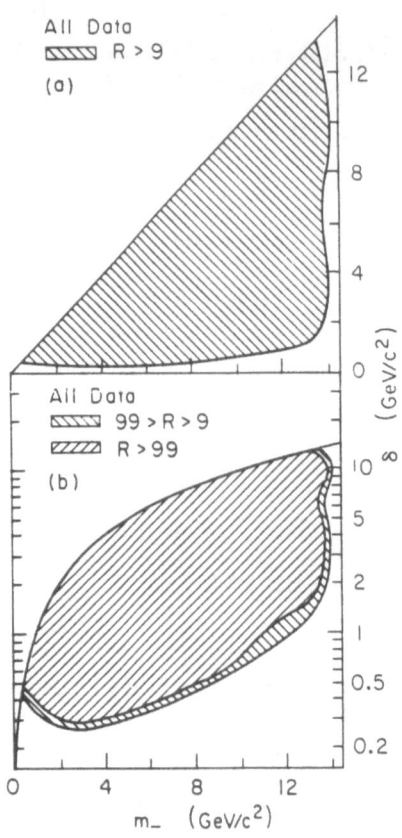

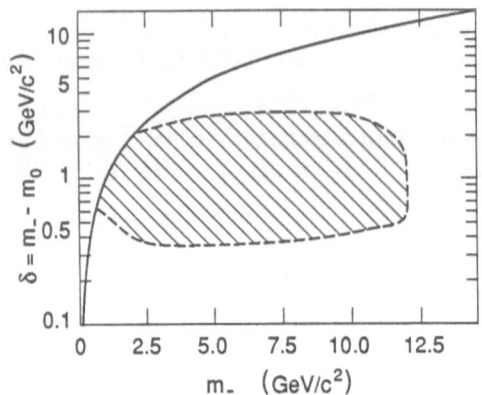

Fig. 6. $L^- - L^0$ pairs are excluded from the hatched $m_- - \delta$ region using 29 GeV e^+e^- data from the TPC experiment at PEP, Ref. 33. $\delta = m_- - m_0$. The boundary gives the 99% C.L.

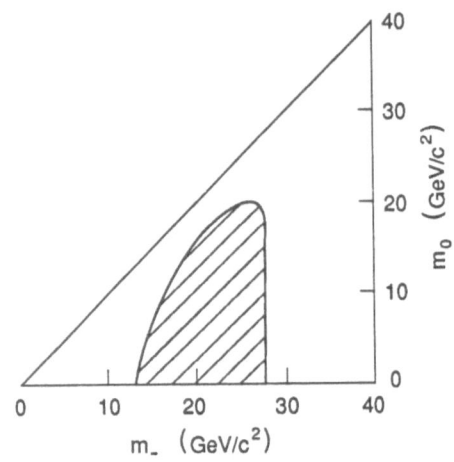

Fig. 5. $L^- L^0$ pairs are excluded from the hatched $m_- \delta$ region using 29 GeV e^+e^- data from the Mark II experiment at PEP, Ref. 29; $\delta = m_- - m_0$. The same results are shows (a) with a linear δ scale and (b) with a logarithmic δ scale. $R > 9$ means about 90% C.L.

Fig. 7. $L^- - L^0$ pairs are excluded from the hatched $m_- - m_0$ region using 56 GeV e^+e^- data from the AMY experiment at TRISTAN, Ref. 34. The boundary gives the 95% C.L.

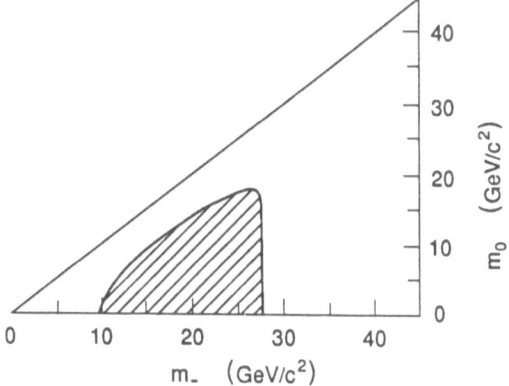

Fig. 8. $L^- L^0$ pairs are excluded from the hatched $m_- m_0$ region using 56 GeV $e^+ e^-$ data from the VENUS experiment at TRISTAN, Ref. 35. The boundary gives the 95% C.L.

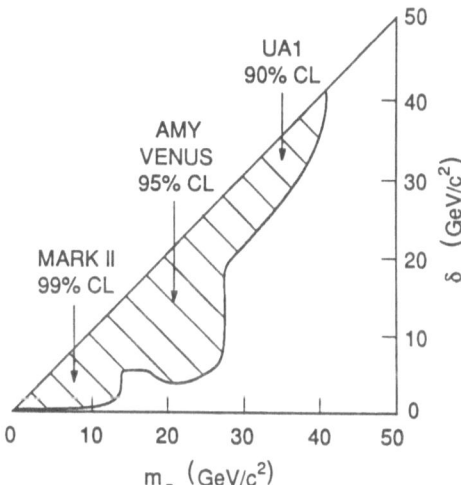

Fig. 9. Composite of $\bar{L} - L^0$ pairs excluded from the hatched $m_- - \delta$ region for: Mark II, Ref. 29; AMY, Ref. 34; VENUS, Ref. 35; UA1, Refs. 31 and 32.

I also note the lower limit on m_- when $m_0 = 0$. The experiments at TRISTAN will explore smaller values of δ as luminosity is accumulated.

These limits are shown in Figs. 5–8 and the combined limits in Fig. 9.

ACKNOWLEDGEMENTS

I greatly appreciated the hospitality of Prof. Antonio Zichichi and the staff of the Centro di Cultura Scientifica ≪ E. Majorana ≫.

Much of the work reported here was carried out with my colleagues K. K. Gan, K. G. Hayes, D. P. Stoker, and B. Efron.

REFERENCES

1. J. J. Thomson, Phil. Mag. **44**, 293 (1897).
2. S. H. Neddermeyer and C. D. Anderson, Phys. Rev. **51**, 884 (1937).
3. M. Bernardini *et al.*, Nuovo Cimento **17**, 383 (1973).
4. S. Orioto *et al.*, Phys. Lett. **48B**, 165 (1974).
5. M. L. Perl *et al.*, Phys., Rev. Lett. **35**, 1489 (1975).
6. K. K. Gan and M. L. Perl, Int. J. Mod. Phys. A3, 531 (1988).
7. W. Bacino *et al.*, Phys. Rev. Lett. **41**, 13 (1978).
8. K. G. Hayes and M. L. Perl, SLAC–PUB- -4471 (1988), to be published in Phys. Rev.
9. H. Albrecht *et al.*, Phys. Lett. **202b**, 149 (1988).
10. Y. S. Tsai, Phys. Rev. **D4**, 2821 (1971).
11. T. N. Truong, Phys. rev. **D30**, 1509 (1948).
12. F. J. Gilman and S. H. Rhie, Phys. Rev. **D31**, 1066 (1985).
13. F. J. Gilman, Phys. Rev. **D35**, 3541 (1987).
14. E. Braaten, Phys. Rev. Lett. **60**, 1606 (1988).
15. B. G. Bylsma *et al.*, Phys. Rev. **D35**, 2269 (1987).
16. C. Akerlof *et al.*, Phys. Rev. Lett. **55**, 570 (1985).
17. K. K. Gan *et al.*, Phys. Rev. Lett. **59**, 411 (1987).
18. H. R. Band *et al.*, Phys. Lett. **198B** 297 (1987).
19. S. T. Lowe, Proc. Int. Symp. on Production and Decay of Heavy Flavors (Stanford, 1987).
20. G. B. Mills *et al.*, Phys. Rev. Lett. **52**, 1944 (1984).
21. M. L. Perl, SLAC–PUB–4632 (1988), to be published in Proc. Les Recontres de Physique de La Valle D'Aoste (La Thuile, 1988).
22. K. G. Hayes, M. L. Perl, and B. Efron, SLAC–PUB–4669 (1988), to be published in Phys. Rev.
23. M. L. Perl, Phys. Rev. **D38**, 845 (1988).
24. P. R. Burchat, SCIPP 88/12 (1988), Proc. of the SIN Spring School on Heavy Flavor Physics (Zuoz, 1988), to be published.
25. R. Stroynowski, CALT–68–1511 (1988).
26. M. L. Perl, submitted to Proc. Physics in Collision (Capri, 1988).
27. M. L. Perl, Proc. XXIII Int. Conf. High Energy Physics (Berkeley, 1986), ed. S. C. Loken, p. 596.
28. D. P. Stoker and M. L. Perl, in Electroweak Interactions and Unified Theories (Les Arcs, 1987), ed. by J. Tran Thanh Van (Editions Frontieres, Gif-sur-Yvette, 1987).
29. D. P. Stoker *et al.*, SLAC–PUB–4590 (1988), submitted to Phys. Rev.
30. K. Riles in Proc. DPF–88 (Storrs, 1988), to be published.
31. B. Adeva *et al.*, Phys. Lett. **152B**, 439 (1985).
32. R. Barnett and H. Haber, Phys. Rev. **D36**, 2042 (1987).
33. L. G. Mathis, Ph.D. Thesis, LBL–25261 (1988).
34. W. Ko, KEK Preprint 88–30 (1988), submitted to Proc. Third Asia Pacific Physics Conference (Hong Kong, 1988).
35. K. Abe *et al.*, Phys. Rev. Lett. (to be published).

Chairman: M. Perl

Scientific Secretaries: G.J. Bobbink and A. Marchioni

DISCUSSION

– *Brandt:*

In the 1-prong decay mode problem of the τ, the sum of the inferred branching ratios in row 4 of your table was less than 2.7% without any errors stated. How well is this upper limit known and could you measure these decay modes in an experiment with a good photon detector?

– *Perl:*

The individual limits are at the 90 or 95 % confidence level and then they are all added up. Thus this upper limit is quite strong. With the present photon detectors you cannot measure these decay modes. For example in the decay mode $\tau \rightarrow \pi^- + \nu + 5\pi^0$, the γ multiplicity is 10 and γ counting does not work with these low energies. What is needed is a good photon detector so one can pair the photons to form π^0's.

– *Skarke:*

The detection efficiencies for the various decay modes are not all the same and are probably not equally well known. Can you comment on this?

– *Perl:*

This is a point that deserves more attention.

– *Bobbink:*

Common systematic effects in the measurements will not average away. What happens for example if the τ neutrino mass is substantially different from zero? Or in the case of most experiments using the same program to calculate radiative corrections?

– *Perl:*

If you have instead of a massless τ neutrino one with a mass of 35 MeV/c^2 (the largest mass allowed by experimental limits) then the 1-prong branching fractions change very little. Indeed, we have made a long search for systematic effects and have not found any. This, of course, does not mean there are not any. For example, a still open question is whether radiative τ decays like $\tau \rightarrow \mu\nu\nu\gamma$ are attributed correctly in the 1-prong sample.

– Calafiura:

When you compute averages of results coming from different experiments, how do you take into account systematical errors, which are possibly correlated?

– Perl:

The statistical and systematic error are added in quadrature. We have tried to estimate the effect of correlated systematic errors in the following way: the smallest systematical error has been subtracted from the error of all measurements. The measurements were then averaged and this systematical error was added back to the final error. The results remained essentially the same.

– Barbieri:

In the searches for new heavy leptons in e^+e^- collisions at LEP and SLC energies, do you see major experimental limitations in going down to low mass difference between the heavy charged lepton and the neutral heavy lepton into which it decays?

– Perl

No, I do not see any major limitations. There are two problems though. Firstly, these events have very little visible energy, and are often rejected in an early stage of the data processing. Furthermore there is background from $\gamma\gamma$ collisions producing e^+e^- or $\mu^+\mu^-$ or $\pi^+\pi^-$ pairs with low visable energy. However, with a good (TPC like) detector one should be able to solve these problems

– Barbieri

Is one using V-A coupling in the analysis to obtain limits on new heavy charged leptons and is this important?

– Perl:

Yes, that is important and we have to look at this more carefully.

– Lederman:

Is there any tenable theoretical speculation which could explain the 1-prong τ decay problem without contradicting other known experiments?

– Perl:

No, there are no successful theoretical explanations

– Colas:

Would not the solution to the 1-prong decay problem be to tag one side of a $\tau^+\tau^-$ event and try to classify the other τ decay?

– Perl:

This has been tried by the Mark II collaboration (see P. Burchat thesis) and a large fraction of the events, about 2/3, do not have enough information so one has to assign them a probability to belong to one of the catagories. When you do that, the electron, muon, and rho mode go up, and take up the difference as must happen. This clearly has to be repeated with a better detector and higher statistics.

– Fordham:

Have you studied the possibility of a low estimate by all experiments of the background in the 1-prong sample due to hadronic events?

– Perl:

This deserves further study.

THE FERMILAB UPGRADE

Leon M. Lederman

Fermi National Accelerator Laboratory
P.O. Box 500
Batavia, Illinois

INTRODUCTION

In 1978, Fermilab set out a goal of building a superconducting accelerator (Energy Saver) which would raise the proton energy to close to 1000 GeV for operation in two modes. Tevatron I would provide proton-antiproton collisions at a total CM energy of near 2.0 TeV to study the particle mass domain beyond 100 GeV. Tevatron II would provide extensive facilities for the programmatic study of Standard Model physics in an upgraded fixed-target program. There was of course the realization that with the right mixture of precision and imagination, the collider could add significantly to Standard Model physics (e.g. W and Z physics, W,Z pairs, B-physics) and that the fixed-target program could explore beyond the Standard Model (e.g., rare K-decays, CP violation). In 1988, we are engaged in setting out the future program of the Laboratory based upon the success of the Energy Saver, TeV I and TeV II construction programs. This future program assures that operation of the TEVATRON facility for physics is the overriding priority between now and perhaps 1993 and it also assumes that the Superconducting Super Collider (SSC) will be funded for construction in 1990 and will begin producing physics by 1999.

A ``brief history'' of upgrades is presented on page 307.

HISTORY

The notion of going to higher luminosity in the Collider and more intensity and quality for the fixed-target program has been around since the start of TeV I and TeV II. The simply stated goal in collider physics is to increase the mass range which can be searched for new

The Superworld III
Edited by A. Zichichi
Plenum Press, New York, 1990

phenomena and in the fixed-target program to enhance the precision and the detail of our Standard Model base. In Laboratory presentations we have proposed a Superbooster (1980), Dedicated 4 TeV Collider (1983), Brightness Enhancer (Jan. 84), and Source Brightener (Sept 84). Upgrade plans and funding profiles were presented in the 1986, 1987 and 1988 institutional plans. Responses from HEPAP have been positive going back to 1982.[*] Experience with the first engineering run of TeV I in 1985 and the 1986 construction year led to a thorough review of the entire accelerator complex. A Collider upgrade plan was submitted (short form 44) with a TPC of $267M in January 1986.

As the first phase, the Linac Upgrade was submitted in January 1987 and resubmitted in February 1988. The plan has emerged into two stages: an adiabatic series of improvements which will bring the peak luminosity of the $\bar{p}p$ collider to about 5×10^{30} cm^{-2} sec^{-1}. This should also make over 3×10^{13} ppp available to the fixed target, an improvement of almost a factor of two. The Collider energy would be 1.0 TeV and the fixed-target energy near 900 GeV. Given reasonable R&D funds and the Linac line item, all of this should be available for a D0 and CDF run of $\int L dt$ > 10 pb^{-1} in 1992.

In the period until 1993, there would be no planned shutdown in excess of several months for installation of upgraded components. This period would also see modest upgrades to Collider Detector at Fermilab (CDF) and some decisions on major new detectors and upgrades for the fixed-target program. In 1993, one can contemplate a 6-10 month shutdown for the second phase of the upgrade. This would be designed to deliver in excess of 100 pb^{-1} per run to the collider detectors and in excess of 4×10^{13} ppp for the fixed-target program. Given enough protons, it will pay to improve the fixed target duty cycle even more - perhaps from 30% to 60%.

There are now several competing elements for the second phase of the upgrade. The purpose of this note is to review these which, at this writing, are evolving out of extensive high-energy physics (HEP) community discussions.

REVIEW OF UPGRADE MOTIVATION

The Fermilab collider is the highest energy machine in the world. Until SSC or LHC or the Soviet 3 GeV x 3 GeV collider turn on and begin to produce physics data, this will remain so. We believe we have a time window that will go to 1999 or so since it will take several years for any of the above machines to go from commissioning to real physics. The window is not only an opportunistic window, it is essential that there be continuity in the production of physics results. Whereas, if SSC is proceeding towards, say, a completion date of 1997/8, a fairly large community will be occupied there by 1992, but one cannot put graduate students, new postdocs and pre-tenure professors on many of the SSC detectors until they are much closer to physics. This is borne out by CDF and DO experience. The Fermilab Collider physics in the period 1994-1999 will also be invaluable as a guide to SSC both from the point of view of collider and detector technology but also from the physics knowledge base. Since a year of SSC is worth $250M (1988), it is terribly cost effective to be as well prepared for the SSC era as one can possibly be. Finally we note that there may well be niches of physics for which TEVATRON energy is well enough above threshold; a vast increase in energy may then only increase backgrounds.

The knowledge base will come from both the Fermilab Collider and the fixed-target program, especially those experiments which illuminate high-rate technology and those which use precision and detail to test and extend the Standard Model.

To present a glimpse of the relative merits of the various upgrade options we present a series of graphs calculated by E. Eichten. See Fig. 1,2,3. We stress that whereas the optimum plan is not yet clear, what is perfectly clear is that the design goals are such as to double the discovery limits, i.e., equivalent to doubling the effective machine energy. Furthermore, it makes possible the collection of huge amounts of data for particles in the W, TOP, e.g., $\leq$ 125-GeV mass range.

A doubling of the mass reach could be compared to building a 400-GeV e^+e^- machine with sufficient luminosity to double the mass reach of LEP II. Another comparison scale is the current attention to B-physics and proposals for electron-position B-factories. An upgraded TEVATRON has impressive capabilities here although the issue is complicated by backgrounds.

The potential for discovery of new physics by our upgrade or for the clarification of discoveries which may be made in the early stage of TEVATRON are very significant. We also stress the important support this kind of data gives to SSC where the parameter $M/\sqrt{s}$ will very rarely reach the Upgrade goal of $\sim$ 0.4.

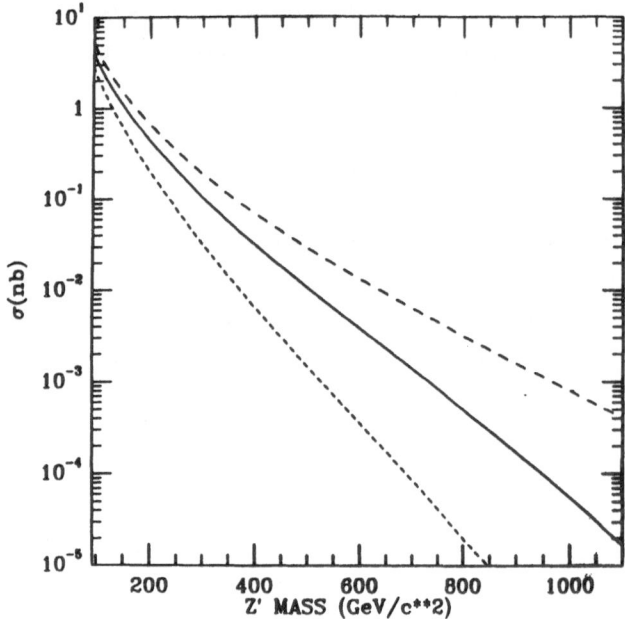

Fig. 1. Production of heavy Z^0. The mass reach is determined by the integrated luminosity and the discovery level required, e.g., 100, produced events. See Chart I for results.

Chart I

Numbers of Produced Events in a 5 Mo (3×10^6 sec) Year Based on Assumed L_p

	No Upgrade	Phase I	Phase II		
L_p	3×10^{29}	4×10^{30}	4×10^{31} A.	2×10^{32} B.	1×10^{31} C.
CM Energy	2 TeV	2 TeV	2 TeV	2 TeV	$\geq$ 3 TeV
Int	$\bar{p}p$ 1pb^{-1}	$\bar{p}p$ 10pb^{-1}	$\bar{p}p$ 100pb^{-1}	pp 500pb^{-1}	$\bar{p}p$ 30pb^{-1}

Mass	Discovery Limits		Z^0	"Factory" Regime	
200	500	5000	50 K	100 K	20 K
400	30	400	4 K	3.5 K	2.4 K
600	--	40	400	200	300
800	--	7	70	15	75
1000	--	--	10	--	30

Mass			TOP		
75	300	3000	30 K	150 K	30 K
100	70	700	7 K	25 K	6 K
125	20	200	2 K	5 K	1.8 K
150	8	80	900	2 K	900
175	3	30	300	500	380
200	1	10	100	250	150
250	--	--	--	25	--

Mass			Gluino		
100	300	3000	30 K	150 K	30 K
150	20	200	2000	10 K	3 K
200	2	20	200	800	500
250	--	3	30	100	90
300	--	--	5	10	30

			W-Pairs		
W^+W^-	5	50	500	1000	240
Z^0Z^0	--	7	70	150	40
W^+Z^0	0.7	7	70	200	30

Mass			Technipions		
50	800	8000	80 K		62 K
100	200	2000	20 K		18 K
150	70	700	7 K		6.5 K
200	30	300	3 K		3 K
250	12	120	1.2K		1.4 K

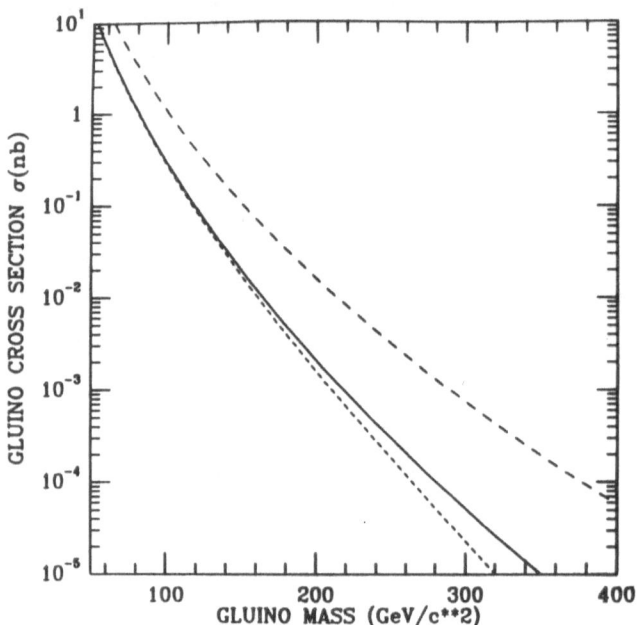

Fig. 2.   Production of gluinos.

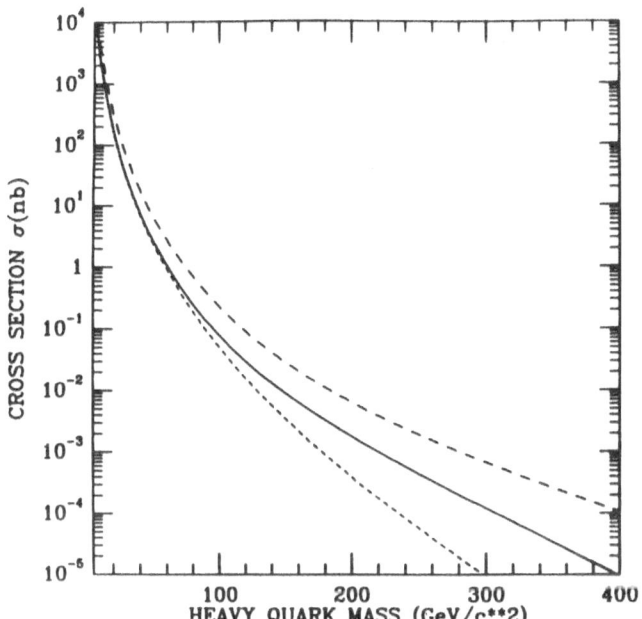

Fig. 3. Production of heavy quarks.

Advancing fixed-target physics will be critically dependent upon advancing the art of detectors. Exploiting higher luminosity in the collider also requires confidence that the detectors are up to resolving signal and background in the high rate environment.

UPGRADE: PHASE I

Goal 5×10^{30} cm^{-2} sec^{-1} and 3×10^{13} ppp 1988 - 1993

The first phase involves a series of steps:

1. Replace Cockcroft-Walton by RFQ, new first tank on linac.
2. Replace last 5 linac tanks by side-coupled cavity type of tank at 800 Mhz (instead of 200 Mhz). This will raise the energy of protons injected into the Booster to 400 MeV. The transverse emittance should go to 12π mm.mr or even as low as 6π at 1×10^{10} p.
3. Strong low-β quadrupoles for D0, CDF; Goal $\beta^* = .25$m.
4. Possible shorter bunches.
5. $\bar{p}$ Source and cooling improvements.
6. Dipole magnet development for separator space; goal is 6.6T dipoles.
7. Cryogenic developments to achieve about 3.9°K TEVATRON for 900-GeV fixed-target and 1000-GeV Collider operations.
8. Electrostatic separators - helical orbits. 50 kV/cm for 2.5cm gap.

These steps, carried out by AIP, R&D and Linac Line Item funding can and should be complete in time for a 1992 run of CDF and D0 with a goal of $\geq 10\text{pb}^{-1}$. Included here are already scheduled improvements in the CDF detector, completion of the D0 detector and new starts on a major fixed-target spectrometer, given PAC approval. Other fixed target activity involves continued upgrades of major existing detector facilities.

UPGRADE PHASE: II

Goal 5×10^{31} cm^{-2} sec^{-1} and 4×10^{13} ppp at > 50% Duty Factor 1994-1999

Introduction

We have looked in some detail at several approaches to this next factor of ten. The luminosity goal is designed to keep the CDF and D0 detectors from melting, but this luminosity will require substantial upgrades to both detectors. These involve replacement of front-end electronics, perhaps central tracking and perhaps some calorimetry improvements. Consideration is also being given to a possible third collider detector, which would be specifically designed to do B-physics. What is also open is whether this gets its own collision region or goes in to alternate with CDF, say. Finally, considerable weight is given to the fixed-target program and how it is benefitted from the various options. Whereas the TEVATRON Collider mode may be supplanted by SSC, the fixed-target program will probably extend well into the SSC era, taking advantage of SSC detector R&D, the almost certain need for more precision and detail, and the continuous need for test beams. We now list the options as currently understood and later indicate some variations and phasing possibilities.

A. $\bar{p}p$ with Superboosters

Here, in order to supply two IR's with 5×10^{31} luminosity we need an improved source fed by an improved Main Ring (MR) and a place to store 3×10^{12} $\bar{p}$'s. Some means of recovering $\bar{p}$'s which have diffused is also useful.

The major devices here are two 20-GeV rings; one, the proton superbooster, injects into the Main Ring at 20 GeV yielding high transmission, small emittance ($\leq 12\pi$), good lifetime and high proton intensity for proton production. The second ring, a $\bar{p}$ ring, is an antiproton depository. This would also involve 8-16 GHz cooling in the $\bar{p}$ accumulator and depository. The total cost including R&D, pre-op is $124M. The technical problems of actually achieving 5×10^{31} are formidable. A more conservative goal is to have a 5-month run (repeated annually) to yield an integrated luminosity of 100 pb^{-1}.

B. pp Option

This suggests a pp option, where more than 5×10^{31} is assured and an overall efficiency of twice the $\bar{p}p$ option seems reasonable. We then assume that we can collect 500 pb^{-1} in a collider portion of one year run. Higher luminosity, i.e., 2×10^{32}, can be achieved for special purposes not including the normal operation of the CDF and DO detectors. Another virtue of pp is the small interaction diamond which benefits all short lifetime experiments, e.g., B-physics. The pp option, as an accelerator project, is not particularly challenging. However, it requires removing the MR from the tunnel (it becomes the Main Injector) and providing a 120- to 150-GeV tunnel into which MR components would go. All overpasses and other TeV-MR hindrances would disappear. The new injector could also be a $\bar{p}$ producer and be organized to provide 150-GeV test beams during Collider operation. MR removal would allow space for a second superconducting magnet string. Longer straight sections would be needed in order to bring beams into collision. This could be done with small displacements of the CDF and DO detectors. The total cost estimate here is about $240M.

C. $\bar{p}p$ High Energy Option

A third option removes the MR and/or the TEVATRON from the old tunnel and replaces it with a ring of 6.6 to 8T superconducting magnets of the SSC/HERA style. This would permit $\bar{p}p$ operation at 3 -3.5 TeV in the CM. Since there would be no superboosters, the luminosity would be only slightly better than 5×10^{30} that was achieved in Phase I. Both CDF and DO detectors would work well here with much less extensive upgrades than option B. The mass reach of such a 3.5 TeV $\bar{p}p$ collider is about that of a 2.0-TeV collider at $> 5\times10^{31}$. The fixed-target program can gain substantially from a higher energy extracted beam and/or the higher intensity of secondary and tertiary beams. The improvement factors come from the benefits of a redesigned Main Ring (Main Injector) and the luminosity gain due to the higher energy (1.5 - 1.8 TeV). We take 10^{31} as the design luminosity and therefore an integrated luminosity of 30 pb^{-1} per year. The energy increase could be a significant help in many fields, e.g., in heavy quark studies, in hyperon research and in structure function data. Open questions here have to do with the removal of the MR and the cost of higher field magnets. This will probably come to about the cost of the pp option.

SELECTION CRITERIA

Which of the three options (or none of them) to choose will depend on a number of criteria:

(1) Physics reach in the collider mass domain ''beyond the W, Z''
(2) Implication for advancing fixed target physics
(3) Cost
(4) Time and downtime to implement
(5) Detector Implications
(6) Technology experience relevant to SSC

(7) B-physics

(8) SSC at Fermilab? If so, it may favor one option more than others.

Many of these criteria are not simple. Physics reach with high luminosity is clouded by backgrounds, pile-up, etc. It may be useful to assume the following about detectors:

1) CDF requires new electronics at ≥ 5 x 10^{30} @ $10M

2) DO requires new electronics at 5 x 10^{31}

3) CDF will require new tracking, vertex, etc. at $\sim 10^{31}$

4) Both detectors will require much more major upgrades at $>$ 5x10^{31}. However, even these upgrades, at an estimated total cost of $25M each, are much less in time, money and people than starting over.

5) With SSC demands, it makes no sense to contemplate a brand new ''standard'' 4π detector. New ideas, however ...

PHASED OPTIONS

As phase II in the upgrade one can consider building a new main ring of 120-150 GeV in its own tunnel with new magnets but, initially, only minimal power supplies. This could be constructed and commissioned without interference with the on-going program. Its objectives: i) excellent injector into TEVATRON; ii) excellent p̄ producer e.g. 2 sec/cycle; iii) provides 150-GeV beams to fixed-target program during collider runs, saving - 2 months of calibration, timing, commissioning of fixed-target experiments; iv) may provide very intense neutrino and k-beams for special experiments. This would also free CDF and DO from Main Ring backgrounds and provide a space for another interaction hall at E-ZERO. Also, it frees space in the existing tunnel for another superconducting ring. When completed and commissioned, there would be a shutdown for tunnel connections, moving of MR power supplies, etc., and perhaps removing MR magnets. This may be ~ 3 months. Options B or C would follow as phase III.

Other phases, as demanded by physics and allowed by resources would be to upgrade from the modest luminosity of the 3 TeV option to perhaps $3-5$x10^{31} using option A devices. Alternatively, if the original TEVATRON ring is still in the tunnel, pp collisions (1.5 TeV x 1.0 TeV) can be contemplated, especially for the B-detector but perhaps for additionally upgraded CDF/DO.

SUMMARY: PHYSICS

Our options as of July 1988 are now recapitulated. We assume a 5 month collider run, 5 month fixed target run and 2 months of changeover, studies, etc.

A. $p\bar{p}$ $\sqrt{s}$ = 2 TeV $\qquad\qquad$ $\mathscr{L}\,dt$ = 100 pb^{-1} /year

B. pp $\sqrt{s}$ = 2 TeV $\qquad\qquad$ $\mathscr{L}\,dt$ = 500 pb^{-1} /year

C. $\bar{p}p$ $\sqrt{s}$ ≳ 3 TeV $\qquad\qquad$ $\mathscr{L}\,dt$ = 30 pb^{-1} /year

The physics graphs and Table take into account the different quark content of pp and $\bar{p}$p.

From the graphs and from the Table, it is clear that the TEVATRON upgrade has two physics benefits. Any of the options extends the discovery potential for a characteristic subset of theoretical speculations by a factor of two in mass: it permits a thorough exploration of the interesting 200-400 GeV mass domain - ''the foothills of the TeV summit.'' Recall that in new Technicolor theories, the crucial parameter is F_π = 246 GeV.

Equally significant, for masses near the lower end, it provides ''factory'' potential. TOP is an excellent illustration. If, as some theorists intimate, the TOP mass is under 125 GeV, then the upgrade makes tens of thousands of TOP quarks per year and thus defines a TOP factory. This applies to many of the potential discoveries - one will be able to exploit the discovery of a GLUINO or TECHNIPION in some detail if the masses are not too high. Perhaps all the theories are wrong - still the exercise indicates that whatever nature has in the 50-400 GeV mass domain, the TEVATRON upgrade will be a powerful tool to guide particle physics on the correct road from the Standard Model toward the ultimate unification.

We have not yet listed some of the obvious ''goodies'' that have been widely discussed elsewhere:

b-quarks: The upgrade will result in of the order of 10^{10} $B\bar{B}$ per year pairs with option B giving 10^{11} $B\bar{B}$'s. Fermilab proposal P-784 has under design a detector which can carry this to the observation of CP violation.

<u>W + Z's</u>: The 100 pb^{-1} luminosity yields 10^6 W's per year and 2x10^5 Z^0's. With precise Z^0 masses derived from e$^+$e$^-$ machines and a highly precise mass ratio of W to Z, one can derive unique values for important radiative corrections which involve the Higgs mass.

Compositeness, Drell-Yan, Fourth Generation and many other processes and issues will also be addressed.

<u>Fixed Target</u>: Although we have stressed the benefits to the Collider, the gains to the fixed target are also important with option C probably having the largest influence. Here even a modest increase in energy gives a very large increase in, for example, photoproduced B's (factor ~ 20). Secondary beams gain in energy and intensity, hyperon beams also gain from the increase in laboratory lifetime.

FUNDING SCENARIOS

In our firm, unalterable 15-year plans we have presented funding profiles which have not noticeably produced cardiac arrest among DOE readers. Table X (Profile I) out of the 1988 Institutional Plan is typical. Below this is an alternative plan which assumes less civil construction and more R&D in the realization of the upgrade program. It assumes we do something between the costs of p̄p and pp or p̄p at high energy. The difference is ± \$10M/year. It includes funds for detector upgrades <u>and</u> fixed-target initiatives.

CONSTRAINTS

In guiding this discussion we have in fact made a number of constraining assumptions:

1. The non-SSC funding level of \$560M will not be increased during SSC construction.
2. SSC physics will be in full swing with first physics publications by ~ 1999.
3. The upgrade over the period 1989-1994 should require increments to the Fermilab budget of less than \$50M/year.
4. No new 4π detector can be contemplated. CDF and DO may be upgraded but not replaced. A special purpose new detector for

Table X
Profile I
Laboratory Funding Summary
($ in Millions)
Fiscal Years

	FY 87	FY 88	FY 89	FY 90	FY 91	FY 93	FY 93	FY 94
DOE Effort	135.4	145.0	151.3	169.9	171.01	174.0	180.0	181.0
Work for Others	3.4	5.7	1.1	0.6	0.6	0.6	0.6	0.6
TOTAL OPERATING	138.8	150.7	152.4	170.5	171.7	174.6	180.6	181.6
Capital Equipment	27.5	25.3	30.8	37.0	34.0	32.0	40.0	42.0
Program Construction	8.2	11.2	4.6	43.4	63.4	30.3	0.0	0.0
AIP/GPP	8.4	9.1	9.5	17.0	14.0	14.0	14.0	14.0
General Purpose Equip	0.0	0.0	0.0	0.0	0.0	0.0	4.0	0.0
TOTAL LABORATORY FUNDING	182.9	196.3	197.3	267.9	283.1	250.9	238.6	237.6
PROPOSED CONSTRUCTION	0.0	0.0	0.0	0.0	0.0	0.0	0.0	0.0
TOTAL PROJECT FUNDING	182.9	196.3	197.3	267.9	283.1	250.9	283.6	237.6
*Outyear Escalation Rates (Base Year - FY 89)				3.6%	3.3%	2.8%	2.3%	2.3%

304

B-physics _is_ conceivable if its cost is modest compared to original CDF/DO costs.

5. The upgrade should begin to produce physics by 1994-5.
6. Until 1993 we plan no shutdowns in excessive of 6 weeks.
7. CDF and D-Zero must have at least 10 pb^{-1} of good data before a long (6-10 mo.) shutdown.

A HISTORY OF UPGRADES

A. Cornell

300 MeV	'49	
1 GeV	54	
2 GeV	64	
10 GeV	68	(SLAC 20 GeV Linac)
8x8 GeV	79	
8x8 GeV Upgrade	88	

B. BNL

AGS 30 GeV Upgrade (linac)
'70 (Fermilab 200 GeV)
AGS Upgrade (booster etc) 88

C. SLAC

Linac	'67	
Spear	73	
PEP	79	
SLC	88	
400 GeV e^+e^-	? 92	

D. CERN

Cyclotron	'58	
PS	60	
ISR	71	
SPS	76	
S$\bar{p}$pS	81	
S$\bar{p}$pS+ ACOL	88	(TeV I Going)
LEP I	89	
LEP II	92	
LHC	?	

E. DESY

DORIS	'74
PETRA	77
DORIS Upgrade	85
HERA	90

F. FERMILAB

400 GeV	'72
TeV I	87
TeV II	84
UPGRADE	→ 93 proposed

RESUME OF UPGRADE VIRTUES

1. Physics is first rate with very large discovery potential and strong programmatic power.
2. This is the highest energy machine in the world. It deserves the full exploitation compatible with realistic costs, time scale and manpower needs. It represents an investment of $500M in R&D, Equipment, construction and AIP funds. The history of upgrades also speaks eloquently to this.
3. HEP must maintain its excitement and its vitality, especially during the long construction schedule for the SSC. Discoveries, press releases, etc., will serve to keep the flow of new students and will insure the attention which is needed to secure a decent SSC funding profile.
4. The learning curve of new physics and of handling collider subtleties alone will pay the upgrade costs. These can modulate SSC detector design and will be relevant up to turn-on and beyond. CDF and D-Zero must learn to cope with subtle signatures at the level of 10^{-10} of the total cross-section. No amount of simulation substitutes for learning by doing. This acquired skill becomes the experience base of the SSC and is terribly cost effective at SSC annual costs of $250M/year. CDF and D-Zero at $> 10^{31}$ luminosity are unique sources of this learning curve.

NOTES

Jan '82   Subpanel on Long Range Planning - Excerpts (p.29) ''The achievement of a luminosity greater than 10^{30} cm^{-2} sec^{-1} will, in our judgement, take some years of operational experience., On the other hand, a number of improvements seem possible. Thus, an ultimate goal of $L = 10^{31}$ appears reasonable to us.''
''The TEVATRON projects will be the focus of a major part of the U.S. program ... they will open up entirely new areas of physics and accelerator development and will be essentially unique in the world.''

306

<u>July '83</u> Subpanel on New Facilities (p.51) ''The viability of the [TEVATRON] facility after about 1992 will depend on the physics interest and the availability of other facilities. If the level of research activity remains high, then an upgrade of the facility and its detectors may be warranted, with a consequent extension of the useful life of the machine for perhaps another five years.''

<u>Sept '85</u> Report of the 1985 HEP Study (p.27) ''Because new phenomena may not conform to our current expectations, it is natural to expect the configuration of these detectors [CDF-DO] to evolve in response to our growing understanding ... A program of detector upgrades and accelerator improvements will be an essential part of the hadron collider physics program.''

In fixed-target experiments ... experiments can be grouped in terms of the physics questions ...

(1) CP violation in Kaon Decays

(2) Rare Kaon Decays

(3) Heavy Quark Physics

(4) Hadron Dynamics Other than Perturbative QCD

(5) Neutrino Oscillation Experiments

(6) Particle Searches with Beam Dump

DISCUSSION

- Brandt:

Are there any running or proposed experiments at Fermilab to repeat the UA4 diffractive experiments or study high mass diffraction like UA8?

- Lederman:

There are total–cross section and diffractive cross–section experiments, but no proposed experiments to do high mass diffraction.

- Marchionni:

Can you comment on B physics experiments at Fermilab and future plans?

- Lederman:

Many of the following experiments expect in the present data sample or in the next run to have B possibilities. E687 is a photoproduction experiment and hopes to get on the order of 100 reconstructed B's in the next run. E690 uses an elaborate processor which can analyze 10^5 events per second on–line; they have demonstrated this capability off-line with data from Brookhaven. They should observe a substantial B signal in the course of doing Charm physics. Again the objective is under 1000 B's. E771 is a specific experiment designed to look at Beauty production with proton beam. Again if you look at the rates, apertures, and reconstruction possibilities, you expect to find on the order of 1000 events. In addition, there are several proposals to do Beauty physics. One of these was based on E772 and instead of dimuons looks for two body B decays with an excellent two particle spectrometer. This is a very high rate experiment–they were taking interactions at the rate of 500 megahertz. Unfortunately, this was rejected by the PAC due to lack of analyzing power. They are considering renewing the proposal in a more powerful way. In addition, there has been a working group under the leadership of Bjorken looking at what we call the ultimate B physics spectrometer. Experimentally this is a fantastic challenge. The cross section for producing $B - \overline{B}$ pairs is of the order of 10^{-6} of the total cross section. Choosing a particular decay mode reduces the cross section by an additional factor of 10^4 to 10^5. Experimental efficiencies cost you another factor of 10. So you clearly need an enormous rate of protons on target per second to overcome these losses. The challenge is to deal with 10^{10} protons per second (needed to get on the order of a million B events per year) and have the detector pick out the B mesons. The problem is very similar to using a high luminosity collider, and in some sense this is easier than looking for Higgs in a collider, so if we fail here we will fail there too. There is another group that is looking at the collider option to study bottom physics. The Bottom Collider Detector group has been studying this problem for more than a year with Monte Carlo simulations. In a $p\bar{p}$ collider the detectors should be largely forward and backward and therefore they need a dipole magnet. There have been many other tentative design decisions such as particle identification and so forth. You can do very good physics with a collider and its likely luminosity of up to 10^{31}, but to solve the problem of CP violation, the luminosity has to be at least 5^{31}. This will be an evolutionary thing; clearly there is a lot of work going on in B physics.

– *Marchionni:*

Is there an interaction region available for such a detector?

– *Lederman:*

One of the upgrade programs would allow such an interaction region to be available. This would not be available before 1992. Another possibility is to retire one of the older detectors.

– *Mincer:*

I am interested in what the total charged particle multiplicity looks like at the collider and are there too many particles going down the beam pipe to tell if the multiplicity scales like log s at lower energies?

– *Lederman:*

CDF looks at charged particle multiplicities and of course you have to correct for what goes down the beam pipe. Roman pots, pioneered at the ISR, show that you can put detectors within a few millimeters of the beam. So you should be able to measure multiplicities with an angular cut of only a few microradians. CDF does have that facility, but the experiment called E735, which is a Quark- gluon plasma search, has already published charged multiplicities.

– *Gourdin:*

What type of accuracy is expected from a measurement of the relative phase between the two CP violating amplitudes η_{00} and η_{+-}?

– *Lederman:*

They have not taken data yet, but I believe it is on the order of 1 degree or better.

THE SOLAR NEUTRINO PUZZLE

Rudolf L. Mössbauer

Technical University of Munich
Dept. of Physics
D-8046 Garching, FRG

ABSTRACT

Terrestrial measurements of neutrinos emitted in hydrogen fusion processes provide information on the nuclear reactions proceeding in the solar interior. The observed deficit by a factor of three in the solar neutrino flux cannot be reconciled with the Standard Solar Model. The GALLEX project presently being built-up in the Gran Sasso Underground Neutrino Laboratory is aiming at the low-energy branch of the solar neutrino spectrum. Any deviations from the expected neutrino flux, which would be measured in the new experiment, can be attributed to particular neutrino properties, such as neutrino mass and neutrino mixing.

1. INTRODUCTION

The sun is the only star which is sufficiently close to provide experimental information on the nuclear reactions proceeding in its interior. The luminosity of a star is the total radiated power, but a time of up to a million years elapses between the generation of the radiation in the stellar interior and its arrival at the surface. Numerous interaction processes destroy all memories of the emerging photons on their origin. Information on the solar interior may be obtained from helioseismology, where Doppler-shift measurements provide solar vibration frequencies, associated with very slight motions of the solar surface. These frequencies depend on the temperature and density distribution within the sun and on its chemical composition. The relation with nuclear reactions, however, is in spite of a wealth of data rather poorly understood. Only measurements of the solar neutrino flux yield at present direct information on the nuclear fusion reactions inside the solar core. Such measurements are of interest both from an astrophysical and from an elementary particle point of view, depending on an interpretation of measured neutrino fluxes in terms of solar models, such as the Standard Solar Model, or in terms of neutrino properties, such as neutrino mixing and neutrino mass parameters[1]. The latter aspect will be exclusively dealt with in this paper.

2. NEUTRINO PROPERTIES

The neutrino, hypothetically introduced some 60 years ago by W. Pauli, was for the first time observed in 1956 via a direct nuclear reaction by Reines and coworkers[2]. Yet even today many of its properties remain a mystery: We do not

The Superworld III
Edited by A. Zichichi
Plenum Press, New York, 1990

know why neutrinos occur in several flavors, nor do we know how many such flavors exist. We do not know, whether flavor conservation is exact or whether neutrinos of different flavor couple to each other. We have little information on
neutrino stability, though astrophysical observations yield neutrino stabili-
ties exceeding the age of the universe. We do not know, whether neutrinos are
massive particles or whether an unknown symmetry principle might give rise to
zero masses. We do not know, whether neutrinos are Dirac or Majorana particles,
i. e. whether a distinction between particles and antiparticles is necessary or
not. The Majorana option is provided by the neutral character of the neutrino,
but a mixture between Dirac and Majorana character is likewise not in contradiction with present experimental results. Neutrinoless double beta- decay would
require a Majorana character of the neutrinos, but thus far there is no conclusive experimental evidence. We furthermore do not know , whether an expected
neutrino background as a remnant from a big bang creation of the universe does
exist. Such a neutrino background radiation should fill the entire universe with
number densities and a peak temperature of its Planck distribution approximately
equal to the electromagnetic background radiation, the differences essentially
being due to the different statistics for fermions and bosons and due to the fact
that neutrinos occur in several flavors. The discovery of such a neutrino background would yield dramatic support for the big bang hypothesis, but such a discovery is presently way out of experimental possibilities, due to the low peak
energy of order of two degree K of the background neutrino spectrum. Neutrinos
are subject only to the weak interaction and the still much weaker gravitational
interaction. Neutrinos are therefore on the one hand the perfect particles for
studying weak interaction processes, while on the other hand their study provides enormous experimental difficulties. Neutrino detection can be performed
in various ways: detection via inverse beta-decays involve threshold energies,
which are of order 0.1 to 1 MeV. The reaction cross sections (for energies sufficiently above threshold) are proportional to E^2 in the energy range of nuclear
physics and are typically of order $10^{-43} cm^2$. Neutrino detection may also proceed
via elastic scattering from electrons, with cross sections of order $10^{-45} cm^2$.
A detection via coherent scattering off nuclei, which would not be hampered by
threshold problems, has not yet been achieved. Neutrino detection in the energy
range of nuclear physics, due to the small cross sections involved, requires neutrino sources of extraordinary strength. Two copious sources of neutrinos are
available for such measurements:

1. <u>Nuclear Fission Reactors:</u> The heavy nuclei undergoing fission are rich
   in neutrons and their fission products are correspondingly left in high
   neutron-rich excited states, which decay to stable ground states by reducing their neutron contents via the reaction

$$n \rightarrow p + e^- + \bar{\nu}_e.$$

A nuclear power reactor with a (thermal) energy of, e.g., 2800 MW is a source
of $5 \cdot 10^{20} \bar{\nu}_e/sec$ with energies of up to 8 MeV.

2. <u>The Sun:</u> The nuclear fusion processes constantly proceeding in the solar interior are likewise a rich source of electron neutrinos. The fusion of Hydrogen into Helium inside the solar core proceeds via the reaction

$$p \rightarrow n + e^+ + \nu_e$$

giving rise to a neutrino production rate of about $2 \cdot 10^{38} \nu_e/sec$. Neutrinos
measured via inverse beta processes, e.g. via the reactions

$$\bar{\nu}_e + p \rightarrow n + e^+ \tag{1}$$

$$\nu_e + n \rightarrow p + e^- \tag{2}$$

with their kinetic energies of order 1 MeV vastly exceeding possible masses
of order 1 eV or smaller, may always be considered as ultrarelativistic particles travelling with roughly the speed of light. Such particles would

travel within about 8 minutes from the solar interior to a terrestrial detector, thus providing nearly instant information on the nuclear fusion reactions proceeding inside the solar core.

3. NEUTRINO-OSCILLATIONS IN VACUUM

Up to this date there is no conclusive evidence that neutrinos are massive particles. Experiments provide only upper limits on neutrino masses. The limit on the mass of electron neutrinos is $m(\nu_e) < 20 eV^3$. Measurements of neutrino masses in the range of eV or below are complicated, because it is very difficult to control possible solid-state or atomic excitations of the same order. Information on neutrino masses can in principle also be obtained via a search for neutrino oscillations, in case neutrino flavor is not a good quantum number, i.e. in case lepton families are coupled. Such a neutrino coupling is suggested by the experimentally observed weak interaction coupling in the quark sector: The weak isospin doublets partaking in the charged current interactions are given by

$$\begin{pmatrix} u \\ d' \end{pmatrix}_L \quad \begin{pmatrix} c \\ s' \end{pmatrix}_L \quad \begin{pmatrix} t \\ b' \end{pmatrix}_L$$

Cabbibo has introduced a coupling scheme, where $u = cos\theta_c d + sin\theta_c s$ in order to explain the possibility of strangeness conserving ($\Delta S = 0$, coupling $\propto cos\theta_c$) and strangeness violating ($\Delta S = 1$, coupling $\propto sin\theta_c$) interactions. The scheme was extended by Glashow, Iliopoulos and Maiani (GIM-mechanism) to explain the absence of strangeness violating neutral currents[4]:

$$\begin{pmatrix} d' \\ s' \end{pmatrix} = \begin{pmatrix} cos\theta_c & sin\theta_c \\ -sin\theta_c & cos\theta_c \end{pmatrix} \begin{pmatrix} d \\ s \end{pmatrix} \tag{3}$$

Here the states on the left side describe weak interaction eigenstates as produced in weak decays, while the states on the right side are mass eigenstates as propagating in space. The unitary coupling matrix involves the Cabbibo angle θ_c. Experimentally one found $sin\theta_c \simeq 0.23$, while a theoretical interpretation is still lacking. The 2 x 2 mixing matrix of equation (3), coupling the first two lepton families, has to be replaced by a 3 x 3 matrix in the case of a three family coupling. This Cabbibo-Kobayashi-Maskawa coupling matrix involves three mixing angles and one phase angle, the latter describing CP violation.

With the family mixing experimentally observed in the quark sector of the weak interactions one may also expect a family mixing in the lepton sector, as long as one does not know of any counteracting principle. Confining ourselves for simplicity to a two-family mixing scheme, we may write

$$\begin{pmatrix} \nu_e \\ \nu_\mu \end{pmatrix} = \begin{pmatrix} cos\theta & sin\theta \\ -sin\theta & cos\theta \end{pmatrix} \begin{pmatrix} \nu_1 \\ \nu_2 \end{pmatrix} \tag{4}$$

where ν_e and ν_μ describe two weak interaction eigenstates, the indices of which can readily be replaced by others, while the states ν_1, ν_2 on the righthand side refer to mass eigenstates. Theory provides no guidance as to the value of the mixing angle θ which appears in the unitary mixing matrix. Neutrino oscillations would be the consequence of such neutrino mixing, because each weak interaction eigenstate becomes a mixture of propagation (mass) eigenstates with different frequencies. This becomes apparent if one writes, e.g. $\nu_e(t) = cos\theta\nu_1(0)e^{-iE_1 t} + sin\theta\nu_2(0)e^{-iE_2 t}$. The beat frequency is apparently proportional to

$$E_1 - E_2 = \sqrt{p^2 + m_1^2} - \sqrt{p^2 + m_2^2} \approx (m_1^2 - m_2^2)/(2p) = (m_1^2 - m_2^2)/(2E) \propto \frac{1}{T_{OSC}} \propto \frac{1}{L_{OSC}}$$

Choosing proper initial conditions, such as e.g. $\nu_e(0) \neq 0$ and $\nu_\mu(0) = 0$, it is easy to work out the probabilities

$$P[\nu_e(0) \to \nu_\mu(L)] \quad and \quad P[\nu_e(0) \to \nu_e(L)]$$

313

for situations, where one starts with a ν_e at the source position and finds a ν_μ or ν_e, respectively, at the detector position a distance L away:

$$P[\nu_e(0) \rightarrow \nu_\mu(L)] = |\nu_\mu(L)^2|/|\nu_e(0)^2| = sin^2 2\theta sin^2[1,27(m_2^2 - m_1^2)L/E_\nu] \qquad (5)$$
$$P[\nu_e(0) \rightarrow \nu_e(L)] = 1 - P[\nu_e(0) - \nu_\mu(L)] \qquad (6)$$

In these equations, $\Delta m^2 = m_2^2 - m_1^2$ is in units of eV^2, L in meters and E_ν in MeV. Neutrino oscillations, apparently, exist only if there is lepton family mixing ($\theta \neq 0$) as well as a difference in mass of at least two mass eigenstates. Equation (5) describes appearance experiments, where a new type of neutrino is generated in the oscillations. The observation of such a new type of neutrino, e.g. a muon neutrino, requires energies in the neutrino beam sufficient for the muon neutrino to generate a muon in a detection reaction. Such appearance experiments, therefore, are not feasible with reactor produced electron neutrinos due to lack of energy. Equation (6) describes disappearance experiments, where one looks for an anomalous reduction in neutrino intensity of one and the same type of neutrino. Appearance experiments being the domain of high energy machines are apparently much more sensitive to mixing angles, while disappearance experiments, being at low energies the domain of reactor experiments, are more sensitive to small mass parameters Δm^2. The experimental search for neutrino oscillations consists in moving a detector sensitive to only one type of neutrinos away from a neutrino source and looking for an anomalous behaviour in the neutrino beam. Table I indicates neutrino oscillation lengths for various situations, differing in mass parameter Δm^2 and in energy E_ν.

Table I.

Δm^2 ↓	Sun 300 keV	Reactor 4 MeV	Mesonfactory 20 MeV	Accelerator 1 GeV
1 eV^2	0.75 m	10 m	50 m	2.5 km
10^{-3} eV^2	750 m	10^4 m	$5 \cdot 10^4$ m	2.500 km
10^{-10} eV^2	$7.5 \cdot 10^9$ m	10^{11} m	$5 \cdot 10^{11}$ m	
$5 \cdot 10^{-12}$ eV^2	$1.5 \cdot 10^{11}$ m			

For illustration we show in Fig. 1 limits on mass parameters Δm^2 and mixing parameters $sin^2 2\theta$ obtained in the Gösgen-reactor experiments[5] and some results obtained by CERN[6]. In each case, the parameter areas to the righthand side of the curves are excluded as a result of the experiments.

4. NEUTRINO OSCILLATIONS IN MATTER

The possibility of neutrino oscillations in matter was pointed out by Mikheyev and Smirnov[7], based on a previous theoretical analysis by Wolfenstein[8]. The MSW-effect is based on the fact, that electron neutrinos in the sun may interact with the solar electrons via charged current reactions in addition to the neutral current reactions, which are possible for neutrinos of all flavors, as indicated in Fig. 2. The additional charged current interaction in matter exclusively confined to ν_e causes in a two-neutrino approximation the following relations between neutrino eigenstates ν_e, ν_μ, mass eigenstates ν_1, ν_2 in vacuum and mass eigenstates ν_{1m}, ν_{2m} in matter:

$$\begin{pmatrix} \nu_e \\ \nu_\mu \end{pmatrix} = \begin{pmatrix} cos\theta_v & sin\theta_v \\ -sin\theta_v & cos\theta_v \end{pmatrix} \begin{pmatrix} \nu_1 \\ \nu_2 \end{pmatrix} = \begin{pmatrix} cos\theta_m & sin\theta_m \\ -sin\theta_m & cos\theta_m \end{pmatrix} \begin{pmatrix} \nu_{1m} \\ \nu_{2m} \end{pmatrix} \qquad (7)$$

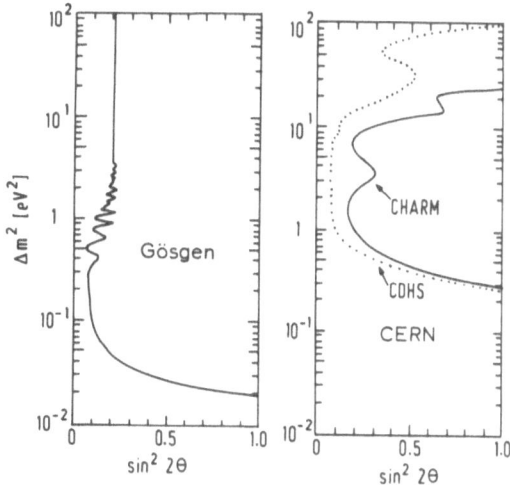

Fig.1. Permitted and excluded ranges for neutrino oscillation parameters $\Delta m^2 = m_2^2 - m_1^2$ and $sin^2 2\theta$, based on a two- flavor model. Excluded are in all cases with 90% confidence the parameter ranges right to the curves. Oscillation parameters left to (or below) the curves would still be compatible with the experimental results. The left side pertains to disappearance experiments with reactor neutrinos performed at the nuclear power reactor at Gösgen/Switzerland. The right side shows some corresponding limits obtained by accelerator based disappearance experiments at CERN.

Alternatively, we may write

$$
\begin{pmatrix} \nu_{1m} \\ \nu_{2m} \end{pmatrix} = \begin{pmatrix} cos\theta_m & -sin\theta_m \\ sin\theta_m & cos\theta_m \end{pmatrix} \begin{pmatrix} \nu_e \\ \nu_\mu \end{pmatrix} = \begin{pmatrix} cos(\theta_m - \theta_v) & -sin(\theta_m - \theta_v) \\ sin(\theta_m - \theta_v) & cos(\theta_m - \theta_v) \end{pmatrix} \begin{pmatrix} \nu_1 \\ \nu_2 \end{pmatrix} \quad (8)
$$

where θ_v and θ_m are the mixing angles applying to vacuum and matter, respectively.

The matter oscillation angle θ_m expressed in terms of the ratio l_v/l_0 is given by:

$$
tan2\theta_m = \frac{sin2\theta_v}{cos2\theta_v - l_v/l_0}, \quad (9)
$$

where the quantity

$$
\frac{l_v}{l_0} = \frac{2\sqrt{2}G_F E_\nu N_e}{m_2^2 - m_1^2}
$$

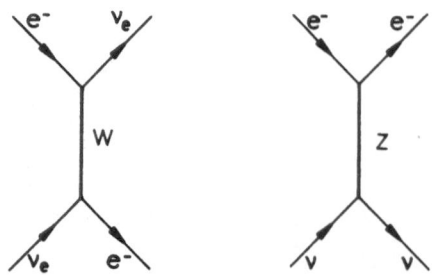

Fig.2. Diagrams for neutrino elastic scattering off solar electrons. The right
side shows neutral current interactions possible for all neutrino flavors; the left side shows charged current interactions possible for electron neutrinos only.

contains in the numerator the relevant charged current interaction, which
is proportional to the electron density N_e in the sun. We may now anticipate the
following succession of events which is illustrated in Fig.3:

1. A ν_e is created in the solar core where it interacts with electrons in a region of high density, which we may approximate by $N_e \to \infty$. Assuming a normal
   mass hierarchy, $m_1 < m_2$, we have

$$cos2\theta_v \ll l_v/l_0 \quad \text{and} \quad \theta_m = \pi/2$$

According to (8) we then obtain for the state vectors

$$\begin{pmatrix} \nu_{1m}(\infty) \\ \nu_{2m}(\infty) \end{pmatrix} = \begin{pmatrix} 0 & -1 \\ 1 & 0 \end{pmatrix} \begin{pmatrix} \nu_e \\ \nu_\mu \end{pmatrix}$$

as illustrated in Fig.3.

2. Neutrinos may pass on their way from the solar interior to the surface
   through a region of electron density, where $cos2\theta_v = l_v/l_0$, corresponding to
   $\theta_m = \frac{\pi}{4}$ (compare Eq. (9)). This resonance condition corresponds to a situation, where the terms in the main diagonal of the interaction matrix, i.e.
   the terms which leave the electron flavor unchanged, just balance. By consequence, a situation of maximum mixing arises, where ν_e mixes most effectively with ν_μ.

3. In continuing their travel, the ν's will finally arrive at the solar surface, characterized be $N_e \to 0$ and a situation arises, where

$$cos2\theta_v \gg l_v/l_0 \quad \text{and} \quad \theta_m = \theta_v$$

According to Eq. (8) we then have:

$$\begin{pmatrix} \nu_{1m}(0) \\ \nu_{2m}(0) \end{pmatrix} = \begin{pmatrix} 1 & 0 \\ 0 & 1 \end{pmatrix} \begin{pmatrix} \nu_1 \\ \nu_2 \end{pmatrix}$$

Fig.3 demonstrates, how this sequence of events would permit, assuming θ_v
being very small, a most effective conversion of the ν_e initially created in the

316

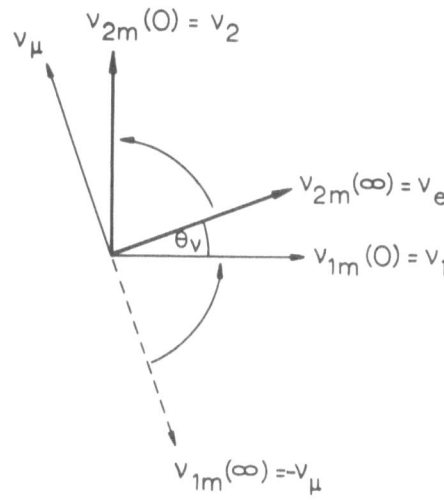

Fig.3. Transformation of the state vectors of the electron neutrinos during their passage from the solar interior ($\nu = \nu_e \approx \nu_{2m}(\infty)$) to the solar surface ($\nu = \nu_2 = \nu_{2m}(0) \approx \nu_\mu$). ν_1 and ν_2 are the neutrino mass eigenstates in vacuum.

solar interior into the mass eigenstate ν_2 at the solar surface. This state effectively resembles ν_μ and travels unperturbed through the vacuum to the earth.

5. THE SOLAR NEUTRINO PUZZLE

The MSW-effect within the sun might be the source of what is frequently called the solar neutrino puzzle. Fig.4 shows the solar neutrino spectrum according to the Standard Model. Experimental studies of this spectrum exist thus far only for the high energy part of this spectrum, involving the ^8B−branch, which has been studied over some 20 years by Davis and coworkers[9]. This radio-chemical experiment, employing the reaction $^{37}Cl + \nu_e \rightarrow ^{37}Ar + e^-$, with a threshold of 0.81 MeV, essentially measures the integral flux associated with the ^8B−branch. The total neutrino event rate above threshold of $R_{obs} = 2.0 \pm 0.3 \; SNU$ has to be compared with the predicted rate of $R_{theory} = 7.9(1 \pm 0.33) \; SNU$, where the solar neutrino unit $(SNU) = 10^{-36}$ events $\cdot$ (target atom) $^{-1}(s)^{-1}$ has been used. The difference between measured and expected values is frequently called the solar neutrino puzzle. To illustrate the achievements of the ^{37}Cl−experiment, we note that a predicted neutrino flux of $6 \cdot 10^6 \nu_e cm^{-2} s^{-1}$ gave rise to approximately one transformation of a ^{37}Cl−nucleus to an ^{37}Ar−nucleus every two days within a detector of 650 tons of C_2Cl_4, which was placed in the Homestake goldmine in South Dakota to protect against cosmic radiation. The few activated nuclei of ^{37}Ar with a halflife of 35 d are periodically extracted from the tank with its $2 \cdot 10^{30}$ nuclei of ^{37}Cl by adding some inactive Ar-carrier and flushing the tank with Helium gas. Following gas separation and scrubbing the Argon was finally transferred to a small proportional counter and its beta-decay served as measure for the number of neutrino capture processes.

The overall efficiency of the ^{37}Cl experiment has never been tested with an artificial neutrino source. Assuming that the noted discrepancies between experiment and expectation do really exist, one might consider the following possible sources:

a) The Standard Solar Model is wrong. This possibility is not favored by theoreticians, the standard solar model being the result of the best physics and input parameters available. It should be noted, however, that the

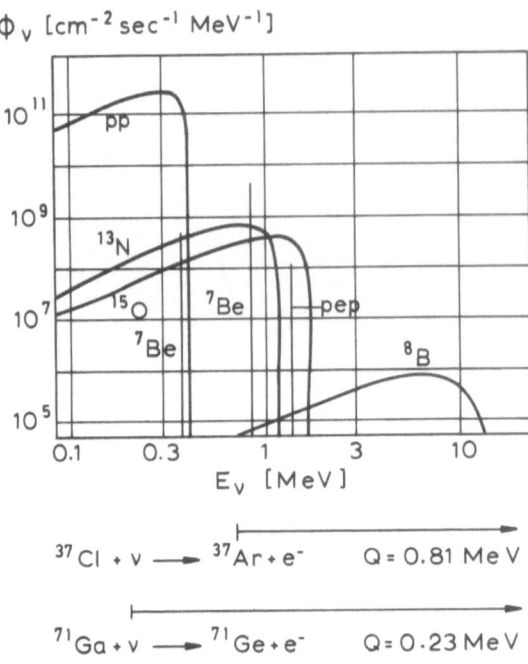

$\Phi_\nu \; [cm^{-2} sec^{-1} MeV^{-1}]$

$E_\nu \; [MeV]$

$^{37}Cl + \nu \longrightarrow \; ^{37}Ar + e^- \qquad Q = 0.81 \, MeV$

$^{71}Ga + \nu \longrightarrow \; ^{71}Ge + e^- \qquad Q = 0.23 \, MeV$

Fig.4. Energy spectrum of solar electron neutrinos. Plotted is the mean terrestrial solar neutrino flux versus neutrino energy for the various neutrino generating reactions occurring in the Sun, according to the Standard Solar Model.

^8B—branch of the solar neutrino spectrum depends very sensitively on the core temperature of the sun.

b) Decay of ν_e on their way from the source to the detector. A decay of ν_e on their way from the solar surface to a terrestrial detector can be excluded on the basis of the photon flux which would be associated with such a decay[10]. The possibility of a decay within the solar interior can be excluded on the basis of recent laboratory studies of neutrino decay at both Gösgen and Bugey[11].

c) Neutrino magnetic moment acted upon by solar magnetic fields. R. Davis has noted a correlation between his measured neutrino events and solar flares. Such a correlation might occur if a neutrino carrying a magnetic moment leaves the sun, and magnetic fields associated with solar flares would act upon the magnetic moment, causing a change in the helicity of the particle. Such a righthanded neutrino appearing in a terrestrial detector would stay unnoticed. This way a reduction in the observed solar neutrino flux could be explained. The required magnetic moment, however, is only slightly below the present experimental limits of some 10^{-10} nuclear magnetons, yielding the suggested explanation marginal. In addition, the neutrino would at least have to exhibit some Dirac character.

d) Neutrino oscillations in vacuum. Complete mixing of an electron neutrino originating in the sun into three neutrino flavors would result in a neutrino flux upon arrival on earth of one third ν_e, one third ν_μ, and one third ν_τ. A neutrino detector sensitive only to ν_e would then measure a flux reduced by a factor of three compared to the expected flux. Such an explanation, however, appears unrealistic, because it would require full mixing of neutrinos, which in all likelihood would already have been noticed in other experiments.

318

e) Neutrino oscillations in matter. The MSW-effect could readily explain the noted deficit in solar neutrinos. For this effect to occur, neutrinos would have to pass through a resonance regime, where the charged current interactions with the solar electrons would just compensate the interaction responsible for the vacuum oscillations. Such a situation might well occur inside the sun, where the matter density changes from some 150 g cm^{-3} in the solar center to zero at the surface.

It is at present not possible to decide whether the solar neutrino puzzle must be attributed to uncertainties in the solar model or to particular neutrino properties, always assuming the ^{37}Cl-experiment to be correct.

6. THE EUROPEAN GALLEX PROJECT

A measurement of the solar neutrino flux associated with the pp-fusion processes with neutrino energies up to 0.42 MeV (compare Fig.4) would remove the uncertainties associated with the solar model, since the pp-neutrino flux can be directly related to the solar luminosity, with 98% of the solar energy emission being related to this pp-branch. The reaction $^{71}Ga + \nu_e \rightarrow {}^{71}Ge + e^-$ with its low threshold of only 0.23 MeV can serve for this purpose. It is instructive to give an estimate of the solar neutrino flux arriving at the earth and originating in the solar fusion process, where ultimately four protons are fused into a ^4He-nucleus, via the reaction $4p \rightarrow \alpha + 2e^+ + 2\nu_e$. The energy liberated in this process obtains from $M(^4H) - M(^4He) = 26.70 \quad MeV$. The total solar luminosity of 2.4 times 10^{45} eV per second, combined with the mean distance between earth and sun of d = $1.5 \cdot 10^{11}$ m, gives rise to an energy flux of about $1.3 kW/m^2 = 8.5 \cdot 10^{21} eV m^{-2} s^{-1}$ at a terrestrial surface oriented perpendicular to the sun. Attributing some 0.6 MeV to the emitted neutrinos (compare Fig.4) and taking into account that two neutrinos are then associated with a photon energy release of 26.1 MeV, we obtain a neutrino flux of $\phi_\nu \approx 2 \cdot 8.5 \cdot 10^{21} eV m^{-2} s^{-1}/(26.1 \cdot 10^6 eV) \approx 6.6 \cdot 10^{10} cm^{-2} s^{-1}$, as compared with the present theoretical value[1] of $6.07 \cdot 10^{10} cm^{-2} s^{-1}$. The actual argument is more subtle and depends on details of the solar reaction cycle: The entrance reaction $2p \rightarrow d + \nu_e + e^+$, which, incidently, determines the lifetime of the sun, is followed by the production of ^3He; the production of alpha-particles is then essentially continuing via ^3He-reactions and the production of one alpha-particle therefore requires the participation of two ν_e, justifying the assumptions made above.

It were essentially financial problems, why a Gallium experiment has not yet been performed. An original collaboration between Brookhaven National Laboratory and the Max Planck Institute for Nuclear Research at Heidelberg was dissolved for financial reasons. These problems have been largely solved by now and two collaborations are presently preparing for a measurement of the pp-branch of the solar neutrino spectrum:

1. The European Gallex project. This project will be performed in the Gran Sasso underground laboratory 150 km east of Rome and will engage 30 tons of Gallium in the form of a highly acidic solution of $GaCl_3$. Measurements are scheduled to start late in 1989. The GALLEX collaboration comprises the following laboratories:
   MPI Heidelberg (low level counting),
   KFK Karlsruhe, WIS Rehovoth, BNL Long Island, N.J. (extraction and chemical processing)
   INFN Milano, INFN Rome (laboratory operations),
   TU Munich (data processing),
   CEN Saclay, CEN Grenoble, NICE (calibration source; astrophysics).

2. Soviet Gallium experiment. This experiment will be performed in a special tunnel digged in the Baksan-valley in the Caucasian mountains and will involve 60 tons of Gallium in the metallic form.

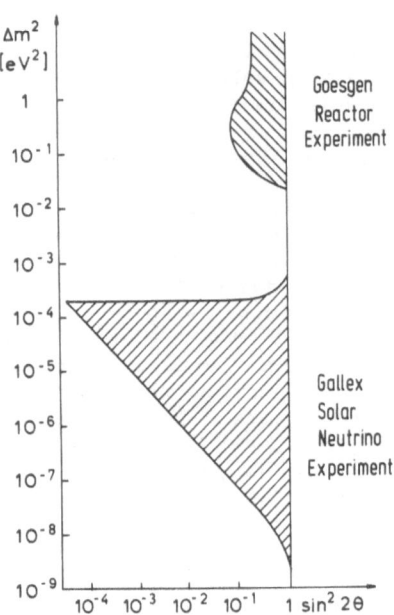

Fig.5. Maximum ranges of sensitivity of the Chlorine- and the Gallium-experiments for oscillation parameters $\Delta m^2 = m_2^2 - m_1^2$ and $sin^2 2\theta$ due to matter oscillations. The horizontal limit follows from the condition for the appearance of a resonance (maximum mixing), assuming a maximum energy of the emitted neutrinos of 14 MeV. The inclined limit follows from the condition of an adiabatic behaviour of the electronic density changes across the resonance for a limiting energy of 233 keV.

Details of the GALLEX project have already been dealt with in another lecture during this Summer School (see paper by L. Votano in this course) and we shall therefore only repeat some crucial aspects: The nuclei of ^{71}Ga in the solution of $GaCl_3$ capturing solar neutrinos will convert to ^{71}Ge, which converts back to ^{71}Ga with a halflife of 11.4 days. Germanium appears in the form of $GeCl_4$ and because of its different valency state can be readily extracted by flushing the target container with gas. The extracted $GeCl_4$ is chemically separated and purified and finally converted into GeH_4, which serves together with Xe as counter gas in a proportional counter with a volume of some 0.5 cm³. This detector counts the decay Auger electrons in the K- and L-peak with energies of 10.4 keV and 1.2 keV[12].

The experiment will be calibrated with an auxiliary neutrino source of 1 MCi strength. The source will consist of reactor produced ^{51}Cr, which decays with a halflife of 27.7 days via the reaction $^{51}Cr + e^- \rightarrow ^{51}V + \nu_e$ and emits neutrinos of energy 746 keV (90%) and 426 keV (10%). This source will be introduced into the center of the Gallium tank and will serve to verify the calculated overall efficiency of the entire neutrino detecting system. The measuring period will be some four years, including two calibration procedures, trying to achieve a statistical accuracy of about 10%, comparable to existing theoretical uncertainties.

In concluding we show in Fig.5 once more the measured experimental limits on mass and mixing parameters obtained in the Gösgen reactor experiment, which excludes parameters within the shaded area. The Fig.5 shows likewise the sensitivity range of the GALLEX solar neutrino project. The horizontal limit arises here from the requirement for a resonance to occur within the sun, with the maximum value of the electron density yielding a maximum allowed value for the mass parameter. The inclined limit arises from the requirement for the neutrinos to pass sufficiently slowly through the resonance, such as to prevent transitions from one neutrino-eigenstate to the other (adiabaticity condition). In other words, the passing through the resonance must be slow compared to a vibrational period at resonance.

In concluding, we should like to emphasize to following points:

1. It should be noted, that the distance between sun and earth does not enter here, because the phenomenon of matter oscillations in confined to the solar interior (with only a small possible reconversion effect upon entering the earth).

2. There exists a gap in mass parameter of about two orders of magnitude between the lower limit of the Gösgen experiment and the upper sensitivity limit of the GALLEX project. It would with reasonable investment be experimentally possible to lower the Gösgen limit by one order of magnitude down to about $\Delta m^2 \approx 10^{-3} eV^2$; further lowering this limit would require enormous experimental and financial efforts.

REFERENCES

1. For a recent review of solar models, neutrino experiments and helioseismology see J.N. Bahcall and R.K. Ulrich, Rev. Mod. Phys. 60: 297 (1988).

2. C.L. Cowan et al., Science 124: 103 (1956)

3. "86 Massive Neutrinos in Astrophysics and in Particle Physics , eds. O. Fackler and J.Tran Thanh Van, Editions Frontières, Gif-sur-Yvette, pp 441-584 (1986)

4. S.L. Glashow et al., <u>Phys. Rev. D</u> 2: 1285 (1970)

5. Munich-Caltech-SIN-Collaboration, G. Zacek et al.,<u>Phys. Rev. D</u> 34: 2621 (1986)

6. Charm Collaboration, F. Bergsma et al.,<u>Phys. Lett. B</u> 142 : 103 (1984); CDHS Collaboration, F. Dydka et al., <u>Phys. Lett. B</u> 134: 281 (1984)

7. S.P. Mikheyev and A. Yu Smirnov, <u>Yad. Fiz.</u> 42: 1441 (1985); <u>Sov. J. Nucl. Phys.</u> 42: 913 (1985); <u>Nuovo Cimento</u> 9C: 17 (1986)

8. L. Wolfenstein, <u>Phys. Rev. D</u> 17: 2369 (1978)

9. J.K. Rowley et al., Article <u>in</u>: "Solar Neutrinos and Neutrino Astronomy", AIP Conference Proceedings No. 126, eds. M.L. Cherry, W.A. Fowler and K. Lande, AIP New York p.1 (1985); R. Davis Jr., Proceedings of 7th Workshop on Grand Unification, ICOBAN, Toyoma, Japan p.237 (1986)

10. G.G. Raffelt, <u>Phys. Rev. D</u> 31: 3002 (1985)

11. L. Oberauer et al., <u>Phys. Lett. B</u> 198: 113 (1987); J. Bouchez et al., <u>Phys. Lett. B</u> 207: 217 (1988)

12. W. Hampel, Article <u>in</u>: "Solar Neutrinos and Neutrino Astronomy" AIP Conference Proceedings No. 126, eds. M.L. Cherry, W.A. Fowler, and K. Lande, AIP New York, p. 162 (1985); T. Kirsten, Article <u>in</u>: "86 Massive Neutrinos in Astrophysics and in Particle Physics, eds. O. Fackler and J. Tran Thanh Van, Editions Frontières, Gif-sur-Yvette, p. 119 (1986)

DISCUSSION

– Rahal:

How can we measure the neutrino mixing angle?

– Mössbauer:

The values for the mixing angles are entirely unknown, and moreover there is no theoretical guidance here.

In our experiment we can set an upper limit on the size of the angles from the errors associated with the measured intensities.

– Rahal:

Could you explain the possible correlation between the neutrino flux and the magnetic field at the surface of the sun?

– Mössbauer:

First one assumes that the neutrino has a magnetic moment. When the neutrinos reach the solar surface they experience strong magnetic fields, which are correlated with solar flares. These fields act upon the magnetic moment and it then becomes possible for a helicity–flip to occur, from a left–handed to a right–handed neutrino. However, the detectors on earth are not responsive to a right–handed neutrino and these particles, therefore, just pass unnoticed through the detector. This gives a reduction in the apparent neutrino flux. Nevertheless, I consider this an unlikely solution to the solar neutrino problem, because a relatively large magnetic moment would be required for the neutrino in order to produce this effect. The experimental limits are presently such as to make this marginally possible. The evidence for this idea comes from an apparent correlation between the flux observed in the Davis experiment and the occurence of solar flares. These experiments, in fact, cover a period of nearly 20 years. There seems to be some counter evidence, since Davis claims to see a jump in the observed flux by a factor of two in recent months, while a recent study at Kamiokande seems to see no such a jump. But let me conclude with caution, that the analysis of the data is very difficult.

I would like now to reply to comments made by Prof. R. Barbieri in his last lecture. He said that maybe there is not really a solar neutrino puzzle. In my opinion, if there is really a shortage by a factor of three or four in the solar neutrino flux, then this is truly a problem. But is there really such a shortage? I should like to point out that in the Davis experiment has never been a calibration and this makes me worry, especially because of the recent jump by some factor of two. There is uncertainty in the cross–section, which, incidentally, is also the case in the GALLEX experiment.

The reason why I mentioned that our collaboration is planning to measure down to 10% of the maximum predicted solar flux, is because we are then hitting the systematic uncertainties in our experiment.

The largest uncertainty in both the Chlorine and the Gallium solar neutrino experiments are the matrix elements for the nuclear transitions. The neutrino energies are relatively high in the experiments and one detects these neutrinos via inverse beta–decay processes. One knows very well the matrix elements for the ordinary inverse beta–decay processes, those with the lowest energy contribution. But there are also higher energy levels, with contributions to the cross–sections which are not so well known. These contributions involve some experimental and theoretical input, whence the corresponding Gamow–Teller type matrix elements are not so well known. We estimate the error for the reaction cross–section in the case of the Gallium experiment to be of order 10%. We therefore believe it crucial to have an absolute calibration for the GALLEX as well as for the Chlorine experiment, because such a calibration would yield to overall efficiency of the experiment and thereby check the expectation. It would provide the combined efficiency for the extraction procedure, for the chemistry and for all the other procedures. In the GALLEX experiment, we shall determine this overall efficiency by means of an artificial neutrino source. Such a calibration has not been performed in the Davis experiment.

– Rahal:

Doesn't the explanation of the problem by interaction with the surface magnetic field assure that only left–handed neutrinos are emitted from the core? Could you justify this?

– Mössbauer:

In the core of the sun the overwhelming production of neutrinos is from ordinary beta–decay processes which are well understood. No one has ever observed a decay into a right–handed neutrino, whence it is pretty clear that practically all neutrinos emitted by the core will be left–handed.

– Barbieri: comment

I agree with the statement that right–handed neutrinos would be produced only negligibly in the core. The magnetic field at the surface required for this solution of the "problem" is only a few kilo–gauss, so I think you cannot rule out this idea. However, there arise other theoretical difficulties in understanding the neutrinos from the supernova 1987 if they indeed have the required magnetic moment.

– *Mössbauer:*

If the experiments on the magnetic moment of the neutrino would lower the existing bound from 10^{-10} down to 10^{-11} Bohr magnetons, then this explanation could already be ruled out. This is why I consider this explanation of the solar neutrino puzzle rather marginal.

– *Rahal:*

If the reaction rate on the experiment is only one nuclear conversion every second day, how is it then possible to isolate this one nucleus out of a total of 600 t of material?

– *Mössbauer:*

The activated material produced in the Davis experiment (Argon) and in the GALLEX experiment (Germanium) is rather volatile. In the case of Germanium, for example, the molecules transform from a trivalent to a tetravalent state and by consequence the molecule becomes very volatile. After two weeks ($\sim$ one halflife for back–decay) several of these molecules have been produced and one then starts extraction procedures in order to avoid a substantial back–conversion of the activated nuclei by beta–decay. Extraction is performed by flushing the tank with gas, adding some inactivated carrier material to avoid losses of the activated nuclei. The amount of this carrier material has not yet been fixed in the GALLEX experiment. After extraction, the wet gaseous mixture is purified, concentrated and in the case of the Gallium experiment converted into GeH_4, which together with Xe–gas will then be used as filling gas of the proportional counters employed in the actual measurement of the decays of the activated ^{71}Ge nuclei back into ^{71}Ga. A sophisticated pulse shape analysis allows to keep the background rate well below the actual signal rate.

– *Skarke:*

What are the possible decay modes of neutrinos?

– *Mössbauer:*

We have examined neutrino decay in a laboratory experiment. In the studied energy range of up to a few MeV, there are two modes $\nu \rightarrow \nu+\gamma$, and $\nu \rightarrow \nu+e^+e^-$.

– *Colas:*

What happens to the electron in the reaction $^{71}Ga + \nu_i \rightarrow^{71} Ge + e^-$, and what is the energy?

– Mössbauer:

The experiment being built–up is a radiochemical experiment. It is capable of looking for only the Germanium, but not for the electron. We are in effect "integrating" over the entire spectrum as well as over large times. The next step would be to do real spectroscopy, but this will be orders of magnitude more difficult. There is a potential for using ^{115}In for neutrino spectroscopy, but this is as yet far from being practical. The Q–value for the reaction is 0.23 MeV; the maximum energy of the pp–process is 420 keV.

– Giannakis:

1) Do you know of other non–standard Solar Model which are not in conflict with experimental evidence, for example models in which you have a different core temperature?

2) Are the pp neutrinos less dependent on the characteristics of the Standard Solar Model than the ^8B neutrinos?

– Mössbauer:

Actually there are some 50 non–standard Solar Models. However from John Bahcall and others, theoriticians are reluctant to follow these models. For example, George Marx from Budapest changes the iron contents of the core and introduces a phase–transition in the iron. Of course, with enough parameters in the model you may explain whatever you want. The Standard Model has very mutually consistent parameters. If you change the core temperature, say, by just a little, then the whole model collapses. With regard to the pp neutrinos, the calculations are particularly safe, since they relate directly to the solar luminosity. More than 98% of the solar energy is associated with the pp–reaction and it is straight forward to deduce the solar neutrino flux associated with the pp–reaction without relying on details of the Solar Model, such as the core temperature. The ^8B neutrinos, on the other hand, depend rather sensitively on the core temperature.

– Brandt:

a) You seem to favour discarding the magnetic moment explanation, is there any alternative explanation of the correlation between the neutrino flux and solar flares, or has this correlation been deemed statistically insignificant?

b) Could you comment briefly on the use of Gallium metal in the Soviet–Los Alamos Solar Neutrino experiment?

– Mössbauer:

1) The statistical significance is indeed the crucial point. In order to obtain the correlation with solar flares the data have to be grouped in a certain way; without this grouping the correlation is much less apparent.

b) The GALLEX project uses $GaCl_3$, because the chemistry of the extraction procedure indicated above is relatively easy. With the use of Ga–metal the extraction becomes more complicated. The use of the metal, however, offers one advantage: in the case of $GaCl_3$ we will employ an artificial neutrino source for calibration, which will be placed in the centre of the Gallium tank. To compete with the sun, this source must provide some ten times more neutrinos than the sun. The same is true in the case of Gallium metal, but with the metal the density is higher and by consequence the neutrino detection efficiency is higher. Thus the calibration experiment is easier to perform with Gallium metal.

– *Gonzales:*

When you were asked about neutrino decay modes, you only mentioned the visible decay modes, but there are some invisible decay modes as well, such as $\nu \to \nu+$ Majoron or $\nu \to \nu\nu\nu$. What about these?

– *Mössbauer:*

We are working here with those modes that can be seen. I agree that my conclusions would be wrong if the modes you mention occur. There is no evidence at all for, however, the presence of e.g. the Majoron.

– *Votano:*

Suppose in the GALLEX experiment that the expected capture rate of the neutrinos is much lower than that predicted by the Standard Model. For what range of Δm^2 for the neutrinos would the experiment be sensitive to?

– *Mössbauer:*

We can at most hope to measure capture rates as 10% of the maximum rate predicted by the Standard Model. The experiment measures only an integrated flux and it is impossible to determine the origin of deviations. Any deviation from prediction might be either due to the sun, or due to oscillating neutrinos. Knowing, however, the luminosity of the sun, we will attribute any measured neutrino deficit to particular neutrino properties such as masses and mixing. I would like to add that if there are oscillations, then the conversion of muon–(or tauon–) neutrinos could produce a "day and night" effect. Furthermore, the distance from sun to earth varies with time, which would yield small seasonal effects.

THE LAA PROJECT: ONE YEAR AFTER

A. Zichichi

CERN, Geneva
Switzerland

The following Physicists, Engineers and Technicians represent the core of the LAA Project:

A. Ali, G. Anzivino, M. Arneodo, F. Arzarello, G. Bari, M. Basile, R. Battiston, U. Becker, J. Berbiers, F. Bergsma, R. Bertin, R.K. Bock, R. Bouclier, G. Bruni, L. Caputi, G. Cara Romeo, R. Casaccia, G. Charpak, M. Chiarini, N.H. Christ, L. Cifarelli, F. Cindolo, E. Colavita, A. Contin, M. Costa, I. Crotty, G. D'Ali, C. D'Ambrosio, S. D'Auria, M. Dardo, S. De Pasquale, R. De Salvo, C. Del Papa, R. Dobinson, J. Dupont, J. Dupraz, T. Ekelöf, J.P. Fabre, P. Ford, F. Frasconi, J. Gaudaen, P. Giusti, K. Goebel, C. Grinnel, B. Guerard, T. Gys, E. Heijne, S. Hellman, M. Hourican, G. Iacobucci, P. Jarron, P. Jenni, L. Jones, W. Krisher, I. Laakso, J.C. Labbé, H. Larsen, G. Laurenti, T.D. Lee, H. Leutz, S. Lone, G. Maccarrone, T. Massam, K.H. Meier, G. Million, R. Nania, Ch. Nemoz, V. O'Shea, A. Oliva, H.P. Paar, P. Pelfer, C. Peroni, E. Perotto, V. Peskov, D. Piedigrossi, S. Qian, J.C. Santiard, G. Sartorelli, F. Sauli, E. Schenvit, J. Schipper, H. Schönbacher, D. Scigocki, P. Sharp, G. Simonet, P. Sonderegger, L. Sportelli, M. Suffert, G.C. Susinno, S. Tailhardat, A.E. Terraneo, L. Votano, T. Weidberg, R. Wigmans, C.H. Yeh, T. Ypsilantis, A. Zichichi and K. Zographos

1. INTRODUCTION

The present status of accelerators and detectors is as follows (see fig. 1).

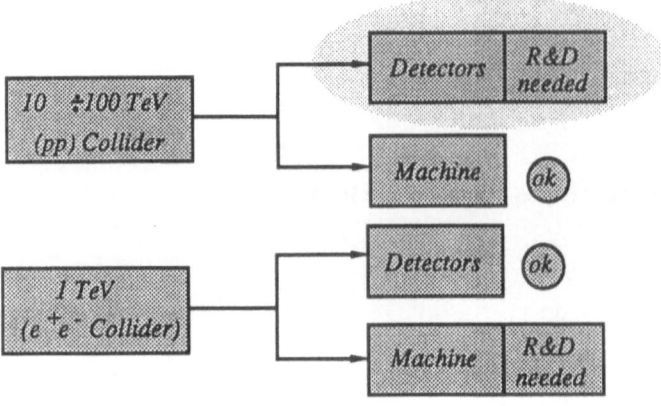

Fig. 1   *The present situation of Accelerators and Detectors.*

No one knows, at present, how to design an (e^+e^-) collider in the TeV energy range, while a conceptual design of ELOISATRON[1] (the 100 TeV (pp) collider) has shown that there are no basic difficulties for the machine to be designed in all details.

The situation is reversed for the case of detectors.

If a multi-TeV (e^+e^-) collider were operative, we could immediately design an experimental set-up. The reason being that, due to the electroweak photon propagator dumping, the cross-section left over is very small.

But, if a multi-TeV hadron collider were operating, no physicist in the planet would know how to perform a meaningful experiment, apart from the trivial "full screened" systems. The basic motivations being the high rate of events and the strong radiation doses to which the detectors are exposed.

As discussed later on, for the first time in the field of High Energy Physics there are basic reasons to justify a jump in energy as big as possible. The multi-TeV energy domain is, at present, the exclusive privilege of proton machines, in spite of the gluon and quark structure of the colliding particles. The 100 TeV level is off-limits for (e^+e^-) colliders. This is why the choice emerging from fig. 1 is R&D for detectors, in proton colliders.

A new multi-TeV collider opens up a new series of problems. These problems are of great general interest for all those who want to do Physics in the next generation of Accelerators.

The urgency of undertaking a series of studies where new ideas and

new instruments could be investigated, has brought us to the present status of the CERN LAA Project[2,3,4,5,6].

The LAA Project consists of ten sub-projects which are open to all physicists and engineers who are interested in participating. The ultimate goal is to prove, on the basis of prototypes, the feasibility of essential components for a detector to operate in a future multi-TeV hadron collider. Special attention is paid to radiation hardness, rate capability, momentum resolution and hermeticity of such a detector assembly.

In what follows, the motivations, the requirements, the choices, and the final achievements of the LAA Project are reported.

2. THE MOTIVATIONS

Is this large R&D programme justified in terms of Physics goals?

Back in 1979 it was clear to some of us that we were entering a new *Era* in Physics.

In spite of the great development of our Physics, little was done to promote the necessary support required by these *new frontiers of advanced research. Why?* Probably because the physics community was not convinced that this new *Era* was really there.

This new *Era* indicated that a big jump was needed in the *energy level*.

The strong feeling that we were entering a new *Era* in Physics initiated the ELOISATRON Project, and LAA is part of this programme. The key point of this new *"Era"* in Subnuclear Physics is that all particles discovered so far, even the very heavy $(W^\pm, Z^0)$, are extremely light.

New ideas, new concepts and new phenomena make physics of just twenty years ago seem as old as millenia. Einstein's quadrimensional space-time seemed to be a conquest beyond which no one would be able to go. This, however, seems to be an incredibly narrow outlook for two reasons, both fundamental: the number of dimensions and the property of those dimensions.

No one had thought, before the sixties, that there could exist space-time dimensions with *fermionic* properties. Those of Einstein are *bosonic*. This is how the new concept of *superspace* was born, and with it, *superparticles* and *supermatter*.

The world in which we live and the matter which we are made of, could have their roots in a *bosonic* superspace with ten dimensions, plus the 32 *fermionic* ones. And this is not all.

The concept of "point" that has held its position for centuries and centuries, falls by the wayside. In its place is the "superstring"[7]: a unidimensional entity with a pointless structure in a 42-dimensional Superspace. More recently, the p-brane[8] (a membrane with p-dimensions) has taken the leading role over the string.

In this extraordinary progress of our knowledge, the winning parameter has so far been and will certainly remain the *energy*. The final goal is the unification of all the fundamental forces of Nature. Figure 2 shows the unification scale, and where the ELOISATRON hadron collider stands. The straight lines and the unification point are just to show the conceptual game. For example, the point at 10^{15} GeV is more like an ellipse. And the straight lines should not be taken as if no basic problems were there.

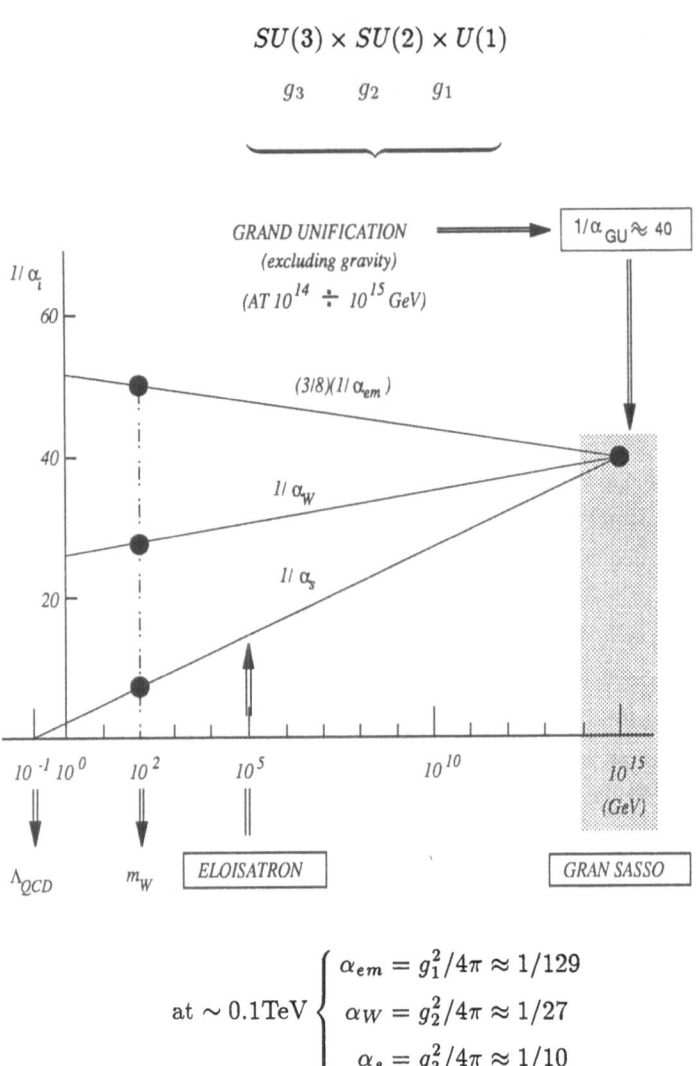

$$SU(3) \times SU(2) \times U(1)$$

$$g_3 \qquad g_2 \qquad g_1$$

$$\text{at} \sim 0.1 \text{TeV} \begin{cases} \alpha_{em} = g_1^2/4\pi \approx 1/129 \\ \alpha_W = g_2^2/4\pi \approx 1/27 \\ \alpha_s = g_3^2/4\pi \approx 1/10 \end{cases}$$

Fig. 2   *All the fundamental forces of nature should be generated by a unique force. The convergence of the three fundamental constants g_1, g_2, g_3, is the basis for GUT.*

There are four basic theoretical problems to be solved in the multi-TeV $(1 \div 100 \, TeV)$ energy domain:

- The *family* problem.
- The *hierarchy* problem.
- The *proliferation* problem.
- The *compositeness* problem.

The questions which arise when translating these theoretical problems into physically observable phenomena are the following:

i) Do *new*, heavier, *quarks* and *leptons* exist?

ii) Are there *other intermediate vector bosons*?

iii) How many *Higgs bosons* exist?

iv) Do *supersymmetric partners* exist?

v) Are *quarks* and *leptons composite*?

vi) Would some *unexpected exotic process* occur?

In order to go from theory to down-to-earth physics the detectors should be able to observe and measure in the multi-TeV range:

 i) electrons and photons;

 ii) muons and other lepton-like long-lived particles;

iii) neutrinos and other non-interacting particles (i.e. missing objects);

iv) leading protons (for hermeticity and new physics);

 v) hadrons and jets, with and without leptons inside.

In addition to all these new effects, there is an impressive series of "expected" phenomena to be studied[9]. The multi-TeV range is overcrowded with short- and long-lived hadrons.

Never before had Physics research such an impressive set of arguments to perform a big jump in energy. From what has been said above, the jump should be as big as possible.

3. THE BASIC DATA

Let me discuss the basic data for a new detector at a future SUPER-COLLI-DER.

The first and main requirement is that a Luminosity as high as possible $(10^{33} \div 10^{34} \div 10^{35} \, cm^{-2} s^{-1})$ should be aimed at. This imposes severe conditions on new detectors. Among them, a vital one is radiation hardness.

A basic feature of new physics is to produce undetectable events. Hermeticity will be essential for the discovery of new phenomena. Therefore the average number of events per bunch crossing ($\langle n \rangle$) must be ONE if the missing energy is to be used as a signature in event selection and analysis[*]

The limiting Luminosity is:

$$L_{pp} = \frac{\langle n \rangle}{\Delta t_b \times \sigma_{pp}}$$

where:

$\langle n \rangle$ = average number of events per bunch crossing,

Δt_b = time between bunch crossings,

σ_{pp} = total (pp) cross-section.

Figure 3 shows the detection limit for rare events (fixed at 10 events per year), and the total minimum bias rate, as a function of Luminosity. Note that a total running time of 10^7 seconds per year, and a total (pp) cross-section of $100\,mb$ are assumed. The "magic" limit of observability, at the $10^{-40}\,cm^2$ level in the cross-section for new physics, is reached if a Luminosity at the level of $10^{34}\,cm^{-2}s^{-1}$ can be achieved.

At present, the following machine parameters:

$$\Delta t_b \sim 100\ ns,\ \text{and}$$

$$L_{pp} \sim 10^{32}\ cm^{-2}\ s^{-1},$$

are well within reach from a technological point of view.

On the other hand, the total (pp) cross-section is expected to be:

$$\sigma_{pp} \cong 100\ mb = 10^{-25}cm^2.$$

These three values together produce $\langle n \rangle \sim 1$. Figure 4 summarizes the present status, including the radiation level.

But what is wanted is $\langle n \rangle = 1$ at higher luminosities. And therefore the corresponding Δt_b reaches prohibitive figures:

$$L = 10^{33} \rightarrow 10^{34} \rightarrow 10^{35}\quad cm^{-2}\,s^{-1}$$
$$\downarrow \qquad \downarrow \qquad \downarrow$$
$$\Delta t_b = 10\ \rightarrow 1\ \rightarrow 0.1\qquad ns$$

---

[*] It must be remembered, however, that $\langle n \rangle = 1$ means that in 37% of the cases the number of events per crossing is 0, in 37% of the cases it is equal to 1, and in 26% of the cases it is greater than 1.

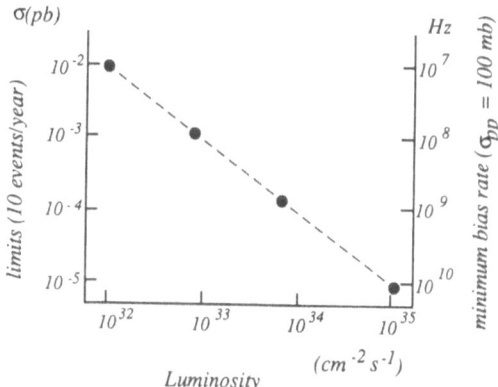

Fig. 3   Discovery limit for rare events and total rate as a function of
         Luminosity.

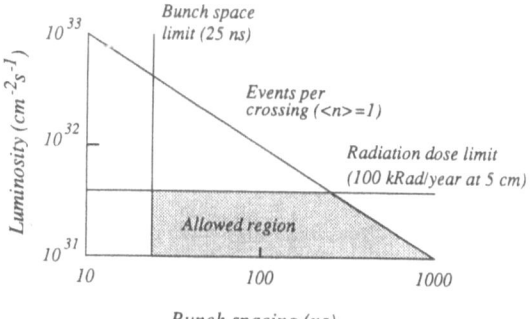

Fig. 4   Luminosity as a function of bunch spacing. The limits correspond
         to the three extreme values: bunch spacing = 25 ns, average events
         per crossing = 1, radiation dose = 100 KRad/year at 5 cm from
         the beam. The grey region is allowed.

High values of Luminosity, such as $10^{34} \div 10^{35}\, cm^{-2}\, s^{-1}$, must be the aim to be reached in future SuperColliders, and Detectors must be ready to cope with this high Luminosity, in order to study the existence of rare phenomena. Conclusion: machine builders[1] and physicists have to solve very difficult problems. This is a great challenge for all of us.

At present, no detector is able to make full use of a (*pp*) collider working with $L > 10^{32}\, cm^{-2}\, s^{-1}$, thus a strong R&D effort on <u>Detectors</u> must be pursued. *This is the aim of the LAA Project.*

4. THE MAJOR REQUIREMENTS

In this section, the major requirements on multi-TeV detectors are outlined, and the R&D necessary to obtain the necessary performances are indicated.

<u>Vertex detector technologies</u>

Requirements:

- Unambiguous and precise determination of primary and secondary vertices;
- Tracking back to the vertex, measuring the charged multiplicities and defining the topology of the event;
- Very high spatial accuracy, double-track resolution and redundancy, because of:
    - large multiplicities (~ 100),
    - high particle densities in jets,
    - high momenta;
- Particle identification.

Superconducting, very high field magnets should be used in association with very high precision vertex detectors.

The above requirements imply R&D on:

- Multidrift gas proportional tubes;
- High resolution scintillating fibres;
- *GaAs* and Silicon microstrips;
- Superconducting magnets ($\sim 10\,T$);
- Current-carrying beam pipe[10]: iron-free space with high toroidal field ($\sim 10\,T$).

<u>Calorimetry</u>

Requirements:

- Large coverage and hermeticity:
    - full detection of showering particles and jets;
    - study of missing p_T (neutrinos, photinos and other "inos");

- High granularity:
    - for electron identification inside jets;
    - increasing when going to more forward regions;
- High rate capability:
    - to cope with the high Luminosity and short time interval between bunch crossings ($\leq 100\,ns$);
- Optimum energy resolution. It depends on:
    - response to electromagnetic and hadron showers (compensation);
    - choice of absorbers and sampling medium;
    - longitudinal segmentation and depth;
    - control of systematics, which become crucial at very high energies.

The above requirements imply R&D on:

- Lead/scintillating fibres;
- BaF_2 scintillators and photosensitive wire chambers;
- Liquid Xenon calorimeters with electrical and optical read-out.

Muon detection

Requirements:

- Good rejection against hadron decays and hadron punch-through;
- Maximum angular coverage for detector hermeticity;
- Good momentum analysis over very large volumes, i.e. high resolution and precision alignment;
- Easy and economic construction;
- Minimum number of electronic channels.

The above requirements imply R&D on:

- Very large area detectors:
    - Limited streamer tubes;
    - Large toroidal and solenoidal magnets (air/iron);
    - Very large, high-precision drift chambers.
- Alignment.

Leading particle detectors

Requirements:

- Large coverage down to the smallest polar angles, needed for:
    - exploring the leading effect and the forward physics domain,
    - studying the longitudinal momentum balance,
    - tagging multiple interactions in the same crossing;
- Capability of facing serious background problems;
- Very high precision in measuring the space points (some μm);

- Stability in position (survey) and response (calibration) over large distances and long periods of time.

Particularly relevant for new heavy flavour research and other expected and unexpected phenomena.

The above requirements imply R&D on:

- *GaAs* microstrip detectors;
- Silicon microstrip detectors;
- Compact calorimeters;
- Ring Imaging Cherenkov counters (RICH);
- Completely new detectors.

Data acquisition and analysis

This is the possible scenario for the trigger and data acquisition system:

- First Level trigger:
  - Reduction in rate from 10^7 to $\sim 10^5 \, Hz$;
  - Decision time: some hundreds of nanoseconds;
  - Simple cuts on energy and p_T;
  - Each detector element provides an independent trigger;
  - Pipelined with the bunch crossing frequency to avoid completely the problem of dead time;
  - Massive use of custom/programmable processors, realized in close connection with industry.

- Second Level trigger:
  - Reduction in rate from 10^5 to $\sim 10^3 \, Hz$;
  - Decision time: $\sim 10 \, \mu s$;
  - Refines the first level trigger results using the digitized signals;
  - Massive use of custom/programmable processors.

- Third Level trigger:
  - Most important part of the whole trigger and data acquisition system;
  - Reduction in rate from $\sim 10^3$ to $1 \, Hz$;
  - Capable of handling event sizes of about $1 \, Mbyte$;
  - Analysis of the full event in a multiprocessor stack of about 1000 CPUs linked by fast data buses;
  - Flexible to explore new energy domains;
  - Implemented in strong connection with off-line (by using the same analysis programmes) to allow a fast implementation of algorithms, coping with new and interesting event topologies.

The above requirements imply R&D on:

- Microelectronics;
- Radiation hardness on the relevant components of the detector;
- Real-time data processing with ultra-high event rates;
- Dedicated Supercomputers.

Theory (QCD lattice calculations for dynamics) and Monte Carlo Simulations

Requirements: Event simulation must be based on QCD calculations.

This implies R&D on:

- Dedicated Supercomputers for lattice QCD calculations.
- QCD perturbative models.

5. THE CHOICE

The previous analysis in terms of physics goals and machine parameters, has brought us to the following choices for the LAA Project: ten basic components (see fig. 5).

1. HIGH PRECISION TRACKING

   Here three parts are needed. The closest to the vertex uses gaseous detectors consisting of multidrift tubes. The surrounding one uses the technology of scintillating fibres. The third one, along the beam, is based on microstrips made of a new type of detector material: Gallium Arsenide ($GaAs$).

2. CALORIMETRY

   Here we follow two lines. One is for the fixed-target mode of operation, and it is based on BaF_2 scintillators coupled to photosensitive wire chambers. The other, for a "collider" mode is based on the so-called "Spaghetti Calorimeter" (lead-plastic fibre calorimeter). Of course, no one can exclude that the first approach becomes so successful as to be extended also to the collider mode.

3. LARGE AREA DEVICES

   Here there are two basic parts. The first studies the problem of constructing large area devices sensitive to charged particles and with high precision. The other part of the R&D refers to the problem of positioning and monitoring the alignment of these large area devices. There is, in fact, no point in constructing a large area device if we do not know how to position it with great accuracy.

4. LEADING PARTICLE DETECTION

This is a crucial component of the LAA project. All, in fact, started with the discovery of the "leading" effect at the ISR[11,12,13,14,15,16]. To work near the beam of a high energy collider presents a series of very difficult, and therefore extremely exciting, problems to be solved. From fast removal of the detectors, to high precision positioning, to high radiation resistance.

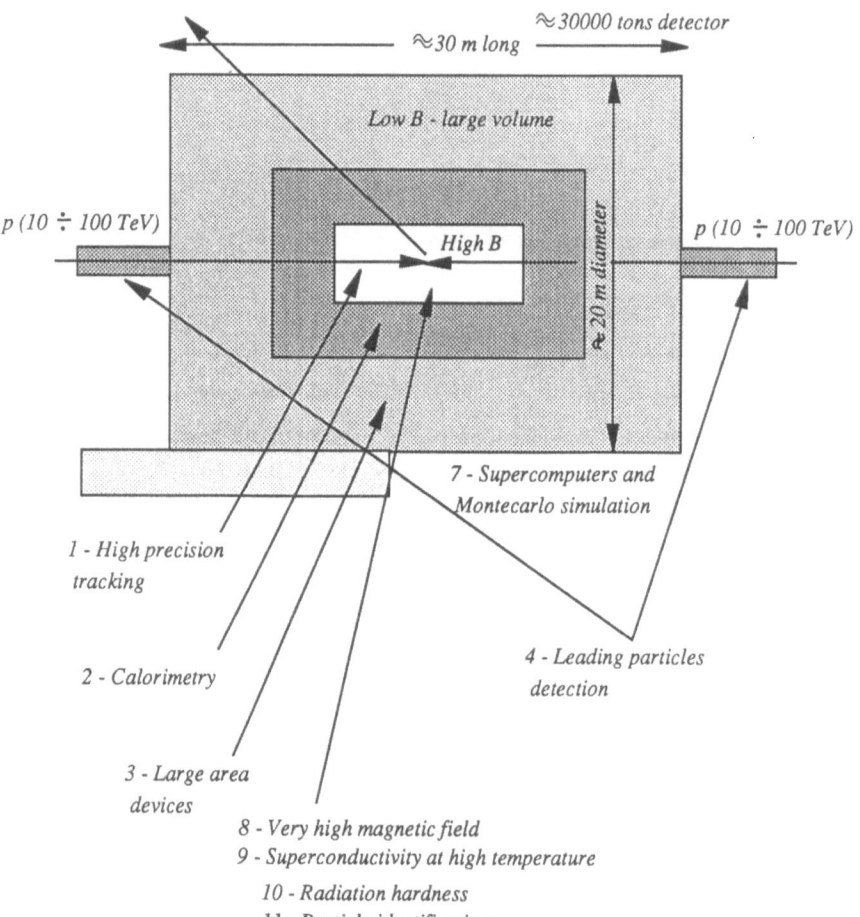

5 - *Subnuclear multichannel integrated detector technologies*
6 - *Data acquisition and analysis*

≈ 30000 tons detector

≈ 30 m long

Low B - large volume

p (10 ÷ 100 TeV)

High B

≈ 20 m diameter

p (10 ÷ 100 TeV)

7 - *Supercomputers and Montecarlo simulation*

1 - *High precision tracking*

2 - *Calorimetry*

4 - *Leading particles detection*

3 - *Large area devices*

8 - *Very high magnetic field*
9 - *Superconductivity at high temperature*
10 - *Radiation hardness*
11 - *Particle identification*

Fig. 5   The eleven components of the LAA Project.

5. SUBNUCLEAR MULTICHANNEL INTEGRATED DETECTOR
   TECHNOLOGIES

   The very high number of channels needed in the detectors for high Luminosity colliders, and, consequently, the problem of compaction of the very many elements, the electric power consumption, etc., lead naturally to the development of integrated electronics. Moreover, the problem of producing radiation resistant electronic components must be solved if these have to be operated very close to the detector. Therefore, this component of the LAA Project consists of two parts, one dedicated to the design of general-purpose integrated circuits with standard technologies (silicon), and the other which will concentrate mostly on new, radiation-hard, technologies.

6. DATA ACQUISITION AND ANALYSIS

   Here, the objective is the real-time processing of the detector signals. Solutions have to be found to the triggering and data compaction problems, filling the gap which exists today between the custom-made electronics and the programmable devices. Specific solutions have to be found for each of the trigger levels, with particular emphasis on the communication between the various components.

7. SUPERCOMPUTERS AND MONTE CARLO SIMULATIONS

   A full understanding of the theoretical predictions on high energy hadron collisions would be absolutely necessary, in order, both to design the detector, and to analyze in a meaningful way the data. QCD non-perturbative calculations (QCD lattice) have never been applied to the dynamics of the hadronic interactions. Powerful supercomputers are needed to perform the calculations in a reasonable time. Moreover, the theoretical understanding of the algorithms is not yet complete. Therefore, a strong effort is needed in this direction. Meanwhile, a unification of the various Monte Carlo programs, based on QCD perturbative models, is necessary. The goal is to produce a single, QCD "Super-Monte Carlo" program, which eliminates all the uncertainties and contradictions still present today in this field.

8. VERY HIGH MAGNETIC FIELDS

   Here the aim is to design a magnet to produce an extremely high magnetic field in the interaction region. Coupled to very precise tracking detectors, this magnetic field will allow the measurement of the momentum and of the charge sign of the particles produced in multi-TeV interactions. The higher the magnetic field, the more compact the full detector will be.

9. SUPERCONDUCTIVITY AT HIGH TEMPERATURE

   The recent discovery of superconductivity at high temperatures opens up a very large range of possibilities in High Energy Physics. Here, most of the problems to be solved concern the technology of materials: for example, how to produce reliable cables, capable of sufficiently high current density to be of practical use. The LAA Project will follow the developments in this field through a series of contacts with the various research centres in the World.

10. RADIATION HARDNESS

Here, all the studies on the radiation hardness of the various materials and electronics, developed by the different LAA components, are co-ordinated.

6. REPORT OF THE FIRST YEAR OF ACTIVITY

6.1 - HIGH PRECISION TRACKING

6.1.a - PROTOTYPE VERTEX DETECTOR USING MULTIDRIFT MODULES

The aim of this component of LAA is to build a vertex detector based on the MultiDrift Modules, whose main present and expected performances are summarized in fig. 6. The expected performances of the complete detector are summarized in fig. 7.

Twenty Mulidrift Modules (MDM) have been built up to now. Some of them are shown in fig. 8. All have the same geometry, but with various improvements.

In particular, different solutions have been tried for the connection of the signal cables to the end plates. Direct soldering of the signal cables to the anode pins gave the most reliable results.

But, the direct soldering is not very convenient for handling, so a new connecting scheme is under development, together with a high density preamplifier block to be directly attached on the module.

The MultiDrift Modules have been tested on the beam for spatial resolution and rate capability. The reconstructed path of a charged particle track, hitting the module perpendicularly, is shown in fig. 9.

The results obtained so far are:
- resolution perpendicular to the wire, measured by drift time: $\sigma_\perp = 60\ \mu m$,
- resolution parallel to the wire, measured by charge division:

$$\sigma_{//} = 3500\ \mu m, \text{i.e.} \quad \frac{\sigma_{//}}{module\ length} = \frac{3.5\ mm}{400\ mm} = 0.9\%$$

Figures 10 and 11 show the difference between the measured and the extrapolated points, for position measurements perpendicular and along the wires, respectively. The extrapolated points have been obtained using a pair of small microstrip detectors with a much higher precision than the MDM ($\sim 10\ \mu m$).

The rate capability and the radiation resistance of the MultiDrift Modules have also been tested.

342

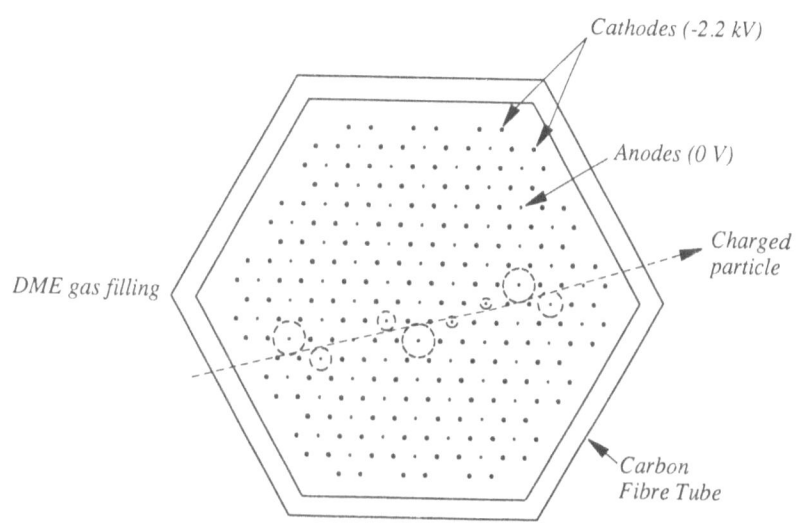

	1987	Achieved 1988	Expected
• cell radius (mm)	1.45		1.45
• modules diameter (mm)	30		30
• number of anodes per module	70		90
• operating pressure (bars)	1		2
• tube length (cm)	80		200
• spatial accuracy transverse to the wire, single wire (μm)	70	60	40
• spatial accuracy along the wire, single wire (mm)	20	3.5 (40 cm length)	2
• two–track resolution (μm)	800		400
• time resolution (ns)	40		35

Fig. 6. Multidrift Module present and expected performances.

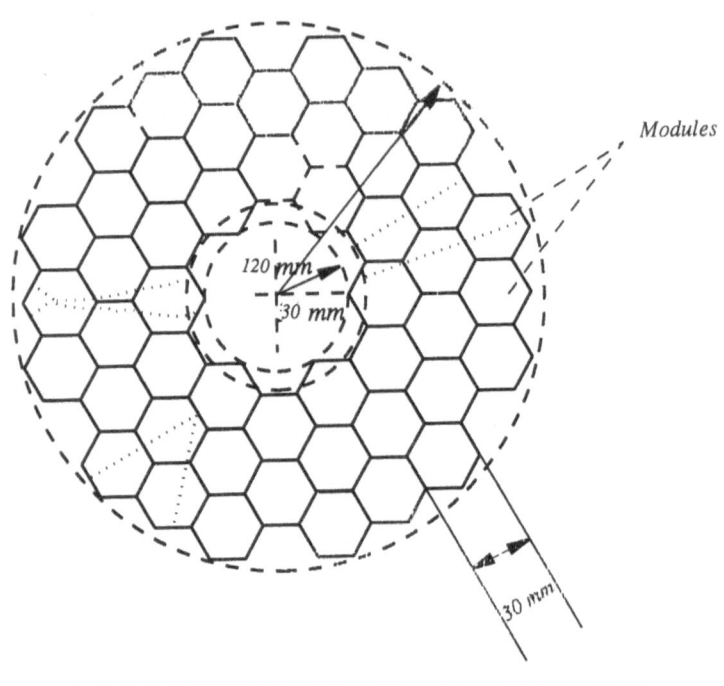

• number of modules	48
• total number of wires	4320
• average number of measurements per track	25
• angular resolution, each track (μrad)	120
• percentage of solid angle covered	99.9

Fig. 7   Vertex detector characteristics

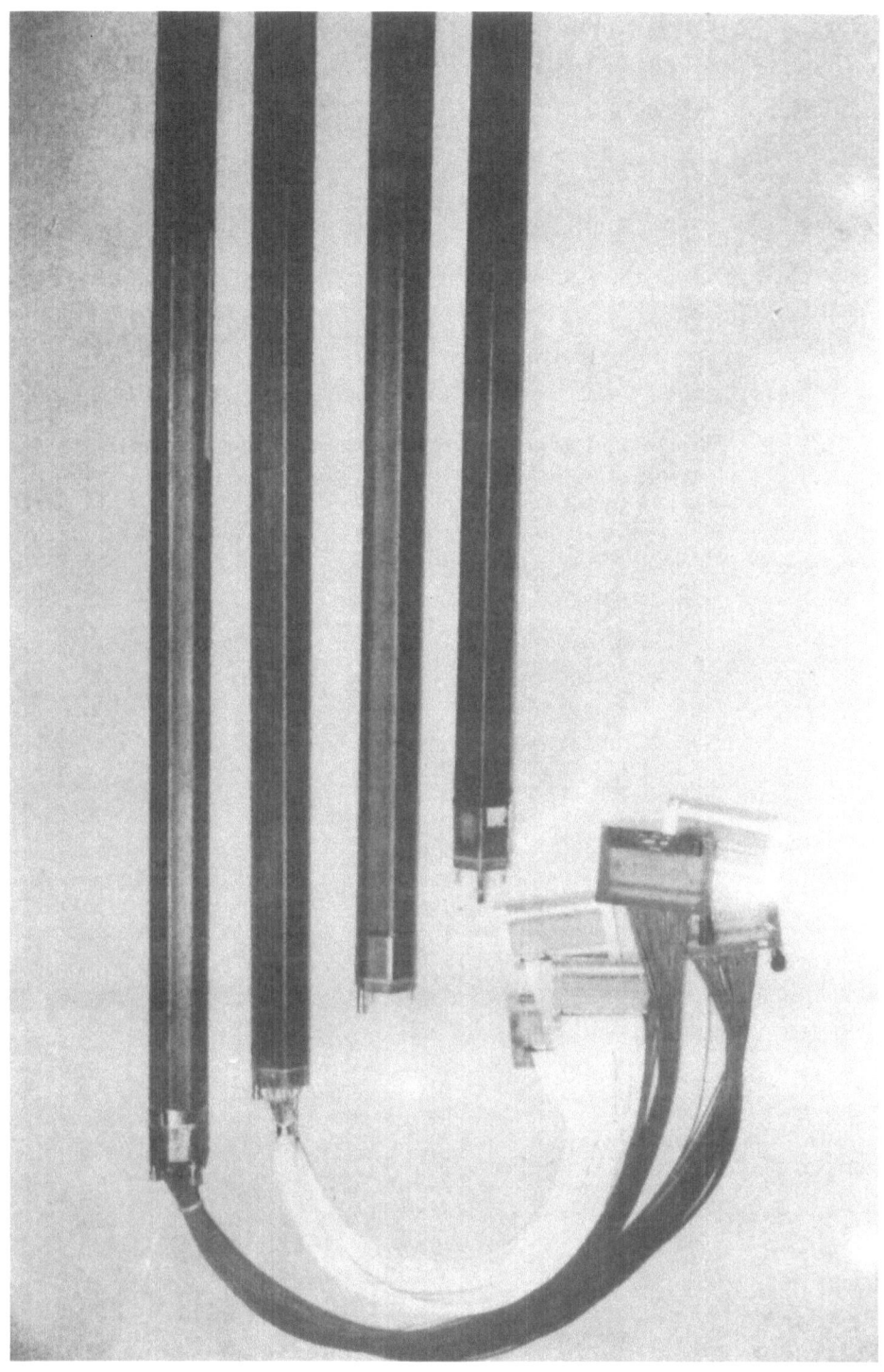

Fig. 8   Some of the MultiDrift Modules built for the LAA Project.

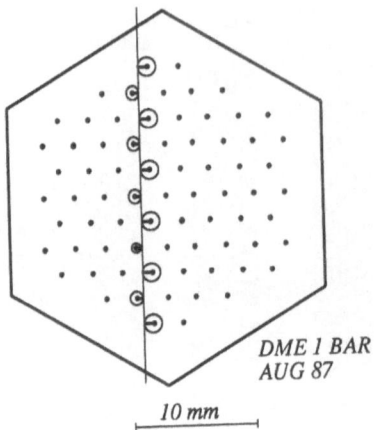

DME 1 BAR
AUG 87

10 mm

Fig. 9   The reconstructed path of a charged particle hitting the module perpendicularly. The circles represent the distance from the anode wire, as
computed from the measurement of the drift time.

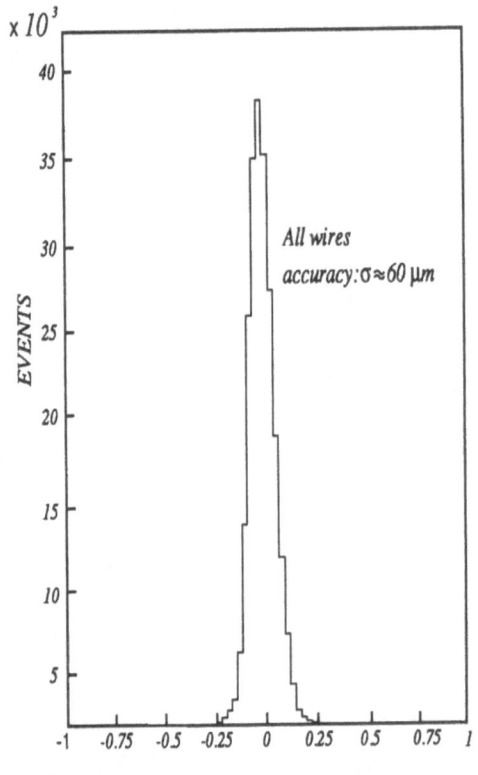

DIFFERENCE MEASURED-EXTRAPOLATED POINT [mm]

Fig. 10   Position measurement perpendicular to the wires: distribution of the
difference between measured and extrapolated points.

Two problems come in when a gaseous proportional counter is exposed to high radiation fluxes.

First, the large amount of slow positive ions released in the avalanches creates a local space charge. This modifies the electric field and therefore the gain above a certain flux (rate limitation). Second, the molecules of the gas, especially if containing an organic quencher, break down and recombine forming heavy polymers that deposit on the electrodes thus permanently modifying the operating characteristics of the module (aging).

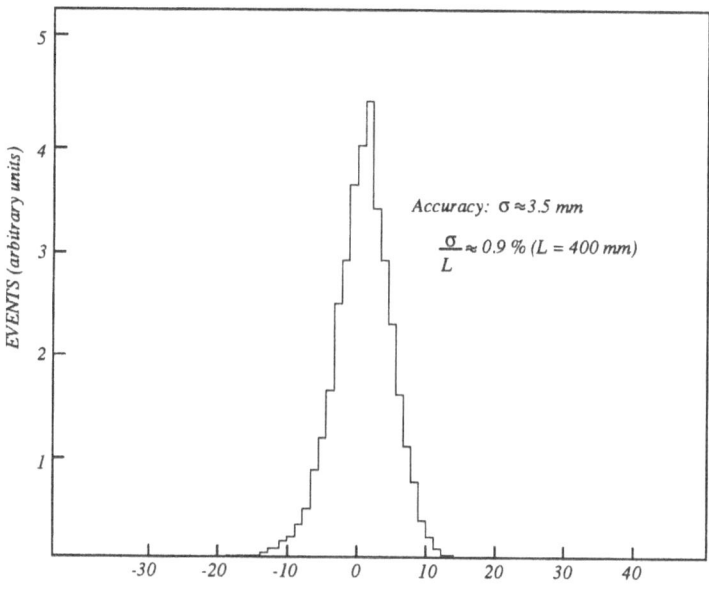

DIFFERENCE MEASURED-EXTRAPOLATED POINT [mm]

Fig. 11   Position measurement along the wires: distribution of the difference between measured and extrapolated points.

Rate limitation

Because of the small drift region (1.5 mm) and the high values of the field (up to 8kV/cm) in the cells of the MDM, we had reasons to believe that the first problem (space charge) would not be very serious. In fact, space charge effects for MDM should be small compared to conventional Multi–Wire Proportional Chambers (MWPC).

The proof is given in figs. 12 and 13, which show the relative gain as function of the rate for a "standard MWPC", and for a MDM, respectively. Full efficiency, i.e. a relative gain at the 100% level, is kept up to a rate of:

- $10^{-3} \mu A/cm$, for a standard MWPC, and
- $5 \times 10^{-2} \mu A/cm$, for the MDM.

Therefore, the MDM is fully efficient up to a rate higher by a factor 50 with respect to the conventional MWPCs.

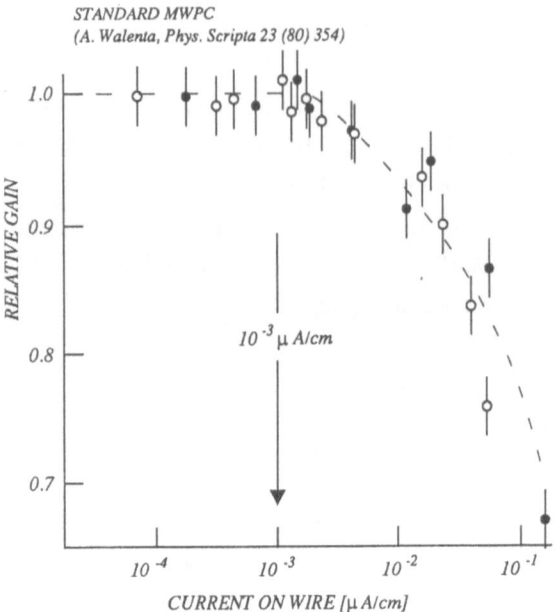

Fig. 12   *Relative gain as a function of the rate on the wire for a "standard"*
          MWPC.

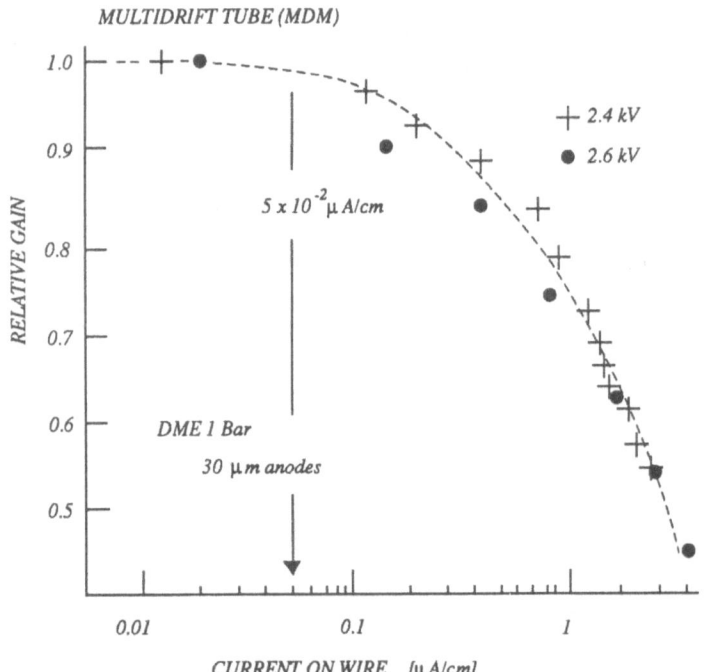

Fig. 13   *Relative gain as a function of the rate on the wire for the MDM.*

Aging

Aging of MWPCs is a very delicate matter. Various groups have obtained substantially different results in apparently similar experimental conditions.

Let me remind you an old direct experience of mine.

I started my activity in cloud chamber technology. The great Blackett was the undiscussed leader in the field. It was a great privilege for me to join his group, where a new big chamber was installed at high altitude. In Manchester I met a student whose Ph.D. was getting postponed. He was literally desperate. He had to study the problem of contamination and there was no way to achieve this goal. No matter what he was doing, the chamber was working very well. It was the glorious Blackett chamber. Other instruments, in principle identical, were incredibly sensitive to the smallest contamination. So much that the cloud chamber was a mixture between exact technology and extra sensitivity to unpredictable effects. I have now the feeling that with MWPCs the story repeats.

It appears that very small traces of unsuspected impurities, such as pollutants in the gas or outgassing of materials, can be the dominant elements in the aging process.

A collaboration with Florida University, Gainsville, has been established to study in a systematic way all the parameters contributing to the aging process.

A complete laboratory for aging and radiation hardness studies has been set–up, with various analytical instruments for trace impurity measurements and surface analysis of damaged wires. Figure 14 shows a design of the experimental apparatus. A single–wire tube is used for these tests, to allow a more accurate control of the experimental conditions. However, the results are directly applicable to the MDM, because the single-wire tube reproduces its cell size and electrical field.

All relevant parameters are monitored by a computer:

- current,
- voltage,
- temperature,
- gas quality,

and kept under control, in order to study the influence of each of them on the MDM performances, separately. As an example, the gas used for aging measurements contained a very small amount of contaminators: less than 18 ppb of Freon11 and 2 ppm of Freon12.

Two typical results of aging are shown in fig. 15. The effect of aging is to decrease the current on the wires, thus decreasing the gain.

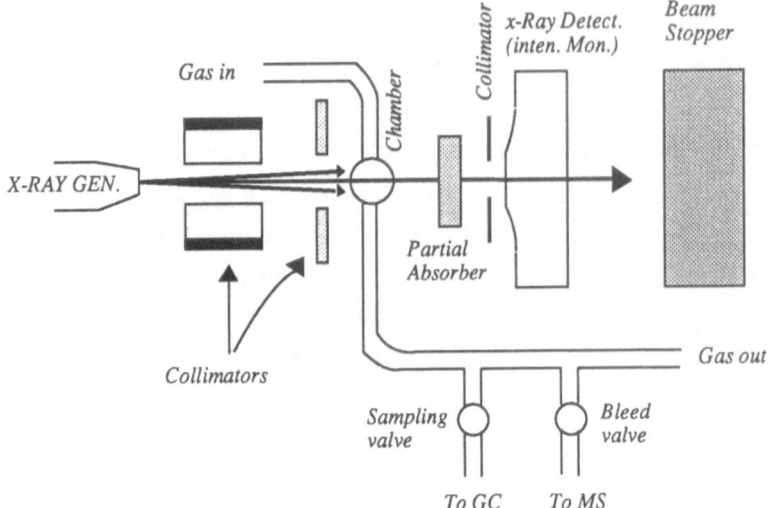

Fig. 14   Radiation hardness test set-up.

The first graph in fig. 15 refers to a NICOTIN (NIckel–CObalt–TIn) wire, which is normally used for position measurements along the wire (by charge division), due to its high resistivity. The NICOTIN wire starts showing aging effects very soon, so it is ruled out for use in a high radiation environment.

The second graph of fig. 15 refers to a Gold–plated Molybdenum wire, which shows no changes in performances with increasing irradiation. However, this type of wire has a low resistivity, thus it cannot be used for charge division measurements.

A third type of wire, made of stainless steel, has been extensively tested for aging effects. The stainless steel wires have a resistivity only 20% lower than NICOTIN wires, so they can be used for charge division measurements, with only a small degradation in performances. Figure 16 shows the results. No aging effects are measured in the stainless steel wire, at least up to an irradiation of 0.62 C/cm, which is the total accumulated up to now. This irradiation corresponds to about 0.62 MRad for our configuration. Therefore, a radiation resistance up to 1 MRad seems well within reach with this kind of wire. *)

Conclusion: using high purity DME gas, the MDM can survive up to 1MRad. The collaboration with Florida University, Gainsville, continues in order to better understand the influence of gas purity and wire surface on the radiation tolerance.

---

*) The latest (22 July 1988) results show that the stainless steel wire continues to behave very well at 1 MRad irradiation.

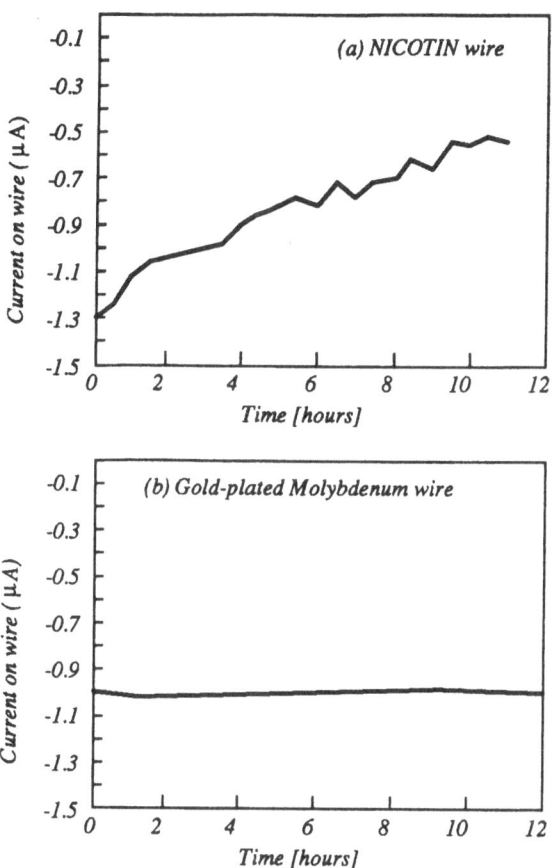

Fig. 15   Current on wire as a function of irradiation time for:
(a) NICOTIN, and (b) Gold-plated Molybdenum wires.

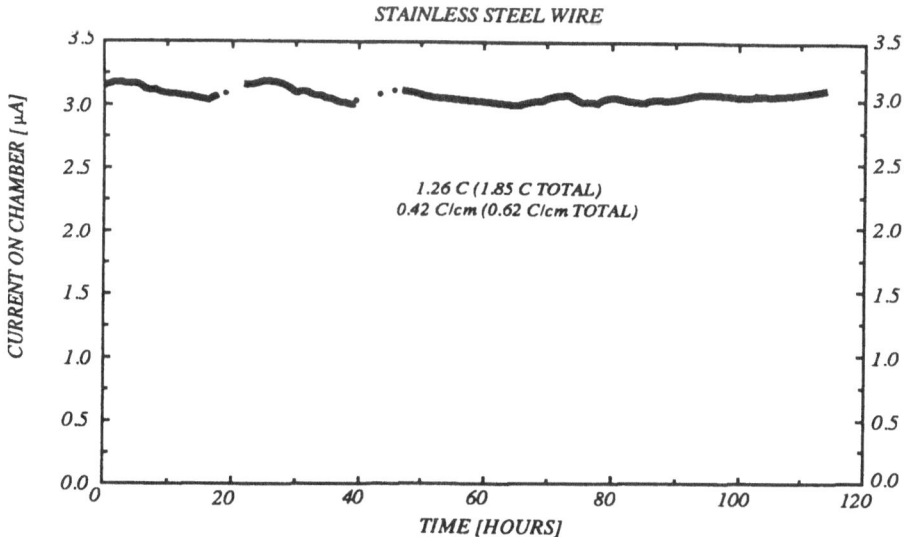

Fig. 16   Current on wire as a function of irradiation time for a stainless
steel wire.

6.1.b – FAST AND PRECISE TRACKING WITH SCINTILLATING FIBRES IN A HIGH MAGNETIC FIELD

This LAA component aims at building a charged particle detector, based on scintillating fibres, to be used, coupled to a very high magnetic field, to measure the values of the charge and the momentum of the charged particles produced in (pp) collisions up to an energy of some TeV.

We recall here the envisaged goals:

- spatial accuracy: $\sim 20~\mu m$ in the (r, ϕ)–coordinates, $\sim 70~\mu m$ in the z–coordinate;
- two–track separation: $\sim 100~\mu m$;
- hit density: ~ 5 per mm;
- time resolution: ~ 10 ns;
- radiation resistance: ~ 1 MRad;
- operation in a 5T magnetic field.

The momentum resolution of the proposed detector, can be expressed as:

$$\frac{\Delta p}{p^2} = \frac{\Delta S}{0.3~BL^2} \left(\frac{256}{N}\right)^{\frac{1}{2}} = Q~\frac{\Delta S}{\sqrt{N}}$$

where:

ΔS = error on the measured points,

B = magnetic field,

L = lever arm,

N = number of points.

For good precision, a small value for ΔS and a large value for N should be aimed at.

But, ΔS small means small fibre diameter (50 μm), and N large means high light emission from the fibres. These two conditions work in opposite directions.

In fact, the smaller the fibre diameter (i.e. ΔS small = high precision), the higher the losses in light transmission (i.e. N small). In fact, the smaller the fibre diameter, the higher the number of total reflections per unit fibre length. As the reflectivity is smaller than 1, this means larger losses of light. If we want a larger N, keeping the diameter small, we need higher doping concentrations. This increases the light absorption in the emission band, and causes a loss of light. The parameters which measure the characteristics of the fibres are:

- Λ_{UV} Ultra–Violet absorption length. This should be much lower ($\sim 1/3$) than the fibre diameter, in order to have a large N.
- Λ_A Scintillator light absorption length. This should be longer than the fibre length (2 m in our case), in order to have sufficient light for particle detection.

The problem to be solved is how to go from the present fibre diameter, 1000 μm, to the diameter of about 50 μm, needed for the position accuracy of the new detector.

The solution stays in finding a new scintillator with a large separation between absorption and emission band, to avoid re-absorption of the emitted light inside the fibre. Two new scintillators have been tested:

- 3HF in Polystyrene:
  A high concentration cannot be reached because of foam production. This solution was rejected.
- PMP in Polystyrene (PS) and in Polyvinyltoluene (PVT):
  Test samples with concentrations ranging between 0.025 and 0.1 Mole ℓ^{-1} have been produced. The results are very good. In fact we have reached the following values for the relevant parameters:

$$\Lambda_{UV} : 6 \div 25 \ \mu m \Rightarrow \ \text{OK for } 15 \div 50 \ \mu m \text{ fibres}$$

$$\Lambda_A : 2 \ m \text{ or longer} \Rightarrow \ \text{OK for our detector}$$

Moreover, the efficiency is comparable with the best available scintillators.

Conclusion: a new scintillator:

$$\text{PMP} = \text{1-Phenyl-3-Mesityl-2-Pyrazolin},$$

never used before in solid materials, has been found.

It has a very good efficiency and its attenuation length is greater than 2 m, in a concentration high enough to produce sufficient light in 50 μm fibres.

The large attenuation length is due to the large separation between the absorption and emission peaks.

This separation is larger in PMP, than in the wavelength shifters commonly used in plastic scintillators. Figure 17 compares the absorption and emission spectra of PMP, with the combination P-TERPHENYL/POPOP, which are the dopants used, at present, in scintillating fibres of much larger diameter. It can be seen from fig. 17 that the overlap between POPOP absorption and emission spectra is much larger than in the PMP case. Therefore, PMP can be used at much higher concentrations, before it starts re-absorbing a sensible fraction of its own emitted light.

Concerning the PMP efficiency, Table 6.1.1 shows that the ratio between the scintillation efficiency of NE110, and PMP-doped samples of PVT (PolyVinyl-Toluene) and PS (Polystyrene) is nearly one. The measurements were made with the apparatus shown in fig. 18. It consists of a $^{90}Sr - ^{90}Y$ radioactive β-source, and a photomultiplier to measure the total light produced in the scintillator sample.

WAVELENGTH (nm)

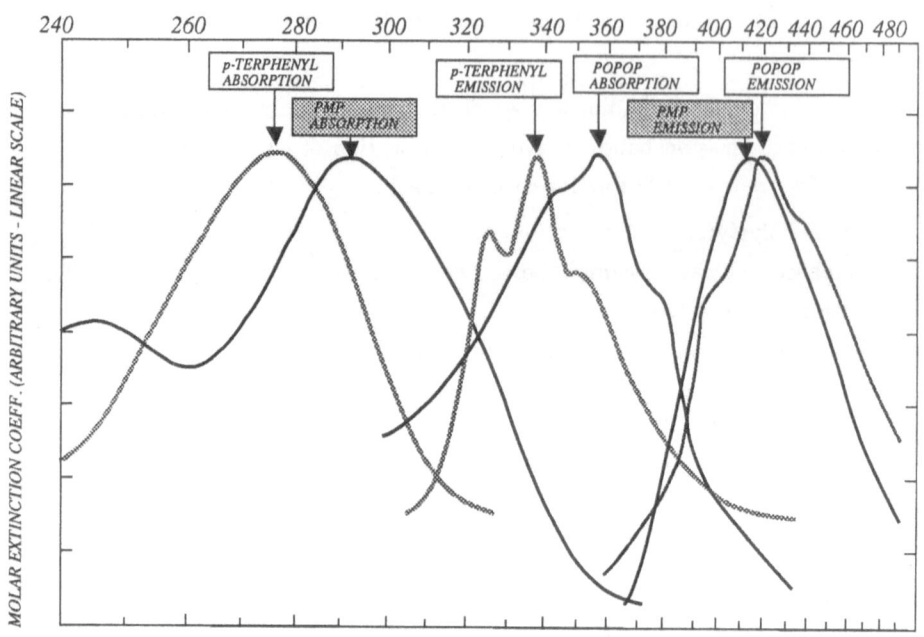

Fig. 17   Absorption and emission spectra of PMP, p-TERPHENYL and POPOP.

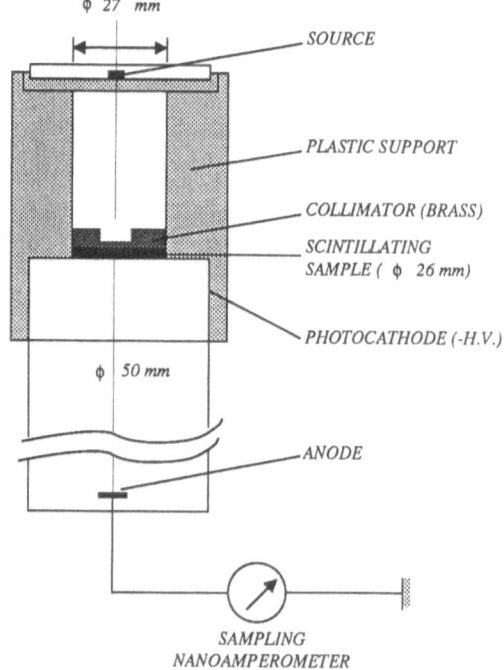

Fig. 18   Test set-up for measuring the scintillation efficiency.

Table 6.1.1   Ratio between the scintillation efficiency of NE110 and PMP–
              doped samples of PVT (PolyVinylToluene) and PS (Polystyrene)

Samples	Ratios		
Thickness:	0.25mm	1mm	12mm
$\frac{PVT+0.1PMP}{NE}$	1.00*	0.83	0.91
$\frac{PVT+0.05PMP}{NE}$	0.90	0.88	0.95
$\frac{PVT+0.025PMP}{NE}$	0.90	0.80	
$\frac{PS+0.05PMP}{NE}$	0.91	0.84	

(*) This is higher because NE becomes inefficient for small thickness.

6.1.c – GaAs MICROSTRIP DETECTORS

The fabrication processes for GaAs microstrips have been completely worked
out at the National Microelectronics Research Centre (NMRC) in Cork (Ireland).

Test structures have been produced in order to measure:

- the leakage current,
- the mean lifetime of the carriers,
- the adhesion of the metal contacts to the substratum,
- the depletion width,
  of the GaAs samples. Some examples of test structures are shown in fig. 19.

The measured leakage current is larger than expected. It ranges between
$400 nA/cm^2/100 \ \mu m$ and $30'000 \ nA/cm^2/100 \ \mu m$.

The structures with higher area–to–perimeter ratios have larger leakage current. Therefore, a high surface leakage current is most probably the main responsible.

With the best structures produced up to now, a leakage current of 40 nA is
estimated for a 50 mm × 50 μm strip geometry. This is already good, in principle,
for particle detection. However, a further reduction of a factor 10 in the leakage
current is aimed for. This will be done by testing different passivation methods to
reduce the surface leakage current.

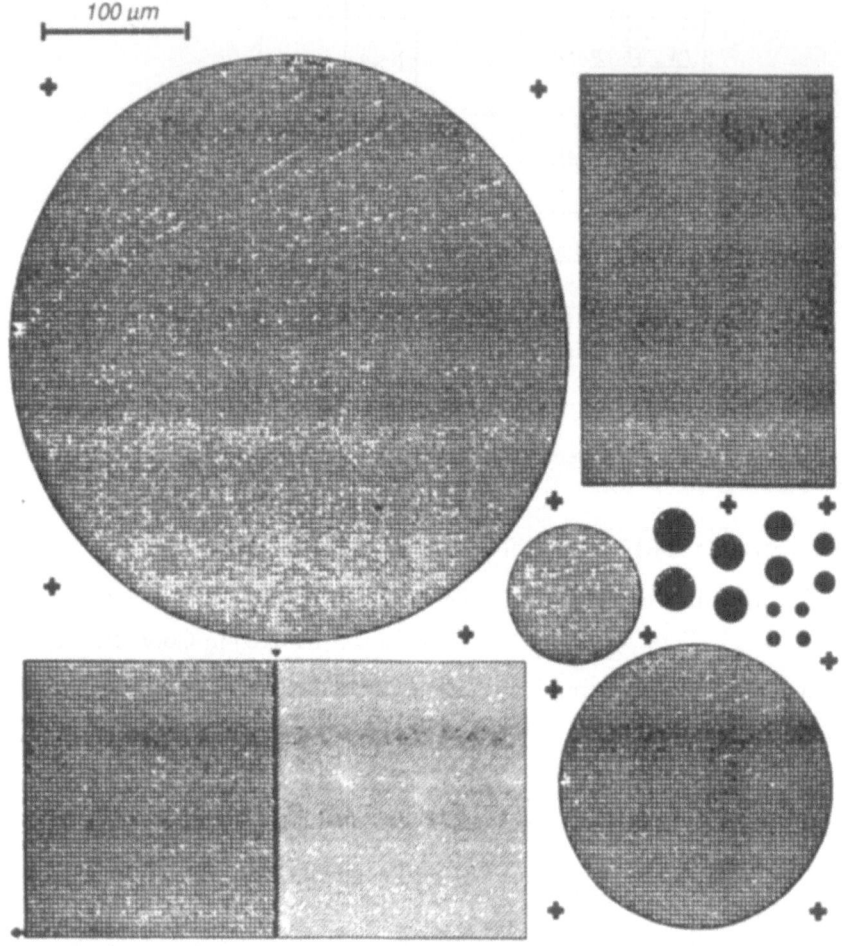

Fig. 19   Example of test structures used to measure the electric characteristics
of the GaAs.

A new bunch of test structures arrived at CERN during the summer. The electrical tests are in progress.

Radiation hardness tests will also start soon.

6.2 – CALORIMETRY

6.2.a – A PROTOTYPE ELECTROMAGNETIC CALORIMETER BASED ON BaF_2 SCINTILLATORS AND PHOTOSENSITIVE WIRE CHAMBERS

The principle design of the electromagnetic calorimeter based on BaF_2 scintillators and photosensitive wire chambers is shown in fig. 20.

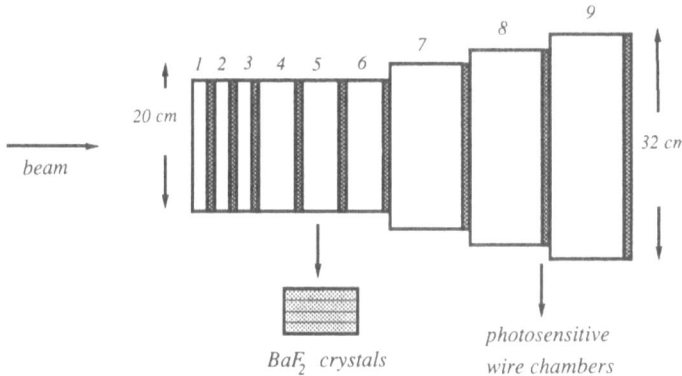

BaF_2 crystal cross–section:	**BaF_2 crystal length:**
slabs 1,2,3: 1×1 cm^2	slabs 1,2,3: 2 cm $(1X_0)$
slabs 4,5,6,7: 2×2 cm^2	slabs 4,5,6: 5 cm $(2.5X_0)$
slabs 8,9: 4×4 cm^2	slabs 7,8,9: 10 cm $(5X_0)$

- transverse shower containment: 10 X_0 (front), 15 X_0 (back) $[1X_0 \cong 2$ $cm]$
- BaF_2 volume: 25 litres
- time resolution: < 1 ns
- energy resolution: $\sigma(E)/E \sim 2\%\sqrt{E(GeV)}$
- length: 51 cm $(25X_0)$
- position resolution: ~ 2 $mm/\sqrt{E(GeV)}$
- e/π rejection: $< 10^{-3}$ up to 100 GeV

Fig. 20   The BaF_2 Electromagnetic Calorimeter prototype to be built by LAA.

The main R&D work performed during last year consisted in testing alternative solutions to Tetra–di–Methyl–Amino–Ethylene (TMAE) as photosensitive gas for the multiwire proportional chambers. Now, a new gas, Ethyl–Ferrocene ($EF = Fe(C_5H_4)_2C_2H_5$), has been found, to replace TMAE.

EF is less sensitive to photons than TMAE, but it has several outstanding advantages:

- Its absorption spectrum is well matched to the fast component (<1ns) of the emission spectrum of BaF_2 (as shown in fig. 21);
- It does not react to Oxygen;
- It is not corrosive to the gases or materials used in wire chambers construction;
- It can be used both as gas in low-pressure chambers, or as a thin liquid layer condensed directly on the BaF_2 crystal.

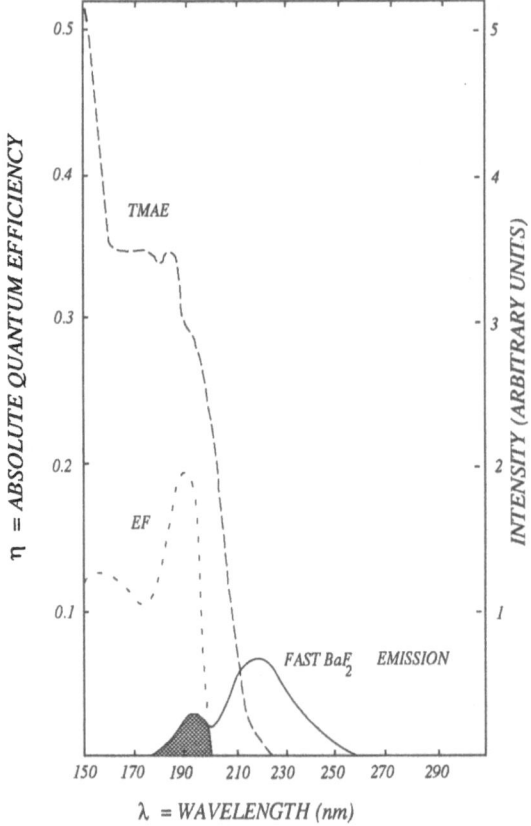

Fig. 21   Ethyl–Ferrocene and TMAE quantum efficiencies.
The fast (600 ps) BaF_2 emission spectrum is also shown. The dashed area of the BaF_2 emission spectrum corresponds to the overlapping component between EF and BaF_2. Notice that the fast component of the BaF_2 emission spectrum is still not understood.

The last point is particularly important, as it allows for a better time resolution and for much easier mechanical operation than TMAE. The efficiency of the condensed liquid layer of EF has been measured with the apparatus shown in fig. 22. Photons emitted from an ^{241}Am γ-source are collimated into a BaF_2 crystal, where a shower develops, emitting UV photons, which are converted into electrons by a thin layer of EF deposited on a copper cathode. A single-wire proportional chamber detects the electrons giving an electric signal which is measured.

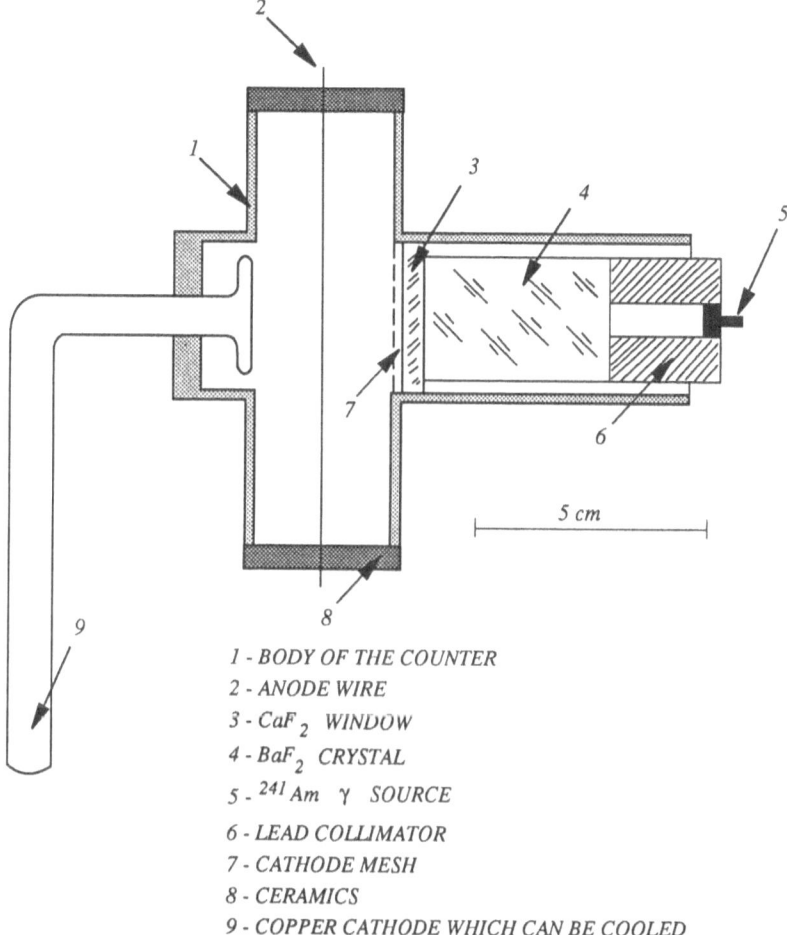

1 - BODY OF THE COUNTER
2 - ANODE WIRE
3 - CaF_2 WINDOW
4 - BaF_2 CRYSTAL
5 - ^{241}Am γ SOURCE
6 - LEAD COLLIMATOR
7 - CATHODE MESH
8 - CERAMICS
9 - COPPER CATHODE WHICH CAN BE COOLED

Fig. 22   The experimental apparatus used for measuring the efficiency of EF to detect the photons from BaF_2.

Figure 23 shows the comparison between the conversion efficiency of EF vapour and of liquid EF. A measurement with a clean cathode was also done for reference. The results show that the EF liquid layer is as efficient as the vapour, so it can be used as photon detector.

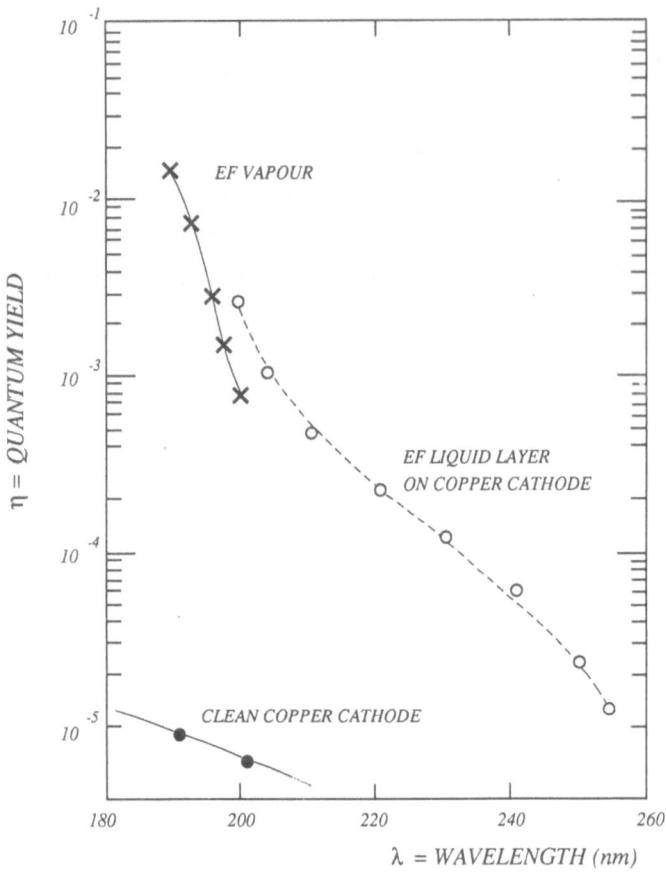

Fig. 23   Quantum yield of liquid EF and EF vapour. The clean copper case is shown for comparison.

The time spread of the photons emitted by the liquid layer, is much smaller than their time spread when a vapour is used. In fact, the liquid layer is some nanometre deep, while 1 or 2 mm of vapour (either EF or TMAE) are needed to achieve the same conversion efficiency. Therefore, a better time resolution can be achieved with the use of EF. The liquid layer of EF is very stable in time. Figure 24 shows the quantum yield of the liquid EF as a function of time. At time t=0, the EF starts condensing on the cathode. After about 5 hours, the thickness of the layer is sufficient to obtain full absorption of photons from BaF_2, and the quantum yield levels off and remains stable.

6.2.b – THE HIGH RESOLUTION SPAGHETTI CALORIMETER

The energy resolution of a hadron calorimeter consists of a component due

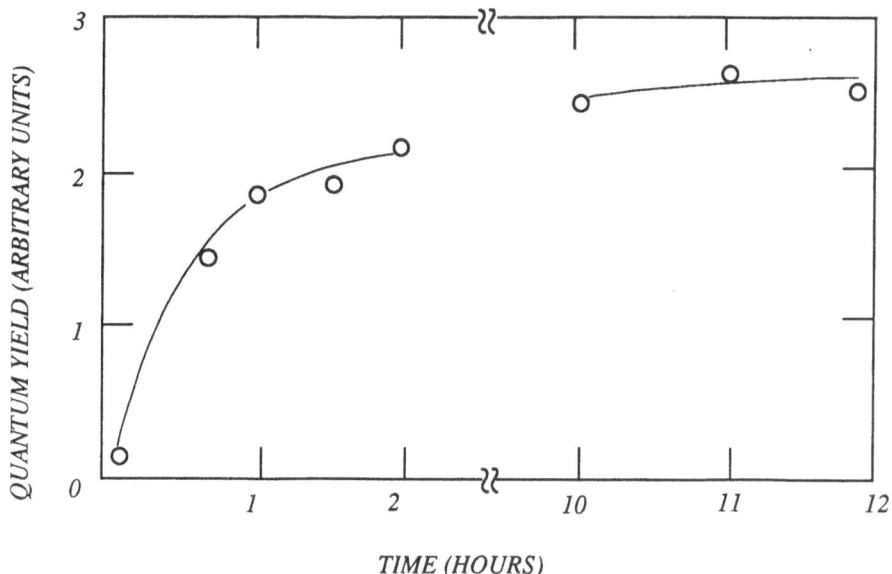

Fig. 24   Long term stability of the liquid layer of EF.

to sampling fluctuations (σ_{sampl}) arising from the fact that only a small fraction of the particle energy, E^p, is deposited in the active layers, and of an intrinsic component (σ_{intr}), due to the fact that the fraction of the particle energy, E^p, transformed into ionization (the visible energy), fluctuates from event to event. The effects of these fluctuations on the energy resolution of the calorimeter add in quadrature:

$$\sigma_{tot}^2 = \sigma_{sampl}^2 + \sigma_{intr}^2$$

The R&D done was primarily intended to separate these two pieces, σ_{sampl}^2 and σ_{intr}^2, in a lead/scintillator sandwich calorimeter, in order to be able to make realistic predictions for the energy resolution of the lead/scintillating fibres calorimeter.

In the fibre case, the sampling is done much more frequently, therefore σ_{sampl}^2 is expected and must be considerably smaller than in the rather crude sandwich case.

The separation between σ_{sampl}^2 and σ_{intr}^2 was done in the standard way: i.e., blocking the light of half the scintillator plates.

In doing so, the contribution of sampling fluctuations in the quadratic sum is doubled, and the comparison between σ_{tot} without blocking and σ_{tot} with blocking allows to separate σ_{sampl} and σ_{intr}.

The results show that the energy resolution of the sandwich calorimeter is completely dominated by sampling fluctuations. The proof is that, when only half of the plates were read out, the resolution for hadron detection increased from:

$$\frac{\sigma_{tot}(E)}{E} = \frac{42.8\%}{\sqrt{E}} \ ,$$

to:

$$\frac{\sigma_{tot}(E)}{E} = \frac{59\%}{\sqrt{E}} \ ,$$

i.e., by a factor:

$$\frac{59.2}{42.8} = 1.38 \approx \sqrt{2}$$

The intrinsic resolution, i.e. the limit for an infinite fine sampling, was found to be:

$$\frac{\sigma_{intr}(E)}{E} = \frac{(12.7^{+4.4}_{-7.5})\%}{\sqrt{E}} \ ,$$

averaged over the energy range $(10 \div 70)$ GeV. Figure 25 shows these results, and the expected performances of the spaghetti calorimeter (SPACAL).

This is significantly better than the intrinsic limit for a compensating uranium/scintillator sandwich calorimeter, which was measured to be:

$$\frac{\sigma_{intr}(E)}{E} = \frac{22\%}{\sqrt{E}} \ ,$$

using the same method.

The author of this important finding is our LAA collaborator Richard Wigmans. He explains this effect as follows.

The intrinsic resolution of a hadron calorimeter is largely dominated by fluctuations in the amount of the incident particle energy used to release nucleons (N) from atomic nuclei in the nuclear spallation reactions: binding energy losses. This energy is, in principle, "invisible" to the sensitive material of the calorimeter.

But, in the case of high–Z materials, most of these N are neutrons, and there is a correlation between these invisible nuclear binding energy losses, and the kinetic energy carried away by the neutrons. An efficient neutron detection, such as plastic scintillator, therefore reduces the effect of fluctuations. Notice that the neutron detection is a crucial ingredient for compensation, i.e. for equal response to hadrons and electrons.

The correlation between the nuclear binding energy losses and the kinetic energy of neutrons is considerably better in Lead than in Uranium. In Uranium, many of the neutrons come from fission processes, and their energy is not correlated at all to the nuclear binding energy losses.

All these considerations strongly support the construction of the lead/scintillating fibre hadron calorimeter.

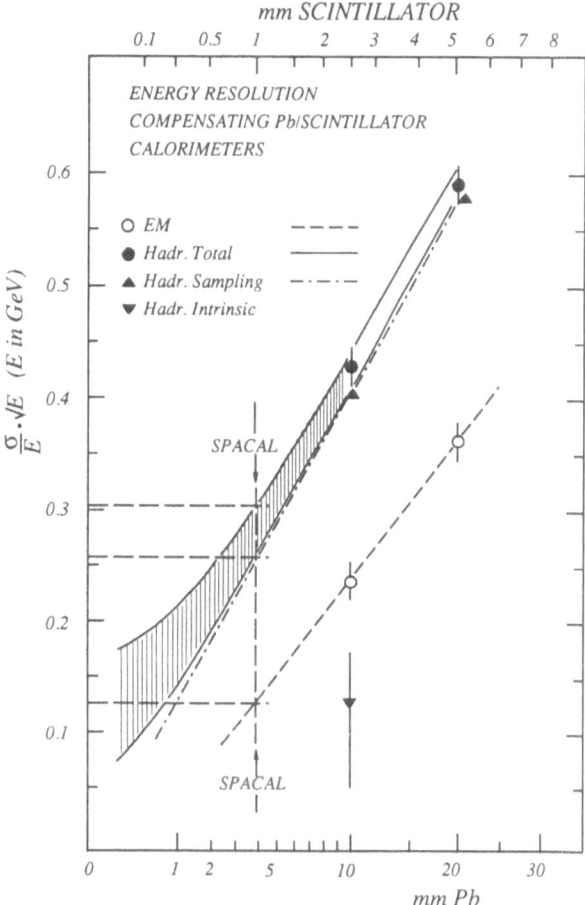

Fig. 25   Energy resolution measurements of the compensating lead/scin-
tillator sandwich calorimeter, as a function of the scintillator (and lead)
thickness. Notice that the ratio lead/scintillator is kept at 4:1, to ensure
compensation. The expected performances of the Spaghetti Calorimeter
(SPACAL) are also shown.

In order to test all the technology involved in the construction of such a detector, a lead/scintillating fibre test module has been built. It is shown in fig. 26. Figure 27 shows the fibres coming out of the module. They are bundled and glued to a plexiglass cylinder. The read–out is done using a photomultiplier optically coupled to the plexiglass.

The purpose of the prototype module was to study the problem of filling the Pb with 1 mm diameter fibres, without damaging the fibres.

The characteristics of the test module are summarized below:

Shape:	hexagonal, 48.7 cm^2 cross–section,
Length:	20 cm,
Number of fibres:	1141,
$\lambda_{electromagnetic}$:	7.5 mm,
$\lambda_{nuclear}$:	21cm

The final prototype calorimeter will have 2 m long modules, with the same cross–section.

One of the main problems which had to be solved before constructing a full–size module of the calorimeter, is how to collect all the produced light with minimal losses, i.e. how to obtain a sufficiently high (> 6 m) attenuation length for the light transmission in 1 mm diameter fibres.

This problem has been solved combining two methods:

- filtering out the short wavelengths of the outcoming light (since the light attenuation is strongly wavelength–dependent, and the short wavelengths are more attenuated, this increases the attenuation length even at the cost of some loss in the total light output);
- putting a mirror at the open end of the fibre (this doubles the light output in a way which effectively increases the attenuation length).

The final result is shown in fig. 28: the fibre response is uniform within ±6% over the full depth of the calorimeter (2 m). This uniformity is better then needed and hoped for.

Another strong point in favour of the use of scintillating fibres is the extreme fastness of the fibre signals. This is particularly important for experiments at high luminosity hadron colliders. The decay time of the light produced in the fibres is 2 ns, and the whole signal does not take more than about 7 ns. Fig. 29 shows the direct and the reflected (by the mirror at the open end) signals as a function of the point z along the fibre, where the light is produced. The direct and reflected light signals are recorded separately, even if the fibre was excited as close as 23 cm from the mirror.

The exceptional time resolution of the fibres might be used for discriminating between electromagnetic and hadronic showers without the need of splitting the detector into separate parts.

Fig. 26 · The Spaghetti Calorimeter prototype module (length = 20 cm).

Fig. 27 – The Spaghetti Calorimeter prototype module. Detail showing the fibres coming out of the module. The diameter of the plexiglass cylinder to which the fibres are glued is of about 5 cm.

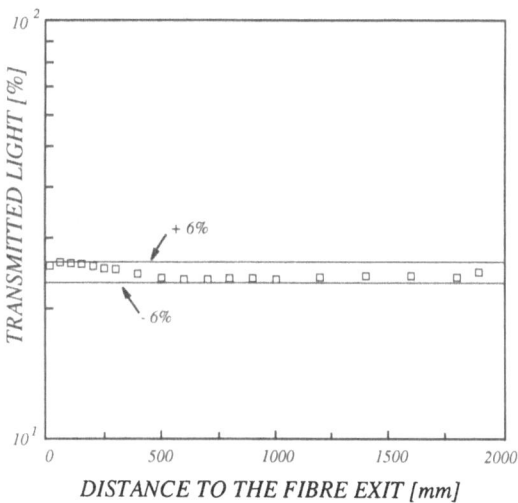

Fig. 28   Attenuation of the light transmitted in the scintillating fibres. Final and best result.

A crucial issue for experiments at a high luminosity hadron collider is the radiation resistance of detector elements. Dedicated efforts to improve the radiation stability of scintillating fibres have been undertaken at Kyowa, Japan, our supplier of fibres. A recently developed fibre is claimed to be an order of magnitude more resistant than the fibres we are using in our prototype.

Figures 30 a) and 30 b) show the attenuation along the fibre for the two samples. The effect of radiation is very much lower on the newly developed fibres (fig. 30 b)).

Preliminary results on the crucial issue of $(\pi - e)$ rejection indicate that the time structure of the showers can indeed be used to discriminate pions against electrons at the level of one per cent.

6.3 –   LARGE AREA DEVICES

6.3.a –   HIGH PRECISION CONSTRUCTION PROBLEMS

A detector module based on Limited Streamer Tubes (LSTs) has been built. Its innovative feature is a polar $(\rho - \phi)$ coordinate read–out. Fig. 31 shows the prototype tested in a 3.5 GeV π beam at the CERN PS.

This coordinate system matches the deflection of charged particles in a toroidal magnetic field, which is a good candidate for forward muon detection in a future supercollider.

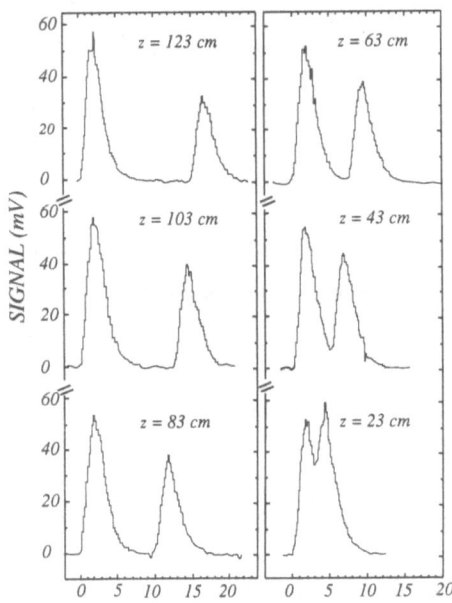

Fig. 29   *Signals from direct (peaks at the left in each histogram) and reflected (peaks at the right) light, as a function of time, for different excitation point z along the fibre.*

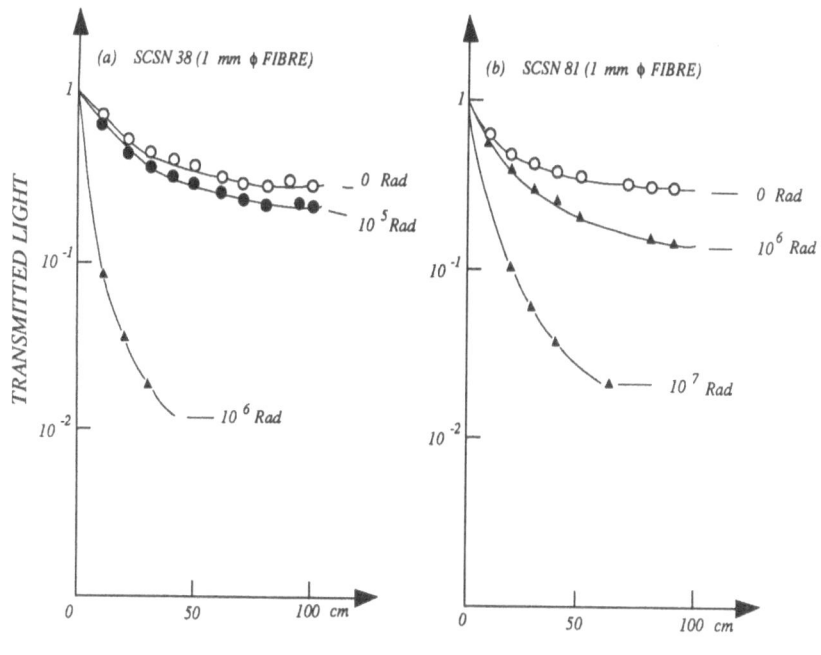

Fig. 30   *Attenuation curve in fibres manufactured by Kyowa, Japan, for different values of the irradiation: (a) presently used fibres; (b) newly developed fibres.*

The granularity of the prototype detector is 2 cm in ρ and 1.8^0 in ϕ.

The read–out electronics, allowing a parallel read–out of all strips, has been especially developed and shown to work as expected.

The main problem of the LST detector is the rate capability. This is, at present, limited to about 100 Hz/cm, as shown in fig. 32. This is a factor 10^2 lower than expected in the $(3^0 \leq \theta \leq 15^0)$ forward range for a luminosity of $10^{33}\ cm^{-2}s^{-1}$.

In order to overcome the problem of high rates, still keeping the $(\rho - \phi)$ read–out possibility, a new type of detector is under development.

It consists of $10 \times 10\ mm^2$ tubes with a blade which acts as anode and low-resistive walls, as cathodes.

Three prototypes have been built. They are shown in fig. 33. The relevant measurements, to be done on this new detector, are:

- Efficiency as a function of blade position and blade edge and shape,
- Time resolution,
- Space resolution:

Fig. 31   The LST prototype detector tested in a CERN PS beam.

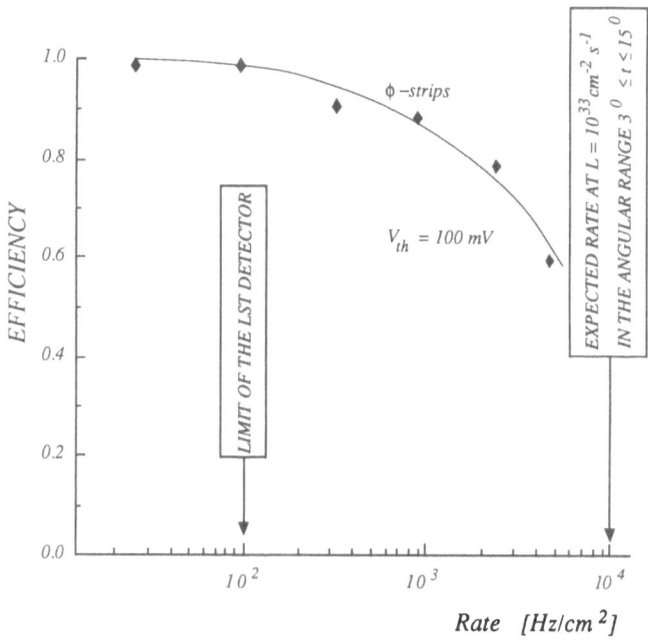

Fig. 32   *Efficiency versus rate for the LST prototype detector. The measurement have been done using ϕ-strips, with an applied threshold $V_{th} = 100 \ mV$.*

Fig. 93 · The three prototypes of blade chambers.

- transverse to the blade (by measuring the drift–time),
- along the blade (by measuring the collected charge, using Vernier–like or delay–line techniques).

6.3.b – HIGH PRECISION ALIGNMENT PROBLEMS

The main aim of this component of LAA is to provide the instruments to allow the measurement of muon momentum at the percent level. The working hypotheses are:

- a magnetic field of about $0.5 \div 0.75$ Tesla, and
- a lever arm for the measurement of muons of about 5m.

With these hypotheses, a sagitta of 1 mm is expected for 1 TeV muons. Therefore, in order to measure momentum in the TeV region with a precision of, say, 2%, a precision in the measurement of the sagitta of 20 μm is required.

At this level of precision, no detector can be better than its alignment. Therefore, a strong effort has to be put into the development of instruments to precisely localize the position of the detector in space.

The following set of specifications has been worked out:

- length: 5 μm in 5m,
- angles: $1 \div 5$ μrad,
- lines: $4 \div 10$ μm in 10m,
- plane flatness: 10 μm in 5×5 m^2.

Instruments to measure length up to these accuracies have been developed:

- one 2 m length standard accurate to 3.6 μm,
- one 2 m length transfer gauge with the same precision.

The length standard will be kept as the reference length in every laboratory building the parts of the final apparatus. For actual measurements on the detector, a transfer gauge is used, which is calibrated from the length standard. Figure 34 shows the production template on which the length standard was manufactured. Figure 35 shows the final accuracy of the length standard, and the contributions due to the measuring instrument (a LASER interferometer) and to the production template. By unfolding the errors due to the production template and to the LASER interferometer, the intrinsic error of the length standard, due to the production procedure, is found to be 3.6 μm (fig. 35).

An electronic level to measure angles with a precision better than 5 μrad (= 1 arc second) has been developed. Three prototypes have been built. Two of them are shown in fig. 36. Optical tests, for the development of a self-centring target made up with photosensors, have been done. An optical macrobench, to carry out 10m alignment and verification down to 2μm, is in an advanced design stage.

Fig. 34   *The production template (on the left) on which the length standards have been manufactured. A prototype is shown on the right.*

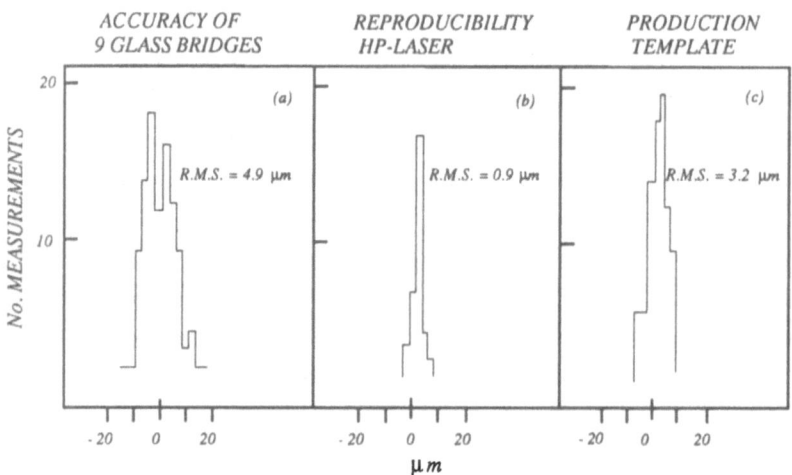

Fig. 35   *(a) Error on a length standard prototype. Also shown are: (b) the error on the measuring device (a LASER interferometer), and (c) the error on the production template.*

Fig. 36   The electronic level prototype.

6.4 LEADING PARTICLE DETECTION

A design study of the principle of the mechanics for a pot holding the detectors inside the beam pipe, has been made.

The solution which has been worked out allows:

- to move the detectors and the electronics far from the beam quickly (∼1 ms);
- to have a good (∼ 10 μm) positioning of the detectors with respect to the beams;
- to shield the detectors from the beam radiofrequency.

Different shapes of the detectors have been studied by Monte Carlo simulation, in order to provide the maximum acceptance without interfering with the beams. Cutting the detectors to elliptical shape is not a trivial job. Test trials with a diamond saw at CERN gave satisfactory results. And this means that cracks below the 50 μm level (fig. 37) have been produced: i.e., 10 times better than the present industrial standard. A LASER cutting machine, also at CERN, has been tested, but with bad results, i.e. with cracks greater than 100 μm. Notice that the silicon diode sensitivity remains undisturbed up to 50 μm level of depth.

In order to study the radiation level and shape expected in a hadron collider, a test run at the CERN SPS has been carried out. The neutron fluxes, and the radiation distribution perpendicular to the beam line, have been measured with a silicon detector.

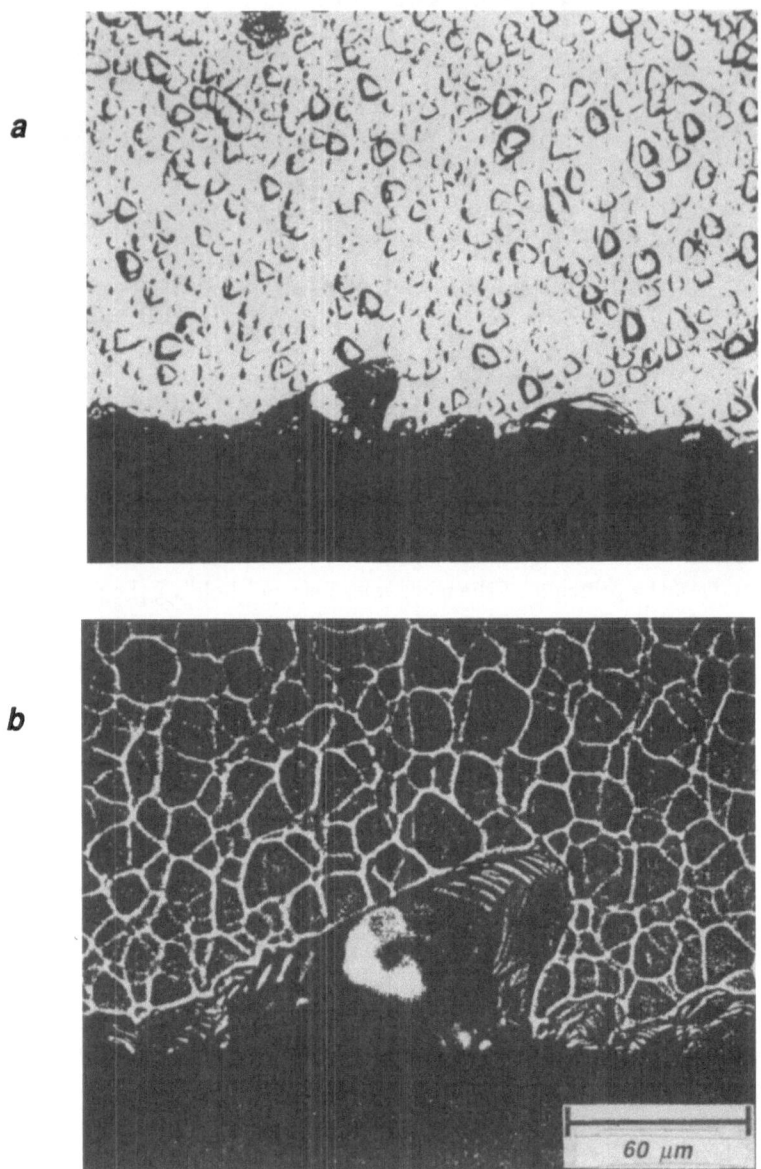

Fig. 37   Edge of the silicon detectors after the cut with a diamond saw.
(a) Magnification: × 200; (b) Magnification: × 500.

The most relevant result is that a factor 1000 variation in radiation doses has been measured between the stable beam conditions, 3 kRad/year, and the machine development conditions, 1 kRad/6 hours. This implies that a real–time monitor of the beam conditions is needed in future accelerators.

The dose rate as function of distance from beam axis under different machine operations was also measured. The results are shown in fig. 38.

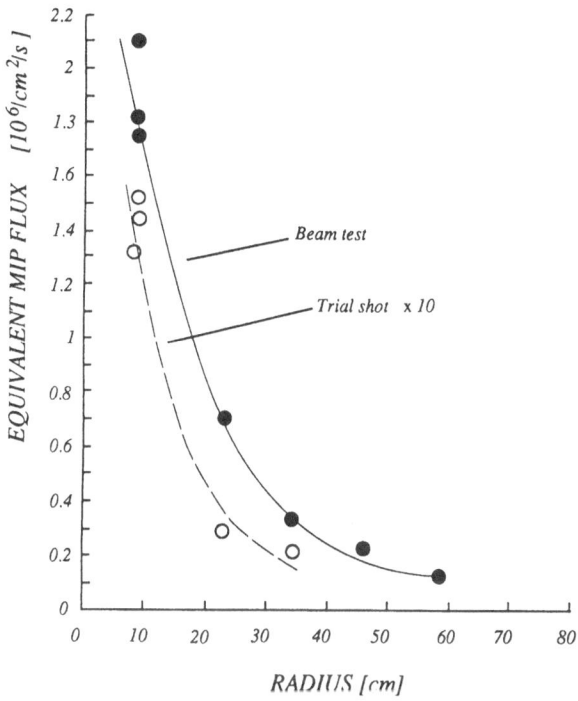

Fig. 38   *Results of the test at the CERN SPS: dose rate as a function of the distance from the beam. 1 Rad corresponds to 3.6×10^7 MIPs/cm².*

At Supercolliders, with Luminosities at the level of $10^{33} cm^{-2} s^{-1}$, we expect 10^4 more radiation: ~ 10 MRad/year with stable beam conditions.

Future colliders will require electronics with:

- High speed read–out.
- Very good radiation hardness.
- Low noise amplification.
- Low power dissipation.

These requirements are satisfied only by a compact electronics, i.e. by Very Large Scale Integration. A collaboration with the University of California, Santa Cruz, has been established to develop such an electronics for the read out of the detector microstrips. Radiation hardening processes will also be tried to reach 10 MRad resistance.

6.5   SUBNUCLEAR MULTICHANNEL INTEGRATED DETECTOR TECHNOLOGIES

6.5.a – SILICON DETECTORS AND MICROELECTRONICS

The first part of the microelectronics development programme is to build a general–purpose, low–cost, read-out chain. As described later, this is already being implemented. Two other parts of the microelectronics development programme are only in the discussion stage, and are listed for completeness:

– Development of a micropattern detector, i.e. a silicon pixel detector, with the electronics integrated in the same substrate. The compatibility between the substrate for the detector and that for the electronics is the main point to be solved.

– Evaluation and improvement of the radiation hardness of the electronics.

Concerning the first part of the microelectronics development programme, this can be done if we succeed to integrate in a single, multichannel, chip all the components shown in fig. 39, i.e. a set of:

• Low-noise amplifiers;
• Discriminators for triggering;
• Fast Analog-to-Digital converters;
• Analog and digital pipelines for reducing the dead time while the first or second level triggers are taking the decisions.

Each of the electronic components mentioned above will be implemented separately on chip, to be used later as "building blocks" in the final chip. At present:

• A 16–channel version of the low–noise amplifier has been designed and mass production has started. An even smaller and faster chip is under test.
• A prototype of the fast discriminator has been done and is under test; the preliminary results follow expectations.
• The digitizers and the analog and digital pipelines are in the design or discussion stage.

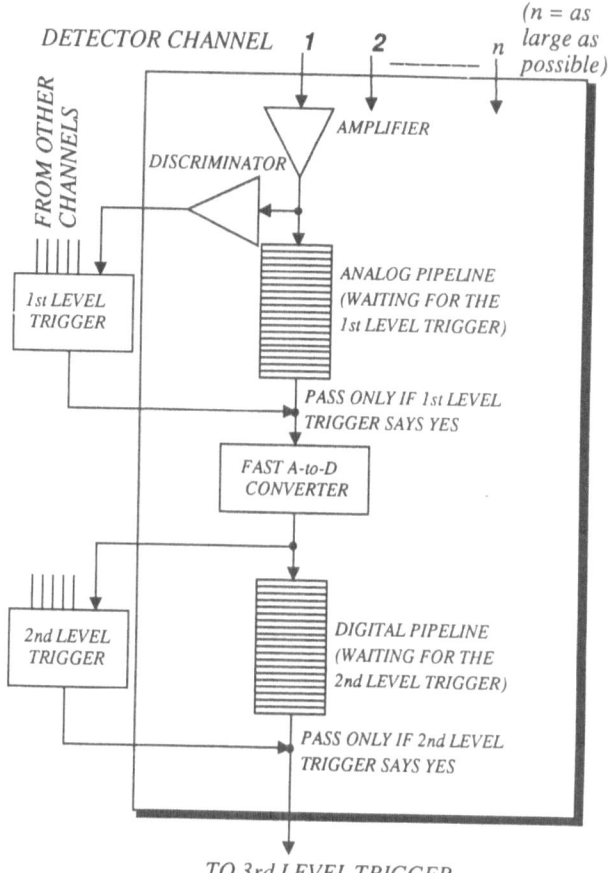

Fig. 39   *Final goal of the microelectronics development programme: to integrate in the same chip all the components enclosed in the box.*

• A block diagram of the 16–channel low-noise amplifier with multiplexed output (acronym: AMPLEX) is shown in fig. 40. The input signal, amplified by the "POTA" (Operational Transconductance Amplifier with a P–channel at the input), filtered and shaped by the "NOTA" (N–channel at the input), is held and stored on capacitor "Ch", at a time defined by the external trigger signal "T/H", which acts opening the switches "SWH1 ÷ 16". The unit–gain amplifier "buffer" protects the signal during the hold stage. If the event is accepted by the next levels of the trigger logic, the signals are retrieved by a multiplexer clocked by an external signal, "clock–in", which acts on switches "SWM1 ÷ 16", one at the time. The digital "clock–in" and "clock–out" lines and the "analog-out" bus can be organized in a daisy–chain configuration of 256 channels.

The gain of the amplifier is 10 mV/fC. This corresponds to an output signal of about 30 mV for a minimum ionizing particle traversing a silicon detector with 25 pF capacitance and releasing an energy of 90 keV, or about 25,000 e–h pairs.

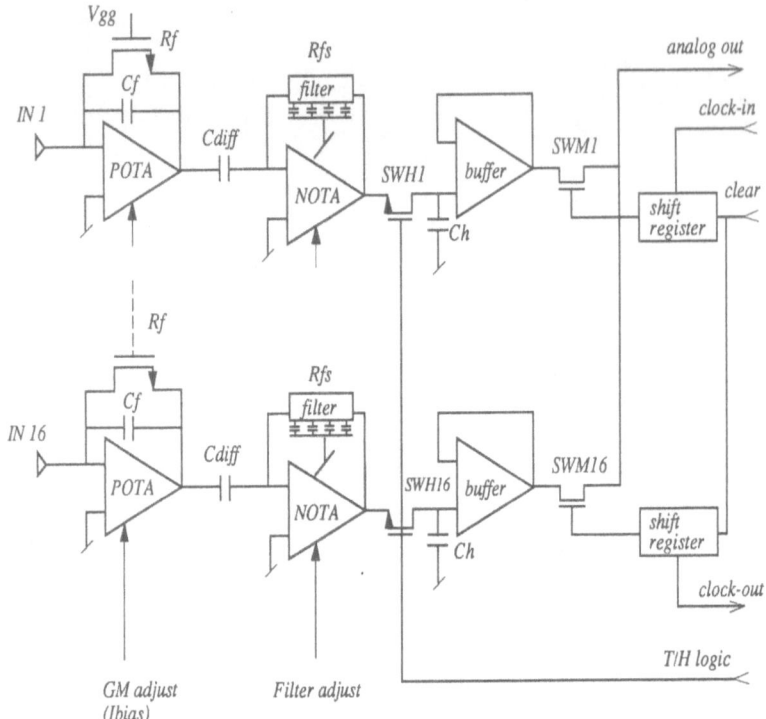

Fig. 40   Block diagram of the 16–channel amplifier (AMPLEX).

Figure 41 shows the result of a measurement performed using 2.3 MeV electrons from a ^{90}Sr β–source incident on a silicon detector. The trigger was provided by a scintillator, so that only the signals from the particles which have completely traversed the silicon were registered. The full width at half maximum of the pedestal is 9 keV. The main spectral peak, which has, as expected, a Landau distribution, corresponds to an energy of 90 keV. It is well separated from the pedestal.

Figure 42 shows a picture of the 16-channel chip mounted on the ceramic chip carrier. The total dimensions are 1×1 cm^2 . The chip has been manufactured in 3 μm n–well CMOS, by MIETEC NV (Brussel/Oudenaerde, Belgium), in close collaboration with the INVOMEC division of IMEC (Interuniversity Micro Electronics Centre in Leuven, Belgium).

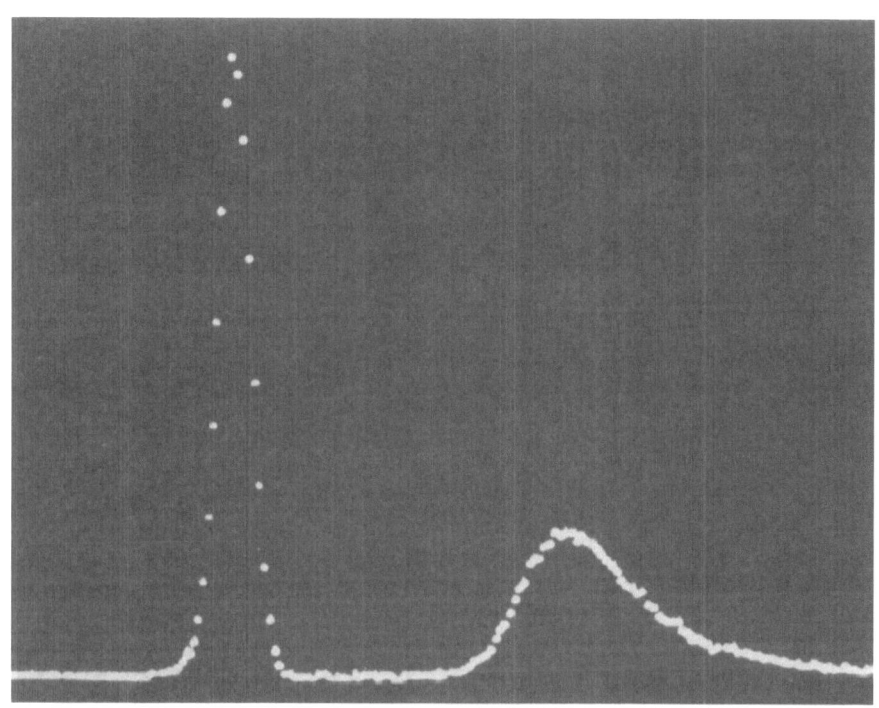

Fig. 41   *Energy spectrum of fast electrons penetrating through a 300 μm thick Silicon detector. The main spectral peak (right) corresponds to 90 KeV energy loss. The pedestal peak (left) is positioned around the zero–energy point. Its width measures the electronic noise: 9 KeV FWHM.*

*Fig. 42 –   Microphotograph of the 16 channel AMPLEX chip mounted on the
ceramic carrier. The total dimensions of the mounted chip are
$1 \times 1 \ cm^2$.*

A new version of this chip, shown in fig. 43, has also been designed, using a
different technology (p–well CMOS). This new version is a factor 2 smaller (100
μm pitch per channel) and a factor 2 faster (300 ns acquisition time) than the
previous version. At present, the first prototypes, with 4 channels on one chip, are
under test.

• The prototype of a fast discriminator is shown in fig. 44. Its key characteristics are:

 – low power: less than 1 mW/channel,
 – high sensibility: ≈ 2 mV,
 – high speed: $20 \div 30$ ns propagation time,
 – high density: 150×150 μm/channel.

The chip was fabricated at MIETEC in 3 μm n–well CMOS technology, on a
design work done at CERN in collaboration with SI, Oslo, Norway.

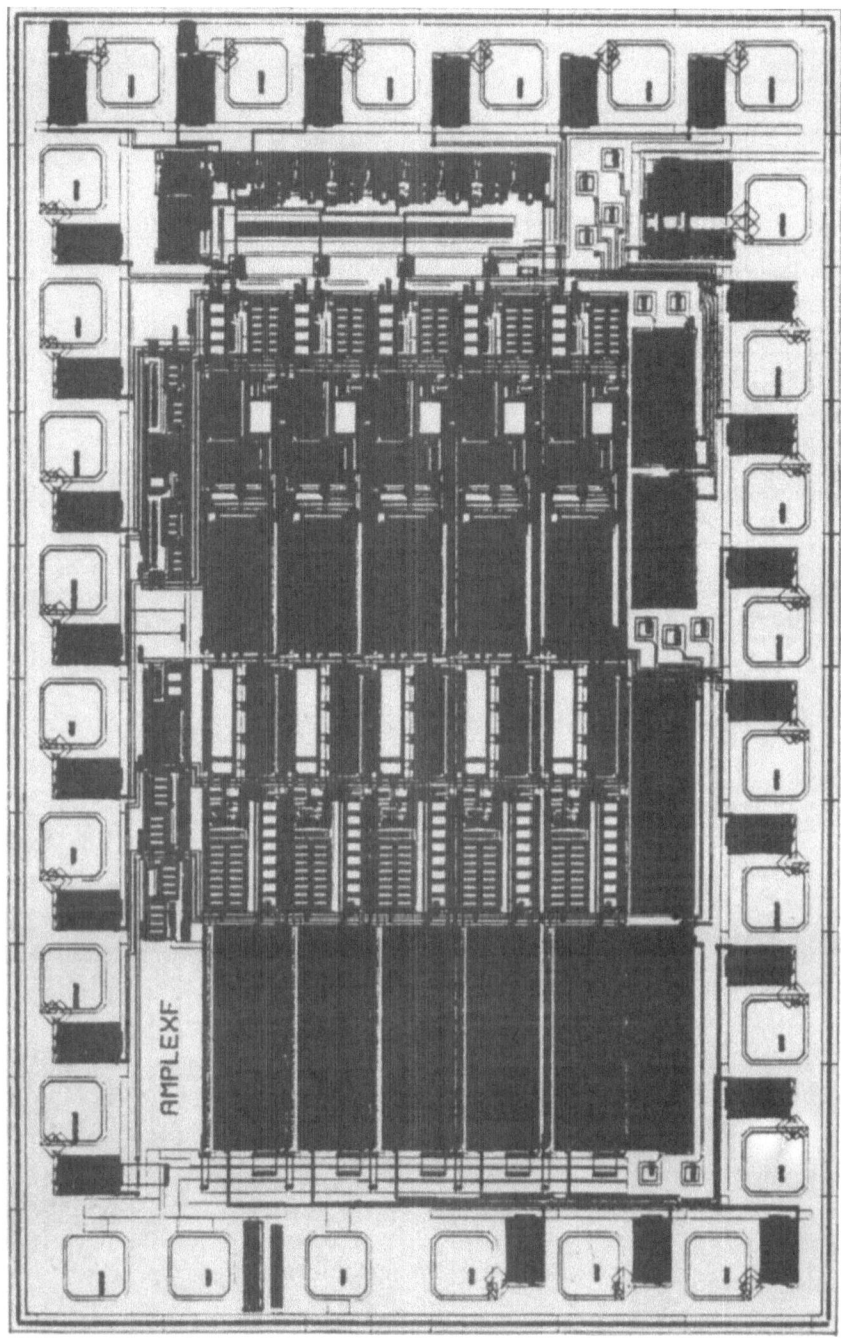

Fig. 43   Design of the new, smaller and faster, version of the AMPLEX
circuit.

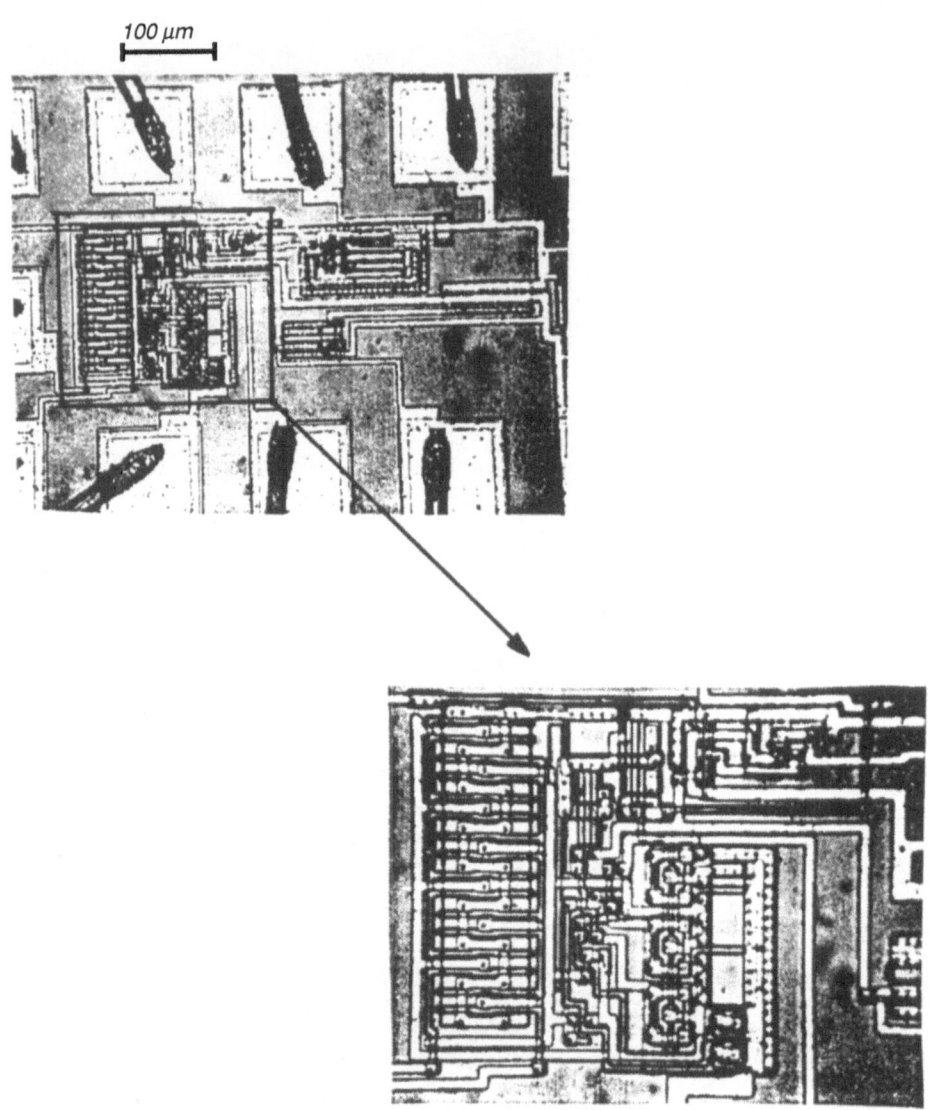

100 µm

Fig. 44 Microphotograph of the fast discriminator chip.

6.6 – DATA ACQUISITION AND ANALYSIS

In order to understand the tasks of the LAA component on Data Acquisition and Analysis, a rough idea is needed of the total data volume expected from each event.

As already mentioned in the introduction, the total event rate in a high luminosity ($L = 10^{33}\,cm^{-2}s^{-1}$) hadron collider, is expected to be $10^7 \div 10^8$ events/s. With at least 10^6 channels/event containing significant data, coded into one byte (i.e., 8 bits), it turns out that the impressive data volume of $10^{13} \div 10^{14}$ bytes/s must be handled by the Data Acquisition System.

Of these, about 10^9 bytes/s participate in the first level trigger, about 10^8 bytes/s in the second level trigger, and about 10^7 bytes/s in the third level trigger. This enormous amount of data must be divided into various pieces, which correspond to the different parts of the apparatus. Figure 45 shows the principle of the implementation of a 3–level trigger, in which information from the full detector is used only at the last level. Levels 1 and 2 use data from the various detector components separately.

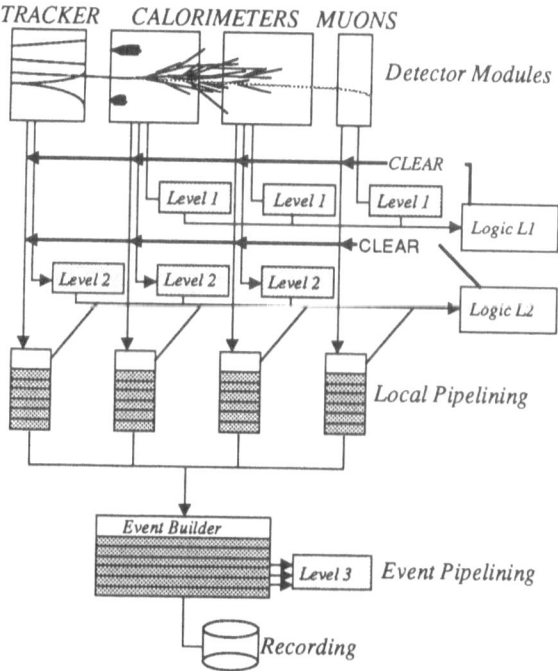

Fig. 45   Principles of the implementation of a 3–level trigger. Levels 1 and 2 are built using the signals coming from the different parts of the detector independently. Only the 3rd level trigger has access to the information from the full detector via the "event builder".

Figure 46 shows the total data volume participating in the trigger (in bytes/s), the decision frequency, and the data volume participating in a single decision (in bytes). The availability of commercial hardware and software decreases going to lower trigger levels, because more and more specific electronics is needed.

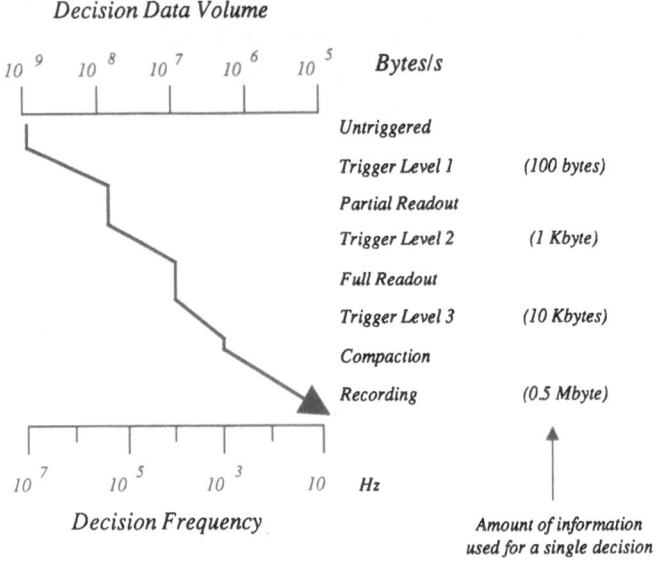

Fig. 46   Schematic multi-level decision making.

The general objective of the Data Acquisition and Analysis component of the LAA project is to find solutions to triggering and data compacting problems that fill the gap existing today between custom-made electronics and programmable devices. In future high–luminosity hadron colliders, fine-grain detectors will allow and need intelligent decisions with extremely demanding constraints on response time: ~ 100 ns at the first level trigger, < 10 μs at the second level trigger (assuming a rate of 10^5 Hz after the first level trigger). Programmable devices of today's technology cannot be considered serious candidates for second level triggering.

A clear message emerges from the initial studies: standard hardware, even if fast enough, will need a serious effort of interfacing to the detector read–out, and of very massive pipelining of data. The development of complex digital VLSI components for this purpose must be a prominent goal of any effort in real–time data processing. This implies training in digital design in close collaboration with industry. This LAA component has therefore initiated a pilot project which will produce an interface VLSI chip between FASTBUS and the future Motorola 88000 produce an interface VLSI chip between FASTBUS and the future Motorola 88000 RISC (Reduced Instruction Set Computer) processor. This will be a learning project for LAA, and produce a highly interesting FASTBUS controller board with large computing capacity for existing experiments.

6.7 – SUPERCOMPUTERS AND MONTE CARLO SIMULATIONS

The needs in dedicated computing power for the Monte Carlo Simulation part of this component of LAA have been defined. The configuration which has been retained, is a VAX 8250 which provides:

- an efficient front-end for network access (to ETHERNET, DECNET, BIT-NET, EARNET, etc.);
- a multi-user (≥ 10) facility;
- large disk space (> 1 Gbyte) as permanent and temporary storage for program editing, building and debugging;
- sophisticated peripherals (graphic terminals, laser printers, etc.) for display.

The VAX 8250 is an optimum computer configuration also in terms of disk space/CPU cost (i.e. power), and peripherals/CPU cost ratios.

This computer is now being delivered to CERN.

In the meanwhile, the implementation of a detailed and exhaustive Monte Carlo simulation program, based on QCD, has been started. It should be used by future experiments in different kinds of very high energy accelerators:

- hadron colliders;
- hadron – fixed target;
- (ep) deep inelastic scattering;
- (e^+e^-) annihilation.

These studies will be centred on the following points:

- a detailed theoretical analysis of all present data;
- a detailed theoretical profile of future high energy experiments;
- the simulation of "new physics", rare decays, oscillations, CP violation, etc.;
- the detailed study of the experimental issues, e.g. detector simulation, performance and resolution studies, signal/background estimates.

The main goal is the implementation of a sophisticated "Super Monte Carlo" program chain using our present best knowledge of theoretical models, and the appropriate and updated software tools. Three powerful computing centres will be used: CERN, DESY, and CINECA (Bologna).

7. CONCLUSIONS

There is no doubt that the feasibility of future experiments will totally rely upon intensive R&D studies on detectors carried out by a large community of researchers. The first results of the LAA Project make us confident that the ambitious goal of operating a detector at the next supercollider, which we hope will be the 10% ELOISATRON, is not out of reach.

8. COLLABORATING UNIVERSITIES, INSTITUTIONS AND INDUSTRIES

Universities, Research Institutions, and Industries from all over the World are collaborating with the different components of the LAA Project. The most important are listed in the following tables VIII.1 to VIII.10.

Table VIII.1   HIGH PRECISION TRACKING

Collaborating Universities and Research Institutions		
1a	University of Florida, Gainesville, USA	
1b	Brussels University, Belgium CNRS, Toulouse, France	
1c	INFN Bologna, Italy Bologna University, Italy	
	Collaborating Industries	**Item**
1a	LABEN, Italy	Microelectronics
1b	EEV, UK	Image Intensifiers
1c	NMRC, Ireland	GaAs microstrips

Table VIII.2   CALORIMETRY

Collaborating Universities and Research Institutions		
2a	Geneva University, Switzerland	
	CRN, Strasbourg, France Academy of Sciences, Serpukov, USSR Bologna University, Italy Institute for Chemical Physics, Academy of Science, USSR	
2b	University of California, San Diego, USA	
	University of Wisconsin,USA FIP, Lisbon, Portugal NIKHEF, The Netherlands	
	Collaborating Industries	**Item**
2a	Harshaw, The Netherlands	BaF_2 crystals
2b	Röhr Stolberg, FRG	Lead modules casting

Table VIII.3   LARGE AREA DEVICES

	Collaborating Universities and Research Institutions		
3a	Bologna University, Italy Calabria University, Italy Palermo University, Italy Perugia University, Italy INFN Bologna, Italy INFN Frascati, Italy		
3b	M.I.T., USA		
	Collaborating Industries		**Item**
3a	LABEN, Italy		Microelectronics
3b	Draper Lab, USA		High precision Measurement tools

Table VIII.4   LEADING PARTICLE DETECTION

	Collaborating Universities and Research Institutions		
4	Bologna University, Italy University of California, Santa Cruz, USA INFN Bologna, Italy INFN Frascati, Italy INFN Torino, Italy		
	Collaborating Industries		**Item**
4	LABEN, Italy AGIP Nucleare, Italy		Microelectronics Precision mechanics

389

Table VIII.5   SUBNUCLEAR MULTICHANNEL INTEGRATED DETECTOR TECHNOLOGIES

Collaborating Universities and Research Institutions		
5	RAL, UK EPFL, Switzerland ETHZ, Switzerland	
	Collaborating Industries	**Item**
5	FASELEC/Philips, Switzerland	Microelectronics
	LABEN, Italy	Microelectronics
	IMEC, Belgium	Microelectronics
	SI, Norway	Microelectronics

Table VIII.6   DATA ACQUISITION AND ANALYSIS

Collaborating Universities and Research Institutions		
6	RAL, UK	
	Collaborating Industries	**Item**
6	LABEN, Italy	Systolic processor

390

Table VIII.7   SUPERCOMPUTERS AND MONTE CARLO SIMULATIONS

Collaborating Universities and Research Institutions		
7	Palermo University, Italy Columbia University, USA INFN Bologna, Italy CCSEM, Erice, Italy COPPE, Rio de Janeiro, Brazil DESY, FRG	
	Collaborating Industries	**Item**
7	Digital Corp., Italy	Minicomputers

Table VIII.8   VERY HIGH MAGNETIC FIELDS

Collaborating Industries	Item
8	ANSALDO, Italy	Magnet design

Table VIII.9   SUPERCONDUCTIVITY AT HIGH TEMPERATURE

Collaborating Universities and Research Institutions		
9	Calabria University, Italy Columbia University, USA CCAST, Beijing, P.R. China	
	Collaborating Industries	**Item**
9	LMI, Italy	Superconductive samples production

Table VIII.10   RADIATION HARDNESS

Collaborating Universities and Research Institutions	
10	Uppsala University, Sweden ENEA, Rome, Italy

Collaborating Industries	Item
9 LABEN, Italy	Microelectronics

REFERENCES

[1] K. Johnsen, in *The ELOISATRON Project*, June 1988.

[2] A. Zichichi, *Report on the LAA Project*, Volume **1**, 15 December 1986.

[3] A. Zichichi, *Report on the LAA Project*, Volume **2**, 25 June 1987.

[4] A. Zichichi, *Report on the LAA Project*, Volume **3**, 15 June 1988.

[5] A. Zichichi, *Report on the LAA Project*, Volume **4**, CERN- LAA/88-1, 25 July 1988.

[6] A. Zichichi, *The LAA Project*, Volume **5**, CERN- LAA/88-2, 19 September 1988.

[7] M. Dine, "Superstring Phenomenology: An Overview", in *The Superworld I* (Plenum Press Inc., New York-London), Erice 1986.
M. Green, "Superstring Theories: a Survey", in *The SuperWorld III* (Plenum Press Inc., New York-London), Erice 1988.

[8] M. Duff, "From Superspaghetti to Superravioli", in *The SuperWorld II* (Plenum Press Inc., New York-London), Erice 1987.
M. Duff, "Classical and Quantum Supermembranes", in these Proceedings.

[9] J.C. Sens, private communication.

[10] A. Zichichi, in *The ELOISATRON Project*, June 1988.

[11] M. Basile et al., Nuovo Cimento **63A** (1981) 230.

[12] M. Basile et al., Nuovo Cimento **66A** (1981) 129,
M. Basile et al., Nuovo Cimento Letters **32** (1981) 321.

[13] K. Alpgård et al., UA5 Collaboration, Phys. Lett. **B121** (1983) 209.

[14] K. Alpgård et al., UA5 Collaboration, Phys. Lett. **B123** (1983) 381.

[15] M. Basile et al., Nuovo Cimento Letters **41** (1984) 298.

[16] M. Basile et al., Nuovo Cimento Letters **38** (1983) 359,
see also: A. Zichichi, "The Problem of New Heavy Flavors: Top and Superbeauty", and "Universality Features in (pp), e^+e^-, and Deep Inelastic Scattering Process", in *How Far Are We From The Gauge Forces* (Plenum Press Inc., New York-London), Erice 1983, 503, 673.

DISCUSSION

– *Calafiura:*

How do you manage to eliminate cross-talk between scintillating fibres of the vertex detector?

– *Zichichi:*

This is not a problem at present. Our present priority is to reduce fibre size from the current limit of 50 microns to 20 microns. The problem is that present scintillators have an attenuation length of the order of centimetres rather than metres. With PMP this problem is solved. After that we will work on the problem of cross-talking.

– *Horne:*

You mentioned a bunch spacing of about 0.1 nanoseconds for very high luminosity. A particle travelling at the speed of light travels only 3 centimetres in 0.1 ns, so how can you distinguish which event the particle came from if the events occur more frequently than the particles leave the detector?

– *Zichichi:*

Good question. At present it is totally science fiction, but you have to ask yourself such questions if you want to make new gadgets. The reason I insist on showing the most impossible luminosities is to call attention to the fact that these problems exist and we have to see if they can be solved. We need to reach the highest possible luminosities to achieve the lowest limits in cross-section. The numbers you see are a result of interaction between physicists and machine builders. The conclusion is that they should not say: there is no point in thinking of high luminosity machines because physicists don't know how to do the experiments. Machine builders should think about machine problems. In the worst case we can always consider fully screened experiments to have a limit on very low cross sections even though this is not what we want in the future machines.

– *Klapisch:*

Ideally, an alternative to increasing the luminosity by increasing the number of bunches would be to increase the density of particles in a bunch. If we could achieve this, it would be much easier because, in addition to what was just said about the distance between bunches, many devices need time in order to process the information. However, for a very high density of particles the space charge effect (the so called intra-beam and beam-beam effect) prevents one from doing this. So the only practical way which has been found so far is to increase the number of bunches. It is worth pointing out that right now in the CERN $p\bar{p}$ Collider there are 6 bunches spaced about 1 kilometre apart. To get to the values of luminosity discussed, one must go to bunches spaced centimetres apart. There is no way of avoiding this.

– *Brandt:*

I noticed that Wigmans was the head of your calorimeter group. Until recently he was a strong proponent of uranium-scintillator calorimetry. I was wondering what has changed his mind to change to lead-scintillator.

– *Zichichi:*

Lead-scintillator was thought to be inferior to uranium because of the intrinsic fluctuations in the lead nucleus. This has been shown not to be the case.

– *Brandt:*

If you just look at the mass of the $q\bar{q}$ pair then you cannot distinguish between a u3 and d4 quark. But doesn't the decay channel allow you to distinguish between the two?

– *Zichichi:*

If you identify the decay products then you can tell the difference. But such an experiment has never been done and is not easy.

– *Mincer:*

Can you explain in more detail how lepton asymmetry relates to effective energy?

– *Zichichi:*

The transverse momentum of the lepton depends on the original mass of the new flavour.

– *Bobbink:*

How many photons are produced when a minimum ionizing particle passes through a scintillating fibre?

– *Zichichi:*

We require enough photons so that the efficiency is greater than zero, approximately 5 photons for example. The number of photons could be increased by adding scintillator, but this increases the attenuation length. These two parameters are used to determine the amount of PMP in the plastic.

– *Rahal:*

Why do you want an average of one event per crossing?

– *Zichichi:*

In order to use the concept of hermiticity: so you can look for missing energy in an event.

– *Lillethun:*

Requiring the collision region in the Eloisatron to be defined within about 3 cm (0.1 ns bunch crossing interval) and having a given beam crossing angle, what will be the limit of the luminosity taking into account the cross section of the bunches and space charge limitations?

– *Zichichi:*

I believe that the limit is above 10^{35}. All beam parameters including the crossing angle are being studied.

– *Bobbink:*

How can you have beams of several ns width separated by less than 1 ns? It appears you have to go to a continuous beam as in the I.S.R. and then you don't know from which bunch it came.

– *Zichichi:*

This is an open question being considered.

– *Servoli:*

What resolution have you achieved with your lead calorimeter prototype?

– *Zichichi:*

About $25\%/\sqrt{E}$.

– *Servoli:*

How do you trigger to go from 10^7 to 10^3 events per second, hardware or software?

– *Zichichi:*

The dream is to do as much selection as possible in the first level trigger in the first 100 ns with the front end electronics. At present this is just in the talking phase, but our aim is to do as much as possible in each trigger level, in order to have real time data acquisition and analysis.

– *Freire:*

Is it possible to obtain information about lepton asymmetry from the collision of identical particles like p-p?

– *Zichichi:*

Yes. In the pp case the lepton asymmetry will be the same in the two hemispheres. In proton anti-proton collisions the asymmetry has opposite sign in two hemispheres.

– *Marchionni:*

Is Eloisatron going to be a proton proton or proton anti-proton machine?

– *Zichichi:*

Eloisatron is a two-ring machine and you can feed it with protons-protons or protons anti-protons.

– *Pietronero:*

You have emphasized that for the first time there is a good theoretical reason to plan more powerful accelerators. It appears, however, that the energy suggested by theory is the Planck mass. That is still much larger than the energy of the planned projects. Could you please comment on this point?

– *Zichichi:*

When the proton synchrotron at CERN was built there was no motivation to go to higher energies, just taste. It was the same for all other accelerators built. The great achievement of present-day theoretical physics is the discovery that the truth of our existence lies at 10^{15} GeV, if you ignore gravitation; at 10^{19} GeV if we include gravitation. Finally, now, and only now, we have a reason to go as high in energy as possible. The Eloisatron is at the limit of where present technology can be extrapolated. History demonstrates that once a machine has operated, what is discovered by far exceeds even the most optimistic forecast.

– *Gourdin:*

I think that the graph that you have just presented is the simplest representation of the desert and I do not personally believe in straight lines between 1 GeV and the Planck mass, 10^{19} GeV. Instead, we expect structures and in particular, in the 10 to 100 TeV region, definite manifestations of new physics. A lot of fundamental problems are still unanswered. Despite the fact that all experimental data in the present available range of energy can be explained using the minimal standard model, most physicists believe in the existence of something beyond the standard model. The only dispute is on what. In particular, the basic problem of the breaking of the gauge symmmetry is expected to be at least clarified by positive manifestations in the 10 to 100 TeV range. This, even only this, would be a great progress.

– *Zichichi:*

The most famous example of a desert is one due to Lord Kelvin who said that there was nothing new to be discovered, 6 months before J.J. Thomson discovered the electron. In our case you need to make a complete analysis of the desert to understand that what M. Gourdin is saying is indeed very true. There is an immense amount of physics to be discovered.

– *Klapisch:*

The basis for believing in a desert was that a plot of the running value of the 3 interactions seemed to indicate a crossing point at an energy corresponding to the Planck mass. The value of such a distant extrapolation was always questionable, but even more so now. In a careful analysis of high precision neutrino and collider determinations of the weak angle, the three straight lines are shown no longer to cross at the same point. Hence there is a need for change of slopes (new physics) if one wishes to restore the crossing point. In conclusion, if the arguments in favour of the desert ever had any value, they have now disappeared.

– *Zichichi:*

There are at least four fundamental problems: the family problem, the proliferation problem, the hierarchy problem, and the structure problem of the so-called fundamental objects, which are outside these straight lines. The desert could be taken seriously if these basic problems were not there. The extrapolation used in my graph is meant to illustrate the value of the unified theories. To take seriously the high precision neutrino data and the determination of the weak angle is, for my taste, not physically justified. No physicist would extrapolate anything to such a number of orders of magnitude. The purpose of the graph is to show how well the 100 TeV for the Eloisatron energy has been chosen.

THE NUCLEUS OF TOMORROW

Denys Wilkinson
University of Sussex, UK

When I began research in nuclear physics in 1943 the "bible" was the great three–volume Bethe–Bacher–Livingston Reviews of Modern Physics article of 1936–37.   Although this article is now an object of veneration rather than of study it is remarkable how much of our later understanding of nuclear structure is fore–shadowed in it. The article may be innocent of the word "parity" but it does envisage the possible relevance of the independent particle, cluster and collective models and it anticipates a fair amount of later nuclear reaction theory.

Of course, even by 1936–37, the nucleus had already changed: the discovery of the neutron in 1932 had converted the nucleus from a collection of protons and electrons into a collection of protons and neutrons, much to the relief of spin and statistics, although even then it was some little time before the neutron was fully recognized as a particle as much in its own right as the proton, before it was no longer thought of as some mysteriously–tight complex of a proton and an electron; so tight that the intrinsic spin of the electron was extinguished in the embrace.

But we knew that another change had to come. Yukawa's pion of 1935, elevated to self–consistent status by Kemmer in 1939 together with the concept of global isospin, demanded for itself an explicit place in our picture of the nucleus of the pions "in the air" between the nucleons whose mutual attraction they are thereby bringing about, must at some stage show itself explicitly and not be wholly absorbable into the nucleonic wavefunctions for which that attraction is responsible.

It is at this stage that we must be very clear as to what constitutes an acceptable intellectual structure for our discussion of the nucleus; how do we know what we know about it? Can we even, with total confidence, assert that it contains neutrons and protons? And here I must make a massive diversion.

We know with total confidence that we can make a complex nucleus by putting together individual neutrons and protons and we know with total confidence that if we, sufficiently gently, bring about the total dissolution of a nucleus we shall be left solely with individual neutrons and protons. But that does not prove that neutrons and protons exist *inside* the complex nucleus: as has been said, the fact that barks come out of dogs does not prove that dogs are made of barks – and, for that matter do not electrons and neutrinos come out of nuclei?

The question as to the degree of validity of the nucleonic starting point for the description of the nucleus is sharpened by our present belief that the nucleon itself has a quark/gluon substructure with the bag–like confinement remaining, despite QCD, a deep mystery of faith/fiat. We also believe, with much less conviction, that at sufficiently high overall densities the nucleonic bags will burst and that the quarks will spill out into a (still colourless) quark/gluon soup. Even assuming QCD as the full and final basis for our ultimate understanding for the strong interaction our ability to handle it in the non–perturbative regime is still little better than anecdotal and we certainly cannot claim quantitatively to understand that under the normal conditions of the tranquil or near–tranquil complex nucleus the bags will not have burst and that the nucleus should not already be a giant 3Aq–bag rather than A 3q–bags.

Of course, the shell model works wonderfully well. Surely that proves that the nucleus is made out of N neutrons and Z protons? But the shell model does not talk about N neutrons and Z protons except for the very lightest nuclei. For most nuclei it talks about just a few valence nucleons which tend to be chiefly in the peripheral orbitals where the density is low and where nucleons might continue to persist as such even though the dense interiors of those same nuclei might have dissolved into quark/gluon soup. And would seem not unreasonable for that quark/gluon soup to have effectively pre–empted the same nucleonic quantum numbers as would have been the case and it not been formed, since the total baryonic number would not have been changed, so that the valence nucleons, active in the shell model sense for our description of the nucleus as a whole and its low–lying excited states, would enjoy the same quantum numbers on either picture.

So we cannot simply quote the success of the shell model as proof of the A–nucleon basis for the nucleus although I shall return later to a further gloss upon this point. Should we then persist with the A–nucleon starting point? What evidence is relevant to the nucleus as a whole or, more particularly, to its depths? The most immediately–appealing work is the beautiful Saclay electron scattering experiments on heavy nuclei differing by a single proton that according to the shell model should be in the (spatially) deeply–lying 3s–state; Tl–205 vs Pb–206. Indeed the subtraction of the overall charge densities inferred for the two nuclei reveals the density distribution just as expected for a 3s–proton, nodes and all, and with a high occupation probability, less than 100% but not unreasonably so in view of the expected configuration mixing. It is difficult to say that the interior of lead is not working pretty well according to the shell model plan although we must note that the 3s proton in question is a valence nucleon in the terms of the present discussion and is not deeply–lying in the energetic sense. Also, beginning many years ago in CERN, and now continuing in CERN and Brookhaven, relativistic collisions, initially of protons with heavy nuclei and now of nuclei up to Si–28 with heavy nuclei, have given results consistent with the A–nucleon picture for the colliding particles. The analyses do not yet have the quantitativity of the electron scattering work, and can probably never achieve it, but they are nicely complementary to it in that they refer to nuclei as a whole; the data can certainly not be interpreted in terms of a model that represents a heavy nucleus as a large bowl of quark/gluon soup on the surface of which float a few nucleonic croutons unless, by some miracle, the shock of the impact effectively converts the soup into born–again nucleons.

I began my diversion already having recognized that the nuclear force, to which I have just referred as, in 1960, simply sticking together unmodified neutrons and protons, in fact cannot do just that because it is itself due to mesonic exchanges that themselves must contribute to overall nuclear properties and that must also engender isobaric excitations so that the A nucleons must immediately be redefined as, at any moment, A baryons, recognizing also that this may be A + a baryons plus a anti–baryons.

And so to my question as to what constitutes an intellectually acceptable structure for our description of the atomic nucleus. The point on which I wish to insist is that we should move away from the neutron/proton description, or supplement it, only when we are forced to do so. We should struggle along with neutrons and protons and effective interactions between them, that we can invent as whimsically as we wish, so long as they are consistent with what else we know, until we encounter phenomena that simply cannot be accomodated within the neutron/proton only framework but which can be systematically accomodated by a supplementation of that framework by mesons, isobars and anti–particles using, in that supplementation, coupling constants and other relevant paraphernalia established independently of the phenomena to the explanation of which that supplementation is being deployed.

More generally, we should not depart from a current framework, neutron/proton–only or neutron/proton plus mesons etc, whatever it might be, unless one of two things happens: (i) new phenomena force us out of that current framework in the sense that I have just indicated or: (ii) the current framework demands such a repulsive complexity of ad hoc rococo elaboration, while an alternative framework displays such an attractive elegance and simplicity, that we switch from one to the other under the slash of Occam's Razor.

So after that supplementary diversion, back to 1960. Those were the heady days of the earliest semi–quantitative working–through of the shell model. Inglis,

Kurath and Lane (1953–56), using what old–timers will recognize as the "a/K" approach, had shown how the 1p–shell could be systematically rationalized; Elliott and Flowers (1955–57) had, similarly and with similar success, extended things into the beginning of the 2s, 1d–shell and had made the first particle–hole descriptions in A=16. Indeed, in the earliest foreshadowing of the latter–day rapprochement between the shell and collective models now finding glorious if still somewhat Delphic expression in the Interacting Boson Approximation (Model), Paul (1957) had showed that F–19 could be as effectively accounted for on the collective (Nilsson) model as on the shell model. As I wrote in 1957: "Indeed the two models, ironically enough, agree better with each other than either does with experiment". Already by 1956 the Chalk River group had pointed out the considerable success of the collective model in the middle of the 2s, 1d–shell, a success reconciled with the shell–model by Elliott through his SU(3) representations of many–particle shell model states in 1958 and, when full effective interaction 2s, 1d–shell calculations became possible in the 70s and 80s, starting with the Oak Ridge group and continuing with Whitehead, Wildenthal and their colleagues, extended to a full and automatic description of the self–evident rotation/vibrational band structure of the shell in independent–particle model terms.

Neutrons and protons are therefore doing very well but, to return to 1960, are they enough? And to return to my criteria: it is not good enough to say that we know that there must be mesons and things in there as well, as Miyazawa had insisted since 1951; where is the compulsion upon us to take them into account? The earliest compulsion came from the simplest nuclear reaction, the radiative capture of thermal neutrons by protons. It was definitively experimentally confirmed by the end of the 1950s that this cross–section is some 10% larger than can be understood simply on the basis of the very–well–established parameters of the neutron/proton system, bound and unbound. The details of the language in which the mesonic and related supplementations to the purely nucleonic contributions are expressed, have changed from time to time since Riska and Brown's first fully credible demonstration in 1972, but the conclusion that these supplementations are quantitatively appropriate to make up the difference between experiment and nucleons–only theory has been robustly independent of that shift of language.

There are now many other places where we must invoke the supplementary contributions of the fleeting virtual components of the total wavefunction. Much work by M. Ericson, Chemtob, Delorme, and particularly Rho has taught us where to look for the most unambiguous signs. We find them particularly, and often in an almost model–independent way, in places where the soft pion contribution is not suppressed; indeed in such cases it seems that this contribution dominates even into regions of momentum transfer where we should not have expected to be able to treat it in so simple a fashion and where it seems that other, "harder," processes associated with heavy–meson exchange and isobar excitation mysteriously cancel as a consequence of, in Rho's terminology, the "chiral filter".

In ordinary nuclear structure experimentation we may expect soft–pion effects to show up most strongly and unambiguously in the time–like component of the axial current; for some years we have recognized such constributions, worth up to a factor 2 or more, in the weak zero–zero transitions with change of parity in A=16 and also, but differently, in B–12 and N–12 beta–decay; they are quantitatively accounted for by the intervention of soft pions.

Again, in the isovector magnetic moment of the A=3 system, the nucleonic description of which is highly trustworthy, the nucleons–only description fails by a large amount while the mesonic–related (chiefly pionic) supplementation quantitatively fills the gap; in the isoscalar moment the pionic contribution falls away, the experimental discrepancy with the nucleons–only expectation is quite small and appears to be consistent with what we should expect on the basis of the rho–pi–gamma contribution: the first suggestion of the need to involve heavy mesons as well as pions in the nuclear structure process.

But undoubtedly the most dramatic manifestations of mesonic currents are found in the electromagnetic properties of very light systems. The classic exploration by Saclay of the electro–disintegration of the deuteron at high momentum transfer but with low energy of relative neutron–proton motion in the final state shows order–of–magnitude discrepancies between experiment and the nucleons–only expectation while the mesonic–related supplementations completely restore agreement. This is a particularly–striking example of the possible relevance of the chiral filter hypothesis since the soft pion contribution by itself gives agreement over pretty well the whole of range of momentum transfer. Comparable dominance of mesonic over nucleonic effects is also found in the magnetic form factors of the A=3 system as measured at Saclay and Bates.

We should note that although the effective nucleon–nucleon interactions employed for the excellent (1980s) fitting to the level schemes of the 2s, 1d–shell, as was the case in the equal successful earlier (1965) and much simpler description by Cohen and Kurath of the 1p–shell, were properly, in accordance with the nucleons–only philosophy, taken as free variables the values that they assumed as a result of the fitting procedure turned out to be quite close to the G–matrix values derived from the empirical free–space nucleon–nucleon interaction plus many–body nuclear matter theory following Kuo and Brown; indeed the Kuo values were directly employed for the less–sensitive 2–body matrix elements. We similarly note that the mesonic–related and other supplementations that we have been discussing were also calculated on the assumption that the various interactions involved inside the nucleus are as they would be in free space apart from standard many–body corrections as appropriate.

There is therefore no doubt that where excellent and tried nucleonic wave functions are available they fail for those processes where we should expect the mesonic–related contributions to be largest and, futhermore, that those supplementary contributions, where they have been properly evaluated, are quantitatively appropriate to describe the experimental values. Virtual mesons are, so to speak, here to stay.

It is important to understand the corollary of this conclusion: although, in accordance with the philosophy that we must not shift our framework until compelled to do so, it has been proper, until this point has been reached, to restrict our description of nuclei as far as possible to neutrons and protons and to regard all parameters quantifying that description, such as nucleon–nucleon and other effective interactions, as freely available for adjustment to optimize the overall fit between theory and experiment, this is no longer the case.

The point is that we now have proof of the relevance of mesonic and related contributions for certain important aspects of nuclear structure, particularly certain moments and transition probabilities, and so should not ignore these. It is no longer the case that we should, as we used to do, struggle to achieve the best overall fit to all nuclear properties on a nucleons–only basis. The framework has changed and we should take mesonic effects etc into account at the same time as we optimize the nucleonic description so as to approach as directly as possible a final nucleonic/mesonic account rather than taking it as a two–step process.

So far I have spoken only of changes that have already taken place or are now taking place, all of a general nature, that had been, or could have been, envisioned for some considerable time. But what of the future? What changes might we see coming about in the ways in which we talk about atomic nuclei? We might yet discover new forms of organization of nuclear matter, in the geometrical sense, and we might discover new modes of excitation of conventional nuclei. We might discover that the shell model is not in fact as nearly literally true as most of us now believe it to be but that nuclei contain major clusterings or substructures, alpha–particle–like or otherwise; these need not be inconsistent with the stupendous success of the valence–nucleon shell model nor with the high (but not complete) population of deeply–lying shell model states such as the 3s in lead. (Would that we could see what the 1s is like...). Nor would it necessarily be inconsistent with the staggering success of density–dependent Hartree–Fock calculations of nuclear charge density distributions throughout the periodic table. But we shall not make a major move in these directions unless we have to; compulsion is presently lacking.

The most obvious possible change for the future comes from the recognition that neutrons and protons are not elementary particles but have a substructure of quarks whose interaction is mediated by gluons probably according to the percepts of quantum chromodynamics, QCD, to which reference has already been made. Reference has also already been made to the possible quasi–deconfinement of quarks, the formation of an extended quark/gluon plasma, that might be brought about by the collision of relativistic heavy ions. This would be most exciting and would powerfully confirm, or modify, our belief in QCD especially if the plasma showed a reasonable match to the previsions of lattice gauge computations when they themselves have reached an appropriate stage of development commanding adequate confidence and respect. But from the viewpoint of the nucleus as such, such investigations are not of direct interest because in them the only function of the interacting nuclei is to bring in the quarks and to provide the compression and high densities that might lead to their quasi–liberation.

The interesting question from the point of view of the nucleus itself is, to continue our basic philosophy, whether there are phenomena to be found in ordinary nuclei in their ground states, or at excitations at which we traditionally talk about them in nucleonic/mesonic terms, that cannot br explained on a nucleonic/mesonic basis as we have it now or with some modest and self–consistent extrapolation, but that can be explained by reference to the quark/gluon substructure of the nucleons. Or, alternatively, are there aspects of the nucleus that, although they can be described in traditional nucleonic/mesonic terms require excessively–repulsive elaboration of that traditional picture to do so whereas the quark/gluon description does the same job with an elegant economy? In this latter event also, although not forced to change our language in the sense that we should be if we were to discover a phenomenon that simply could not be described in the old language, we should not hesitate to make the change. Elegance, beauty, economy and Occam's Razor are the invariable touchstones; the progress of science is the story of switching horses.

I want once more, however, strongly to insist that we should not be tempted to move into quarks because of fashion: "We have done nuclear physics for fifty years without quarks; why do we need them now?" in Herb Anderson's telling words. Most particularly, we should not move into quarks without a thorough understanding of those fifty years and of what we have learnt in them and of the splendid edifice of the many–body shell model and of its refinements through configuration mixing of all kinds (going back to Blin–Stoyle in 1953; Horie and Arima in 1954) and through the mesonic and related effects that we have been considering. We must not say, as has, alas!, been said, that because C–13's magnetic moment is not that expected of a single neutron in the upper 1p–shell we must move to a quark–based picture in which the change of the neutron's structure in the process of quark/gluon exchange with its neighbours appropriately changes its single–particle magnetic moment to fit experiment. Undoubtedly the magnetic moment of a neutron in C–13 is modified by its immersion in the nuclear force field, because the neutron is a complex structure and must respond to that force field, but I want to be compelled to move to that conclusion, not to take it, by ansatz, as a starting point that ignores the knowledge and experience of a lifetime.

It must be said that at the moment we do not know of any effects in ordinary nuclear physics that demand reference to quarks and gluons for their understanding. This is not, of course, in any way to deny our belief that the effective constituents of the nucleus namely nucleons, isobars and mesons of all kinds, are QCD–based, but just to say that we know of no specifically nuclear phenomena that require us to say that. We do not, indeed, understand why the effective constituent model of the nucleus works so wonderfully well but it does so work and so long as it does we have no right to pretend that we can see quarks and gluons in the nucleus.

We do not understand the extraordinary resiliency of nucleons in the nucleus, for the reasons that I have already indicated, nor do we understand the rapid convergence of the effective constituent model but we have no present evidence that the quarks in the nucleus are doing anything other than minding their own business inside the constituent particles for which they are responsible. I hope very much that such compelling evidence will emerge; I await it eagerly; that surely will be the nucleus of the future.

Probably the closest encounter at present with quarks in the complex nucleus is through the EMC Effect in which, crudely speaking, the global quark momentum distribution of a nucleus as a whole is probed by deep inelastic lepton scattering, chiefly, so far, with muons and electrons with the vitally–important complementary data from neutrino sorely needed. What is clear is that the quark momentum distribution in a complex nucleus is softened in the medium momentum range relative to that in a free nucleon. What is not clear is to what this is due. Many effects of a relatively trivial nature such as those associated with binding and with Fermi motion must be allowed for and this allowance is not unambiguous. Discussion of possible residual more fundamental effects must start from the remark that the EMC Effect refers to the nucleus as a whole and not just to the nucleons within it. That remark lead to the two extremes of a whole spectrum of viewpoints. The one extreme places the onus for the effect on the individual nucleons which therefore must in some way be modified in their quark substructure by their exposure to the nuclear environment so that their quark momentum distribution is correspondingly changed. Since the sense of the EMC Effect is, pace the "trivial" effects, a softening of the momentum distribution, the simplest visualization in the change to the nucleon structure would be a swelling; figures as high as 15% in radius – 50% by volume – have been put into play. The other extreme viewpoint regards the nucleons themselves as being unchanged and places the onus on the mesonic exchange currents that bind them together: crudely speaking, a contribution to the overall momentum distribution comes from the quarks and anti–quarks in the mesons that are "in the air" between the nucleons and this contribution is softer than that from the quarks within the nucleons themselves. This picture may well be consistent with our traditional nuclear structure ideas about intra–nuclear mesonic behaviour.

The matter is very far from definitive resolution. Separation of quark and anti–quark contributions, in which neutrino–based information would be most valuable, is badly needed as also quark–counting information such as would come from Drell–Yan studies.

In the absence of any "smoking gun" phenomenon signalling the explicit role of quarks in nuclei, in the absence of any clear indicator impelling us to describe atomic nuclei directly in QCD terms, it seems that the most urgent task is to accept that the effective constituent model in fact works and to try to derive it from QCD complete with its forces and currents.

As to the general question of traditional meson exchange versus quark/gluon exchange as the better basis for discussion of the medium–range and short–range part of the nucleon–nucleon interaction we are still feeling our way. The traditional meson exchange picture (between "point" nucleons) has been remarkably successful most particularly in the highly–elaborated Bonn potential. Remarkably good account is given of the nucleon–nucleon force using known mesons and isobars in a wide range of combinations and using coupling constants either taken from experiment or with guidance from various models. Although it is not a fully dead–reckoning account it succeeds without doing major violence to other received notions.

Quark/gluon based descriptions of the nucleon–nucleon force are at a relatively early stage but are already showing themselves clearly able to account in a very respectable way for many important features, most particularly the "hard core".

Any note about the changing nucleus must have regard for the skyrmion. Skyrme's remarkable observation (1961) was that something very like a baryon could be constructed as a topological soliton in a pion field, effectively, if somewhat arbitrarily, stablized by a quartic term.

This idea, probably in future form with the stabilization to be provided by the omega–meson etc, has had a remarkable recrudescence, not least because it not only accounts remarkably well, in quite a quantitative way, for many aspects of the nucleon–nucleon interaction, for baryon structure and so on, but also because it can be put into an interesting correspondence with QCD and therefore may even turn out to be, for some purposes, an alternative and more convenient representation of it.

It should, however, immediately be said that there is no near–term prospect of writing QCD in skyrmion terms. At very low energies, much less than the QCD characteristic energy scale of perhaps 200 MeV or so QCD is indeed itself equivalent to a theory of Goldstone bosons which we may take to be pions for the purposes of argument. However, this does not establish the QCD/skyrmion connection because the energy of a quark in a baryon is about the same as, not much less than, the QCD characteristic energy.

However, the possibility of a well–defined connection at finite energies emerged from the observation ('t Hooft; Witten) that, assuming confinement, QCD is equivalent to some theory of mesons and glueballs with coupling constant inversely as the number of colours N. However, the meson theory is not a simple one such as underlines the skyrmion; it is complicated and, futhermore, unknown. For the fictitious case of large N simplifications arise and some predictions can be made, for example Zweig's rule, that are independent of the unknown equivalent meson theory. Furthermore if N is large the related solitons can be handled semi–classically. This leads to the frequently–made remark that in the limit of large N, QCD and the skyrmion picture are equivalent but this is as oversimplification. However, since the connection between QCD and meson theory can be made, if only in an unknown way in detail, for any value of N there is some ground for hoping that for N=3 baryons are solitons in some equivalent meson theory.

But those solitons connected to N=3 QCD are certainly not just skyrmions as we now use that term. We might nevertheless hope that the overlap between the QCD Lagrangian and the "simple" skyrmion Lagrangian would be large enough for us to write the QCD Lagrangian, after we have expanded it in terms of our array of effective nuclear constituents, nucleons, isobars, familiar mesons... as the skyrmion Lagrangian plus a modest afforcing by residual effective constituents. However, there is as yet no sign that this may be a profitable exercise in the sense that the afforcement will be either modest or rapidly converging. But that is no reason for not continuing to try.

What might be the relevance of the skyrmion for nuclear structure physics? A major effect might be that it could largely disembarrass us of the quarks. Quarks are seen in nucleons only in high momentum transfer processess, such as deep inelastic scattering, that are not relevant for nuclear structure physics. The skyrmion in an intrinsically more appropriate language for the low momentum transfer processes of the nucleus and if the QCD/skyrmion equivalence could be established (skyrmion now in a generalized sense) we should not need to concern ourselves with the details of nucleon structure for our discussion of nuclear structure. (We do not need to ask the question "where are the quarks in the skyrmion"; that is taken care of by the putative QCD/skyrmion equivalence; the quarks are seen only when we look at the nucleon in the appropriate high momentum transfer way).

An aspect   of this discussion is   the Cheshire Cat   phenomenon promoted by
Nielsen and Rho that arises from the grafting of a quark bag onto an external
skyrmion. It is rigorously proved in one dimension, and plausibly conjectured in
three, that low energy phenomena are independent of the size of the bag which
could shrink to zero without affecting the nuclear physics.

Already, with just the "simple" pion–based skyrmion, remarkable agreement
is found with many aspects of the nucleon–nucleon interaction. It is welcome that
there are no singularities at the origin since the skyrmion itself provides a kind of
natural form factor. There are obvious deficiencies, for example in the important
"sigma meson" region, but these may well be remedied by taking into account
higher order effects. A natural description is also given of what in constituent
language we call meson exchange currents of various kinds.

Even less than in the case of a quark/gluon base for the nucleus are we in a
position to switch to a skyrmion base. Such may come but we must not rush it:
the skyrmion is dangerously seductive.

I have left out a vast amount of changes and possible changes in the atomic
nucleus. I have not mentioned the coming of relativity in our description of nucleonic motions and, in some sense yet to be stabilized, the replacement of the
Schroedinger equation by the Dirac equation. Nor have I mentioned the coming
of relativity in the deeper sense of attempting a nuclear relativistic field theory
(Wilets; Walecka). Nor the vast expansions for our understanding that have come
from studies of all sorts of responses of the nucleus to stress: its resonant reactions
to a host of stimuli and its anxious reorganization of its internal structure as it is
spun towards destruction.

THE END OF SUPERWORLD III

Sheldon Lee Glashow

Lyman Laboratory of Physics
Harvard University

1. Introduction

Once upon a time, the life of the high-energy physicist was dominated by frequent astonishing and spectacular discoveries: new effects or new particles that were unexpected by the lumpen proletariat appeared with great regularity. Recall the spirit of the times a century ago: with the discoveries of the inert gases in 1894, X-rays in 1895, radioactivity in 1896, and the electron in 1897. Here are a dozen more contemporary examples:

(1948) Strange Particles	(1964) Cosmic Background Radiation
(1954) The Pion-Nucleon Resonance	(1967) Partons
(1955) Parity Violation	(1973) Neutral Currents
(1961) Two Neutrinos	(1974) The J/Ψ Particle
(1962) Zillions of Hadrons	(1975) The Tau Lepton
(1964) CP Violation	(1977) The Upsilon Particle

Our list of experimental surprises came to a crashing end over a decade ago. For the longest period of time in a century, there simply haven't been any theory-shaking discoveries. The W and the Z were found, but hardly anyone doubted that they would be. Supernova neutrinos were seen, but that only proved that supernovae are what we believed them to be. The lack of any real surprises is certainly not for want of experimental effort. Consider a partial list of some things that have been looked for but not found:

No fractional electric charge nor magnetic monopoles.

No zeta particles nor monojet nor high-y anomaly.

No neutrino mass nor oscillation nor magnetic moment.

No neutrinoless double β-decay.

No majorons, axions, gluinos, squarks nor sleptons.

No proton decay nor neutron-antineutron oscillation.

No confirmed fifth or sixth forces.

Forbidden decay modes seem to be forbidden.

No evidence for a light Higgs boson or a fourth fermion family.

In short, there is not the least indication of any flaw in the Standard Model that could be exploited to build a better theory. The standard model separates all the questions we may ask into two categories: those which are answered in principle (like the details of hadron spectroscopy or the branching ratios of the W), and those to which no answer at all is offered (like the number of families or the quark masses). Nothing shocking showed up at at the last generation of great machines: neither at the ISR, PEP, PETRA, TRISTAN, nor at the CERN and Tevatron Colliders. Not even the high precision or deep underground frontiers have been bountiful. Can it be that high-energy physics has come to its natural end? Will further progress in fundamental physics be purely deductive with no further experimental clues to guide us? I am convinced that Nature's bag of tricks is not yet empty, yet where is the new physics hiding?

A) There may remain many more *conventional* particles that await discovery. There could be a fourth, even a fifth, family of quarks and leptons. There may be particles with novel transformation properties under $SU(3) \times SU(2) \times U(1)$: quixes or queights of fermions, or perhaps, scalar mesons with charge and color. It is true that no indications of such particles have shown up in e^+e^- interactions up to a collision energy of over 50 GeV. Will this unfortunate situation persist up to the highest LEP energies? I doubt it.

B) Perhaps new interactions and new kinds of force inhabit an energy scale that has not yet been explored. After all, proton decay at a definite level was predicted by $SU(5)$, by far the simplest and most appealing of Grand Unified Theories. Yet, proton decay has not been seen. Something important is missing from our theory. It may well be a new strong force such as technicolored physicists advocate. Let us hope that such a new force lies closer to the Fermi scale than the forever inaccessible Planck scale, and that the technicolor analogs to pions and rho mesons lie just around the experimental corner. Again, there is not yet the least shred of empirical evidence for any new interaction, but we are duty bound to push the envelope by another few powers of ten.

C) Neutrino masses and oscillations have been a fruitless obsession of experimenters for years. At most, neutrino masses *should* be of order M_{Fermi}^2/M_{Planck}, or about 10^{-5}eV — barely enough to generate MSW oscillations, let alone measurable laboratory effects. My guess is that the solar neutrino problem has more to do with our inadequate model of the sun than with the intrinsic properties of neutrinos. Clarity should emerge when the results to GALLEX and other second-generation solar neutrino experiments become available.

D) How forbidden is forbidden? We can easily jiggle the standard model so as to generate otherwise forbidden decay modes. For example, a triplet of scalar bosons (T^{++}, T^+, T^0) of mass M and with Higgs-like couplings can generate the decay mode $\mu \to 3e$. However, the rate is anticipated to be very small:

$$\frac{\Gamma(\mu \to 3e)}{\Gamma(\mu \to e\nu\nu)} \leq \frac{m_\mu^2 m_e^2}{M^4}$$

If the T^{++} is heavier than 100 GeV, the branching ratio for this not-quite forbidden mode can be no more than 3×10^{-17}. Clearly, the experimenters have got a long way to go. Indeed, it would be far easier to search for and discover the charged and doubly-charged T particles in high-energy e^+e^- collisions than to see its tiny indirect effect upon μ-decay. Conversely, the observations of any such allegedly forbidden decay modes as $\mu \to e\gamma$, $\tau \to 3\mu$, and $K \to \pi e\mu$, unlikely as they may be, would have revolutionary impact. Such experimental searches are mandatory, though we cannot count on positive results.

However certain we may be that exciting surprises await the next-but-one generation of accelerators (SSC, CLIC, LHC and UNK), it is our modest hope that new physics can and will show up at the *current* crop of new machines: The Tevatron and CERN Colliders have yet to complete their explorations of an uncharted energy domain. LEP (perhaps even SLC and HERA) have a good chance to encounter something delightfully inexplicable. How strange we particle physicists be — we yearn to demolish our hard-won standard theory to make room for an even better scheme.

In the remainder of this talk, I shall discuss some of the exciting experimental possibilities that are associated with a minor extension of the standard model: the notion of chiral color. We shall learn what wondrous new phenomena could present themselves in real time at LEP and at the Tevatron.

2. Chiral Color

What is it? Chiral color is a consistent alternative to the standard model in which the fundamental gauge group is extended to become

$$SU(3)_L \times SU(3)_R \times SU(2) \times U(1).$$

This group, like its conventional predecessor, breaks down to its exact charge and color subgroup at or around the Fermi mass. Thus, we imagine that the scale of chiral color breaking is comparable with the scale of electroweak symmetry breaking. The rôle of the chiral color subgroup is natural and intuitive: all left-handed quarks are triplets under $SU(3)_L$, whilst all right-handed quarks are triplets under $SU(3)_R$. Thus, chiral color is a precise implementation of some old ideas of Jogesh Pati and Abdus Salam in the modern context of a successful standard model. In its current form, renormalizable, free of flavor-changing neutral currents, potentially unifiable, and compatible with experiment, it has been re-introduced by Paul Frampton and me.

At low energies, chiral color theory is indistinguishable from conventional theory. In particular, the theory may be made anomaly free and renormalizable by an appropriate choice of the fermion representation, a choice that is not unique, but inexorably implies the existence of colored fermions with curious properties such as have not yet been found.

Why are we doing this? Who needs a new model that is manifestly more complex than the standard model yet not particularly more profound? There are two answers. Chiral color can serve as a *foil*. Its predictions at high but accessible energies differ from those of the standard model. To know just how successful is the standard model, experimenters need to have a consistent alternative scheme with which to compare it. Chiral color can also serve as an *indicator* to suggest the possible existence of novel phenomena that are beyond the ken of the standard model. In this sense, it may act as a guide to the experimenter who might otherwise fail to search for things sufficiently bizarre.

Axigluons. Much of the physics of chiral color is independent of the details of the fermion representation. For example, there must exist an octet of massive axial-vector fields. The axigluon mass is anticipated to be larger than the W and Z masses simply because the color coupling constant is relatively large. Indeed, if the dominant symmetry-breaking mechanism is the one giving quarks their mass, we may compute the axigluon mass to be about 320 GeV. Now, axigluons cannot hide, for they must have about the same coupling strength to quarks as do the gluons. They must be produced at a calculable rate in hadron collisions. And, they can be detected as a Jacobian peak in one-jet inclusive distributions, or as a two-jet enhancement at the axigluon mass. No evidence for such effects have yet been reported. Not yet, anyway. Certainly, the appearance of axigluons is the central test of the new theory, a test for which the Tevatron Collider is the ideal examiner.

The Higgs Sector. The scalar multiplets responsible for quark masses must transform as $(\overline{3}, 3, 2)$ under $SU(3)_L \times SU(3)_R \times SU(2)$. Indeed, there must be two such multiplets with different charge assignments, one to fatten the up quarks, and another to fatten the down. This means that a variety of interesting and not-too-heavy scalar bosons must survive symmetry breaking to remain as observable particles, including electrically charged and neutral color octets. They will decay semiweakly into quark pairs. If these particles are kinematically accessible in Z decay, they will lead to identifiable four-jet decay modes. (Another set of Higgs bosons generates lepton masses: these may appear as Z decays into *four* charged leptons.) Again, LEP is the ideal instrument for the exploration of chiral color's necessarily rich scalar sector. Chiral color can provide something interesting for almost any new facility, except, perhaps, for HERA.

The scalar sector can lead to exotic decay modes of axigluons. Since the $(\overline{3}, 3, 2)$'s contribute both to electroweak and chiral color symmetry breaking, there are trilinear couplings linking axigluons to electroweak bosons and scalar octets. This could lead to axigluon decay to W or Z plus a pair of jets at the scalar octet mass, with a branching ratio of several percent. Such events would be spectacular.

The beasts of chiral color may offer an explanation of why the top quark has not been found, and how it may never show up in conventional searches for semileptonic top decays. Perhaps one of the charged scalars (either a color octet or a singlet) is less massive than the top quark. In that case, top particles will decay semiweakly, almost never producing a hot muon or electron. This possibility, recently pointed out by Elizabeth Jenkins. will surely be tested at LEP. None of the many new states required by chiral color can lie much beyond the Fermi mass, which (not coincidentally) corresponds to LEP's ultimate reach.

Lots more Fermions. Let us turn to a simple possibility for the fermion representation. Aside from the usual anomalies, which are to be cancelled in the usual way, there are two new threats in chiral color: anomalies associated with traces of the following combinations of generators:

$$A_1 = Tr(\lambda_L \lambda_L \lambda_L) - Tr(\lambda_R \lambda_R \lambda_R),$$

and,

$$A_2 = Tr(\lambda_L \lambda_L Q) - Tr(\lambda_R \lambda_R Q).$$

Each conventional fermion family contributes to each anomaly: thus, unconventional fermions are needed if all the anomalies are to be cancelled. A minimal fermion supplement consists of a chiral *quix*, a color sextet carrying electric charge 1/3, and a dichromatic $(\overline{3}, 3)$ which is electroweakly neutral. Such a system of exotics, together with *five* conventional fermion families is entirely anomaly free. No simpler pattern will do. Furthermore, troublesome flavor-changing neutral currents do not appear in tree approximation with this choice of the fermion representation. They appear neither in the couplings of Z's, nor axigluons, nor among the various Higgs boson couplings.

Could there really be five families, with five more quarks and two charged leptons lying just beyond reach? And, a quix and a queight to boot? It does seem unlikely, but who can say it's impossible? Some astrophysicists argue that our cosmology seems to disfavor more than four light neutrinos. Curiously, chiral color satisfies this constraint.

Sure enough, five families involve five neutrinos. One of them, however, is involved with the fermion dichromat, and should develop a sizeable mass. The $(\overline{3}, 3)$ of fermions is most easily given mass by a neutral scalar multiplet with identical transformation properties to it. The quone becomes twice as heavy as the queight. However, there is a possible Yukawa coupling of the Higgs $(\overline{3}, 3, 2)$ linking the fermion dichromat to a linear combination of the various lepton states. In particular, mass mixing is induced between the quone and one of the neutrinos. The lighter eigenstate, which is mostly neutrino, develops a mass by means of a *see-saw mechanism*. If this *neutrino* is to be lighter than the queight, and therefore reasonably stable, the quone-neutrino mixing angle cannot exceed 30 degrees. The heavy neutrino will eventually decay by weak interactions, since it is connected by Kobayashi-Maskawa-like angles to the lighter lepton families. We presume that the neutrino state chosen by the quone is mostly, though not exclusively, of the fifth and heaviest family. Thus, our theory suggests the existence of four massless neutrinos plus a fifth unstable but long-lived neutrino-like state which may have a considerable mass.

A key test of chiral color with the above fermion assignment is the LEP neutrino count. It is likely that the fifth neutrino is kinematically accessible to Z decay, though it is suppressed by phase space and by quone mixing. Thus, the neutrino count must yield a result larger than four, smaller than five, but in no case an integer. Now, that would be a surprise!

Last of all, there is a quix: a fundamental fermion transforming as a six-dimensional representation of color $SU(3)$. A mechanism must be introduced to assure that the quix develops a mass, and that it can manage to decay. This is not so easy, and it may be that the quix decay is suppressed by one reciprocal power of the grand-unification mass. There is at least the possibility that the lightest quix state has a lifetime of seconds, or perhaps even centuries. We suggest a search for the production of heavy long-lived stable hadrons at the Tevatron Collider or in cosmic rays.

Monojets again? As one last indication of the rich phenomenology demanded by chiral color, consider the return of the monojet — once the figment of overzealous experimenters' imaginations, but perhaps nonetheless a real phenomenon awaiting discovery. For the putative axigluon may decay into queight plus quone — and our panoply of fermions with all channels 'go' yields a branching ratio of 1/27 for the mode. Sometimes this decay mode can lead to neutrino production — not more often than a quarter of the time, neutrino-quone mixing yields a final state consisting of a queight and a neutrino. The former is hadronic and very unstable, the latter is neutral and long lived: As many as one percent of our axigluons can lead to events in which half the decay energy is unbalanced and unseen: monojets.

Unification. A neat aspect of the standard model is how easily it may be embedded within a simple unifying group. Thus may we understand the appearance of families of fermions with the curious structure they are seen to have. Is there an analogous possibility for unification of our chiral color model? It turns out that the simplest unifying group has 90 generators, compared to the mere 24 of $SU(5)$ or the 45 of $SO(10)$. The group consists of the direct product of *six* $SU(4)$ factors, together with a nonabelian discrete symmetry group to assure that the unifying group is simple, and that it is characterized by a single gauge coupling constant. Details are available elsewhere, so that I mention only a few aspects of the Grand Unification of Chiral Color.

The fermion representation of the unifying group is 252-dimensional. It consists of a dozen $(\overline{4}, 4)$'s, each of which transform nontrivially under two of the $SU(4)$'s, together with six (10)'s, one for each of the factor groups. When the group breaks down to its low-energy subgroup, just 92 of the left-handed fermions survive: five families, the quix, and the dichromat. When depicted as a *moose* diagram, the representation bears a striking resemblance to the chemists' illustration of a benzene molecule. Though the unifying group and its fermion representation are large, perhaps absurdly so, we have been able to construct no simpler alternative. While the unifying group can, in principle, break down to the chiral color theory, we have no idea of why the pattern of symmetry breakdown is as it is.

None of the gauge bosons of the unifying group can mediate proton decay; such a process would depend upon the detailed properties of the scalar sector. On the other hand, because of the richness and arbitrariness of the scalar sector, no unambiguous determination of $\sin^2 \theta_W$ emerges.

3. Conclusion

I have no reason to believe in chiral color — it is merely one fascinating alternative to the standard theory. Surely, Nature will prove herself to be more imaginative. The theory does, however, offer many possible experimental signatures at presently accessible energies. These should be targets of opportunity for workers at the Tevatron Collider and at LEP. The search may well be futile, but it may lead, serendipitously, to the Truth.

The alternative route, the path leading downwards from the Planck mass, has not brought us to any recognizable treasures. String thoughts offer no quick answers to our problems. The time may come for us to find a unified theory of everything, but it is not now. First, we had better be sure that we have experienced all of nature's wonders. The ball is once again in the hands of the experimenters.

DISCUSSION

– Bobbink:

Would the additional quarks in systems like the ψ and Υ, with γ and pion transition, behave unusually?

– Glashow:

No, they would behave normally–there are just five more quarks, but nothing else is unusual. Of course, for quarks heavier than the W the situation is different, since they decay instantaneously semiweakly, and the experimenters are very much prepared for that.

– Bobbink:

How are the parameters and the unitarity of the 3×3 KM matrix affected in your model?

– Glashow:

The KM matrix is just bigger, so there are many more parameters to determine. There is not much impact on the KM matrix for the first three families except perhaps for the top-bottom coupling. There is plenty of room for the five new quarks, as far as we know.

– Volkas:

Below the chiral color breaking scale, I think that there are enough coloured particles to not allow the (left)×(right) colour force decrease with increasing q^2, if one ignores the fact that some of the coloured particles are heavy. Introducing mass thresholds could change this conclusion. Now at the chiral colour scale $SU(3)_{colour}$ is enhanced to $SU(3)_L \otimes SU(3)_R$, and since more or less half the particles go in one $SU(3)$ and half in the other, I would imagine that the chiral colour forces are asymptotically free. Is this correct?

– Glashow:

Yes, chiral colour remains asymptotically free, as a simple counting will show, at least as far as the fermions are concerned. There is a danger with the bosons. For example, how would you give the quix a mass? The simplest way is to invent a $(6, \bar{6})$ scalar, but that is a disasterously heavy representation which would undo the asymptotic freedom. It is possible to avoid this and stay asymptotically free.

This brings up the question of what you get for $\sin\theta_W$ in this model, and the answer is anything you want, simply by varying the scalar sector.

– Volkas:

I have a speculation concerning models that be half-way between the chiral color model and the technicolour chiral colour models of Georgi that you alluded to. Would it be interesting to invoke an old idea of Marciano (and I do not know whether this has been shown to be wrong in the interim) whereby $SU(2)$-doublet colour-sextet fermions are used to dynamically break the electroweak group? Clearly the anomaly cancellation scenario would be different, but maybe a model like this can exist.

– Glashow:

I have no comment.

– Klapisch:

Is there place in your scheme for a fifth neutrino that would be so massive as to not affect the decay width of the Z^0?

– Glashow:

Yes, from Z^0 decay one can get anywhere from 4 to 5 neutrinos.

– Horne:

Is there any way to prevent the many scalar particles from getting very large masses?

– Glashow:

This theory suffers from the same naturalness problem as the standard model in this respect. Perhaps one could supersymmetrize the model to avoid this problem.

– Warr:

Do the multiple Higgs bosons in this theory tend to be weakly or strongly coupled to each other – if they were moderately strongly coupled their widths would be comparable to their masses and they would be undetectable.

Glashow:

I agree, and I see no way to constrain the coupling to be strong or weak, so perhaps these scalars are unobservable.

– Polychronakos:

Is there a reason why a theory of chiral colour should be confining?

– Glashow:

Chiral colour breaks down at a high energy, leaving the ordinary colour which we haven't changed, so it's still confining; thus at low energy the theory is identical to ordinary QCD.

CLOSING CEREMONY

The closing ceremony took place on Thursday, 14th August 1988. The Director of the School presented the Prizes and Scholarships as specified below.

PRIZES AND SCHOLARSHIPS

Prize for **Best Student**

    awarded to Jim HORNE, Princeton University, USA.

The Scholarships open for competition among the participants were awarded as follows:

Patrick M.S. Blackett Scholarship
    Gerjan BOBBINK, NIKHEF, Amsterdam, The Netherlands

James Chadwick Scholarship
    Concepcion GONZALEZ–GARCIA, Universidad de Valencia, Spain

Amos De-Shalit Scholarship
    Brian WARR, University of Texas, USA

Paul A.M. Dirac Scholarship
    Zygmunt A. LALAK, Warsaw University, Poland

Gunnar Källen Scholarship
    Raymond VOLKAS, University of Melbourne, Australia

André Lagarrigue Scholarship
    Carrie FORDHAM, SLAC, Stanford, USA

Ettore MAJORANA Scholarship
    Allen MINCER, Calthech, USA

Giulio Racah Scholarship
    Uri SARID, Harvard University, Cambridge, USA

Jun John SAKURAI Scholarship
    Ioannis GIANNAKIS, The Rockefeller University, USA

Antonio STANGHELLINI Scholarship

    Alexios POLYCHRONAKOS, University of Florida, USA

Worl Laboratory Scholarship

    Bani MITRA, University of DELHI, India

One **EPS Scholarship** was also awarded:

    Vladimir BALEK, Bratislava, Czechoslovakia

Prize for **Best Scientific Secretary** – awarded ex–aequo to:

    Carrie FORDHAM, SLAC, Stanford, USA

    Concepcion GONZALEZ–GARCIA, Universidad de Valencia, Spain

Two participants received **Honorary Mention**

    Paul COLAS, Centre d'Etude Nucléaires de Saclay, France

    Alberto MARCHIONNI, INFN, Firenze, Italy

The following participants gave their collaboration in the scientific secretarial work:

Gerjan BOBBINK	Alberto MARCHIONNI
Andrew BRANDT	Allen MINCER
Luigi CAPPIELLO	Bani MITRA
Paul COLAS	Günter POLLAK
Ornella DI ROSA	Vlado RAHAL
Carrie FORDHAM	Nuria RIUS–DIONIS
Filipe FREIRE	Sigurd SANNAN
Ioanas GIANNAKIS	Uri SARID
Cencepcion GONZALEZ–GARCIA	Harald SKARKE
Zigmunt LALAK	Brian WARR
Thomas MANNEL	

INTERNATIONAL SCHOOL OF SUBNUCLEAR PHYSICS
August 7-15, 1988, in Erice, Italy

PARTICIPANTS

Carlo ALCADE

University of California
Department of Physics
405 Hilgard Avenue
LOS ANGELES, CA 90024, USA

Francesco ANSELMO

IFCAI CNR
Via Mariano Stabile, 172
90139 PALERMO

Fabrizio AVERSA

INFN
Laboratori Nazionali
C.P. 13
00044 FRASCATI, Italy

Vladimir BALEK

Comenius University
Department of Theoretical Physics
Mlynska Dolina
842 15 BRATISLAVA, Czechoslovakia

Riccardo BARBIERI

Istituto di Fisica dell'Università
Piazza Torricelli 2
56100 PISA, Italy

Vincenzo BARONE

Massachusetts Institute of Technology
Department of Physics
CAMBRIDGE, MA 02139, USA

Alessandro BELLINI

Università di Firenze
Dipartimento di Fisica
Largo E. Fermi, 2
50125 ARCETRI (Firenze), Italy

Gerjan BOBBINK

NIKHEF-H
AMSTERDAM, The Netherlands
and
CERN, 1211 GENEVA, Switzerland

Andrew BRANDT

CERN
EP Division
1211 GENEVA, Switzerland

Paolo CALAFIURA

INFN
Via Livornese 582/a
56010 SAN PIERO A GRADO, Pisa, Italy

Luigi CAPPIELLO

INFN
Sezione di Napoli
Mostra d'Oltremare – Pad. 20
80125 NAPOLI, Italy

Paul COLAS

Centre D'Etudes Nucléaires de Saclay
DPhPE/SEPh
91191 GIF–SUR–YVETTE, France

Ornella DI ROSA

IFCAI CNR
Via Mariano Stabile, 172
90139 PALERMO, Italy

Thomas DOMEIJ

Chalmers Tekniska Hogskola
Institute of Theoretical Physics
41296 GÖTEBORG, Sweden

Michael DUFF

Imperial College
Physics Department
Prince Consort Road
LONDON, SW7 2BZ, UK
and
CERN, TH Division
1211 GENEVA 23, Switzerland

Anne EALET

CPPM
Case 907
70 Route Léon Lachamp
13288 MARSEILLE Cedex 09, France

Sergio FERRARA

University of California
Department of Physics
405 Hilgard Avenue
LOS ANGELES, CA 90024, USA
and
CERN, TH Division
1211 GENEVA 23, Switzerland

Carrie FORDHAM

Stanford Linear Accelerator Centre
P.O. Box 4349
STANFORD, CA 94309, USA

Filipe FREIRE

Centro de Fisica Nuclear
Av. prof. Gama Pinto, 2
1699 LISBOA, Portugal

Maurizio GASPERINI

Dipartimento di Fisica Teorica
Corso Massimo D'Azeglio, 46
10125 TORINO, Italy

Ioannis GIANNAKIS

The Rockefeller University
Box 129
1230 York Avenue
NEW YORK, NY 10021, USA

Giorgio GIAVARINI

INFN
Dipartimento di Fisica
Viale delle Scienze
43100 PARMA, Italy

Sheldon L. GLASHOW

Harvard University
Physics Laboratories
CAMBRIDGE, MA 02138, USA

Concepcion GONZALEZ

Universidad de Valencia
Departamento de Fisica Teorica
46100 BURJASSOT, Valencia, Spain

Michel GOURDIN

Université Pierre et Marie Curie
4, Place Jussieu
75230 PARIS Cedex 05, France

Michael GREEN

Queen Mary College
Physics Department
Mile End Road
LONDON E1 4NS, UK

Oscar HERNANDEZ

Harvard University
Department of Physics
CAMBRIDGE, MA, 02138, USA

Jim HORNE

Princeton University
Jadwin Hall
PRINCETON, NJ 08544, USA

Ralf KAISER

Karlsruhe University
Institute of Theoretical Physics
Kaiserstr. 12
7500 KARLSRUHE, FRG

Robert KLAPISCH

CERN
1211 GENEVA 23, Switzerlad

Zygmunt A. LALAK

Warsaw University
Unstitute of Theoretical Physics
ul. Hoza 69
00-681 WARSAW, Poland

Leon LEDERMAN

Fermi National Accelerator Lab.
P.O. Box 500
BATAVIA, IL 60510, USA

Endre LILLETHUN

ICTP
Strada Costiera, 11
P.O. Box 586
34100 TRIESTE, Italy

David MacFARLANE

McGill University
Department of Physics
3600 University Street
MONTREAL, Quebec H3A 2T8, Canada

Michele MAGGIORE

Institut für Theoretische Physik
Sildestr. 5
3012 BERN, Switzerland

Michel MAKOWSKA

Université de Lausanne
Institut de Physique Théorique
1015 LAUSANNE, Switzerland

Thomas MANNEL

Technische Hochschule Darmstadt
Institut für Kernphysik
Schlossgartenstr. 9
6100 DARMSTADT, FRG

Robert MARCH

University of Wisconsin–Madison
Departmen of Physics
1159 University Avenue
MADISON, WI 53706, USA

Alberto MARCHIONNI

INFN
Largo Fermi, 2
50125 FIRENZE, Italy

Allen MINCER

CALTECH
Mail Stop 356–48
PASADENA, CA 91125, USA

Bani MITRA

SGTB
Khalsa College
Delhi University Campus
DELHI 110009, India

Rudolf MOSSBAUER

Technische Universität München
Department of Physics
8046 GARCHING, FRG

Giuseppe MUSSARDO

SISSA
Strada Costiera – Grignano
MIRAMARE – TRIESTE, Italy

Hans Peter NILLES

CERN
TH Division
1211 GENEVA, Switzerland

Martin L. PERL

Stanford Linear Accelerator Centre
P.O. Box 4349
STANFORD, CA 94305, USA

Luciano PIETRONERO

Università di Roma "La Sapienza"
Dipartimento di Fisica
Piazzale Aldo Moro
00185 ROMA, Italy

Irwin PLESS

Massachusetts Institute of Technology
Laboratory for Nuclear Science
CAMBRIDGE, MA 02139, , USA

Günther POLLAK

Tecn. Univ. Wien
Institut für Theoretische Physik
Karlsplatz 13
1040 WIEN, Austria

Alexios POLYCHRONAKOS

University of Florida
Department of Physics
234–E Williamson Hall
GAINESVILLE, FL 32611, USA

Vlado RAHAL

The Rockefellr University
Department of Physics
66th Street York Avenue
NEW YORK, NY 10021, USA

Riccardo RATTAZZI

Scuola Normale Superiore
Classe di Scienze
Piazza dei Cavalieri, 6
56100, PISA, Italy

Nuria RIUS

Universidad de Valencia
Departamento de Fisica Teorica
46100 BURJASSOT, Valencia, Spain

Sarabjit SABHARWAL

University of California
Department of Physics
405 Hilgard Avenue
LOS ANGELES, CA 90024, USA

Nick SAMIOS

Brookhaven National Laboratory
Physics Department
UPTON, NY 11973, USA

Sigurd SANNAN

NORDITA
Blegdamsvej 17
2100 COPENHAGEN, Denmark

Uri SARID

Harvard University
Lyman Laboratory
CAMBRIDGE, MA 02138, USA

Julian SCHWINGER

University of California
Department of Physics
405 Hilgard Avenue
LOS ANGELES, CA 90024, USA

Leonello SERVOLI

Dipartimento di Fisica
Via Elce di Sotto
06100 PERUGIA, Italy

Qi–Xing SHEN

Academia Sinica
Institute of High Energy Physics
P.O. Box 918
BEIJING, China

Harald SKARKE

Tech. Univ. Wien
Institut für Theoretische Physik
Karlsplatz 13
1040 WIEN, Austria

Gemma TESTERA

INFN
Sezione di Genova
Via Dodecaneso, 33
16146 GENOVA, Italy

Samuel C.C. TING

Massachsetts Institute of Technology
Department of Physics
CAMBRIDGE, MA 02139, USA

and

CERN
1211 GENEVA 23, Switzerland

Raymond VOLKAS

University of Melbourne
School of Physics
PARKVILLE, 3052 Victoria, Australia

Lucia VOTANO

INFN
Laboratori Nazionali di Frascati
00040 FRASCATI, Italy

Brian WARR

University of Texas
RLM 9202
AUSTN, TX 78712, USA

Sir Denys WILKINSON

University of Sussex
Falmer
BRIGHTON BN1 9QH, UK

MACRO experiment, 263-265
  detection of monopoles, 263-265
  specification, 263
Magnetic monopoles, 260-263
  astrophysical limits on flux, 262
  detection techniques, 262-263
  MACRO experiment, 263-265
Majorana fermion, 19
Majorana neutrinos, dark matter, 172
Majorana world-sheet spinors, 19
2π-Meson modes, isotopic spin
      analysis, 187-188
π-Meson, two-photon decay, quantum
      field theory, 1-9
Minkowski world-sheet metric, 18
Monojets, 415
Monte Carlo simulations, 339, 341
MSW effect, solar neutrino puzzle,
      317-319
Multichannel integrated detector
      technologies, silicon
      detectors, 378-384
Multidrift modules
  aging, 349-351
  LAA project, 345
  performances, 343
  rate limitation, 347-348
  vertex detector technologies,
      342-351
Multiloop amplitudes, string Feynman
      diagrams, 24-25
Muon detection, LAA project, 337
Muon events, expectation of numbers,
      165

Nambu action, 17
Neutralinos
  as cosmions, 173
  cosmologically interesting,
      defined, 160
  dark matter, 160-165
  Hubble parameter, 160
  as LSPs, 171
  searches in e^+e^- collisions,
      158-160
  trapping on sun and earth, 165
Neutrino events, expectation of
      numbers, 165
Neutrinos
  Gran Sasso laboratory, 249, 251
  mass, Gran Sasso approach, 265
  oscillations with atmospheric
      neutrinos, 266-267
  solar neutrinos, 256-258, 260-261
  see also Solar neutrino puzzle;
      Solar neutrinos
NICOTIN wire, 350-351
NOTA (Operational transconductance
      amplifier with N-channel),
      379
Nuclear structure
  A-nucleons, 400-401
  gluons, 405-407
  mesonic currents, 403
  nucleon-nucleon interactions,
      403-404
  quarks, 405-407

Nuclear structure (continued)
  shell model, 399-400
  soft pion effects, 403

O^2 orbifold, 131-133
Open string field theory, Witton, 29
Orbifold compactifications
  effective Lagrangians, 81-84
  S and T fields, 84
  untwisted sector, 84
Orbifolds
  Coxeter, symmetric Z_N, 94
  degenerate, 139-145
  O^2 orbifold, 131-133
  Z_3-orbifold, nonstandard
        embedding, 142
  Z-orbifold, 133-135

2π Meson modes, isotopic spin
      analysis, 187-188
π-Meson, two-photon decay, quantum
      field theory, 1-9
PC violation
  B^0 system, 220-244
  discussion, 245-248
  introduction, 185-186
  K^0 system, 186-197
  status, 185-244
Peccei-Quinn symmetry, 100, 102, 104
Penguin diagram
  K^0 system, 205-210
  references, 219
Perturbation expansion, superstring
      theory, 29-31
Perturbation theory, s-model, 79
Pion decay, 1-9
  discussion, 10-11
Planck temperature, defined, 30
PMP
  absorption and emission spectra,
      354
  new scintillator, 353
PMP-doped PVT and PS, 355
Polyakov-Zamolodchikov metric, 107,
      113
Polyakov method, string theory
      amplitudes, 17
POTA (Operational transconductance
      amplifier with P-channel),
      379
Proton decay, trapping rate, sun and
      earth, 165
Pseudoscalar coupling
  action term, 7
  slowly varying fields, 6-7
Pseudovector coupling
  action term, 7
  fermions, 1-9

Quantum chromodynamics
  4-dimensional, high temperature
      phase, 30
  nuclear structure, 405
  skyrmion terms, 408-410
  sum rule, 234